REVIEWS IN PHARMACEUTICAL & BIOMEDICAL ANALYSIS

Edited By

Constantinos K. Zacharis and Paraskevas D. Tzanavaras

eBooks End User License Agreement

CONTENTS

CHAPTERS

FOREWORD

Analytical chemistry is playing a critical role in many scientific disciplines and certainly pharmaceutical and biomedical sciences are among the most important. Just a quick glance in the international literature can easily prove this statement.

On this basis it is a pleasure to introduce you to this book under the title "Reviews in Pharmaceutical and Biomedical Analysis". The topic of this book is so wide that it would have been a utopia to expect to cover all aspects of pharmaceutical and biomedical analysis in one single volume. However, the reader can find ten very interesting chapters that cover important fields ranging from sample preparation to metabolomics.

The authors of the chapters of the book are distributed in a large number of countries and they certainly are well-respected and experienced researchers. I strongly believe that this e-book will be a valuable assistance to a variety of scientists and of course to students that are involved to the field of Analytical Chemistry and I strongly recommend it.

Demetrius G. Themelis
Aristotelian University of Thessaloniki, Greece

PREFACE

The scope of this e-book entitled *"Reviews in Pharmaceutical and Biomedical Analysis"* is the coverage and review of new trends and applications in all areas of pharmaceutical and biomedical analytical chemistry.

Our intention was to cover all instrumental analytical methods that are applied to the analysis of compounds with pharmaceutical and biomedical interest, including liquid and gas chromatography, electrophoresis & related techniques, mass spectrometry, hyphenated techniques, automated analytical techniques, spectrometry, luminescence, electroanalysis etc.

We were pleased to see that many authors accepted our invitation to contribute to this e-book covering a wide spectrum of fields. The topics in the order of appearance in the e-book include: an insight on the methods used in the metabolomics analysis (LC-MS, GC-MS, HPLC-DAD, NMR) of several natural matrices with protective health potential (Chapter 1); demonstration of the ability of Artificial Neural Networks (ANN) in successfully predicting the response of an Enzyme-linked Immunosorbent assay (ELISA) (Chapter 2); a discussion of the most recent progresses in bioinformatics tools useful in mass spectrometry-based proteomics (Chapter 3); a review of current methods used in quantitative analysis of pharmaceutical compounds from whole blood matrix by liquid chromatography mass spectrometry (Chapter 4); a discussion of how biosensor-based platforms can be used in conjunction with microbial cells for monitoring, environmental and industrial applications (Chapter 5); electroanalytical methods as tools for predictive drug metabolism studies (Chapter 6); a review of sample preparation methodologies prior to the chromatographic determination of benzodiazepines (Chapter 7); a presentation of the most common analytical techniques for the control of the level of apoptosis (Chapter 8); a review on the various analytical methods designed to meet the requirements for cytokinin analyses in complex matrices (Chapter 9); a review on the applications of pressurized liquid extraction technique in phytochemical analysis in last decade (Chapter 10).

We sincerely believe that the book will prove to be a useful contribution to analytical science. We express our appreciation to all of the contributing authors to Bentham Publishers and their team members for the opportunity to publish this volume. Lastly we thank our family members for their support, encouragement and patience during the entire period of this work.

Constantinos K. Zacharis & Paraskevas D. Tzanavaras
Laboratory of Analytical Chemistry
Department of Chemistry
Aristotelian University of Thessaloniki, Greece

CONTRIBUTORS

Name	Address	e-mail	Telephone
Andrade P.B.	*REQUIMTE/Serviço de Farmacognosia, Faculdade de Farmácia, Universidade do Porto, R. Aníbal Cunha, 164, 4050-047 Porto, Portugal*	pandrade@ff.up.pt	+351 222078934
Arora S.	*(1) School of Biotechnology, Dublin City University, Dublin 9, Ireland. (2) Biomedical Diagnostics Institute (BDI), Dublin City University, Dublin 9, Ireland. (3) National Centre for Sensor Research (NCSR), Dublin City University, Dublin 9, Ireland.*		
B. Blankert	*(1) University of Mons, Faculty of Medicine & Pharmacy, Lab. of Pharmaceutical Analysis, Belgium (2) Université Libre de Bruxelles – Institute of Pharmacy, Laboratory of Instrumental Analysis and Bioelectrochemistry, Belgium*	bertrand.blankert@umons.ac.be	+32 65 373592
Barry Byrne	*(1) School of Biotechnology, Dublin City University, Dublin 9, Ireland. (2) Centre for Bioanalytical Sciences (CBAS), Dublin City University, Dublin 9, Ireland. (3) National Centre for Sensor Research (NCSR), Dublin City University, Dublin 9, Ireland.*		
Brunetti P.	*Institute for Biomedical Technologies, Proteomics and Metabolomics Unit - CNR, Via Fratelli Cervi, 93, 20090 Segrate (Milan), Italy.*	pietro.brunetti@itb.cnr.it	+39 0226422724
Dąbrowska M.	*Jagiellonian University, Collegium Medicum, Department of Inorganic and Analytical Chemistry, 9 Medyczna Str., 30-688; Cracow, Poland*	mtylka@cm-uj.krakow.pl	
Daminelli S.	*Institute for Biomedical Technologies, Proteomics and Metabolomics Unit - CNR, Via Fratelli Cervi, 93, 20090 Segrate (Milan), Italy.*	ra1d83@gmail.com	+39 0226422724
Di Silvestre D.	*Institute for Biomedical Technologies, Proteomics and Metabolomics Unit - CNR, Via Fratelli Cervi, 93, 20090 Segrate (Milan), Italy.*	dario.disilvestre@itb.cnr.it	+39 0226422724
Dotsikas Y.	*Division of Pharmaceutical Chemistry, Department of Pharmacy, University of Athens, Panepistimioupoli Zografou GR - 157 71, Athens, Greece*		
El-Shourbagy T.A.	*Abbott Laboratories, Department of Drug Analysis, 100 Abbott Park Road, Abbott Park, IL 60064-6126, U.S.*	tawakol.a.elshourbagy@abbott.com	+ 1-847-938-4494
Ferreres F.	*Research Group on Quality, Safety and Bioactivity of Plant Foods, Department of Food Science and Technology, CEBAS (CSIC), P.O. Box 164, 30100 Campus University Espinardo, Murcia, Spain*	federico@cebas.csic.es	+34 968396324
Ge L.	*Natural Sciences and Science Education Academic Group, Nanyang Technological University, Singapore*	lge@pmail.ntu.edu.sg	
J.-M. Kauffmann	*Université Libre de Bruxelles – Institute of Pharmacy, Laboratory of Instrumental Analysis and Bioelectrochemistry, Belgium*	jmkauf@ulb.ac.be	+32 2 6505215

Kousoulos C.   *Division of Pharmaceutical Chemistry, Department of Pharmacy, University of Athens, Panepistimioupoli Zografou GR - 157 71, Athens, Greece*

Li P.   *Faculty of Life Science and Technology, Kunming University of Science and Technology, Kunming, Yunnan, P. R. China*

Li S.P.   *Institute of Chinese Medical Sciences, University of Macau, Macao SAR, P. R. China*   lishaoping@hotmail.com

Loukas Y.L.   *Division of Pharmaceutical Chemistry, Department of Pharmacy, University of Athens, Panepistimioupoli Zografou GR - 157 71, Athens, Greece*   loukas@pharm.uoa.gr   +30 210 7274224

Marsili E.   *(1) School of Biotechnology, Dublin City University, Dublin 9, Ireland. (2) National Centre for Sensor Research (NCSR), Dublin City University, Dublin 9, Ireland.*

Mauri PL.   *Institute for Biomedical Technologies, Proteomics and Metabolomics Unit - CNR, Via Fratelli Cervi, 93, 20090 Segrate (Milan), Italy.*   pierluigi.mauri@itb.cnr.it   +39 0226422728

O'Kennedy R.   *(1) School of Biotechnology, Dublin City University, Dublin 9, Ireland. (2) Biomedical Diagnostics Institute (BDI), Dublin City University, Dublin 9, Ireland. (3) Centre for Bioanalytical Sciences (CBAS), Dublin City University, Dublin 9, Ireland. (4) National Centre for Sensor Research (NCSR), Dublin City University, Dublin 9, Ireland.*   Richardokennedy@gmail.com

Papadoyannis I.N.   *Laboratory of Analytical Chemistry, Department of Chemistry, Aristotle University of Thessaloniki, Thessaloniki, GR-541 24, Greece*   papadoya@chem.auth.gr   +30231997793

Pastorella G.   *(1) School of Biotechnology, Dublin City University, Dublin 9, Ireland. (2) National Centre for Sensor Research (NCSR), Dublin City University, Dublin 9, Ireland.*

Pereira D.M.   *REQUIMTE/Serviço de Farmacognosia, Faculdade de Farmácia, Universidade do Porto, R. Aníbal Cunha, 164, 4050-047 Porto, Portugal*   david.ffup@gmail.com   +351 222078935

Rieser M.J.   *Abbott Laboratories, Department of Drug Analysis, 100 Abbott Park Road, Abbott Park, IL 60064-6126, U.S.*   matthew.j.rieser@abbott.com   + 1-847-937-5505

Samanidou V.F.   *Laboratory of Analytical Chemistry, Department of Chemistry, Aristotle University of Thessaloniki, Thessaloniki, GR-541 24, Greece*   samanidu@chem.auth.gr   +30231997698

Skuciński J.   *Jagiellonian University, Collegium Medicum, Institute of Emergency Medicine, 12 Michałowskiego Str., 31-126; Cracow, Poland*   jerzy.skucinski@wp.pl

Starek M.   *Jagiellonian University, Collegium Medicum, Department of Inorganic and Analytical Chemistry, 9 Medyczna Str., 30-688; Cracow, Poland*   mstarek@cm-uj.krakow.pl

Tan S.N.   *Natural Sciences and Science Education Academic Group, Nanyang Technological University, Singapore*   Sweengin.tan@nie.edu.sg

Uddin M.N.   *Department of Chemistry, University of Chittagong,*   nasiru_cu@yahoo.com   880-31-716552-58
                  Chittagong-4331, Bangladesh

Valentão P.   *REQUIMTE/Serviço de Farmacognosia, Faculdade de*   valentao@ff.up.pt   +351 222078934
                  Farmácia, Universidade do Porto, R. Aníbal Cunha, 164,
                  4050-047 Porto, Portugal

Xu R.N.   *Abbott Laboratories, Department of Drug Analysis, 100 Abbott*   raymond.xu@abbott.com   + 1-847-938-8158
                  Park Road, Abbott Park, IL 60064-6126, U.S.

Yong J.W.H   *Natural Sciences and Science Education Academic Group,*   jean.yong@nie.edu.sg
                  Nanyang Technological University, Singapore

CHAPTER 1

Metabolomic Analysis of Natural Products

David M. Pereira[a], Patrícia Valentão[a], Federico Ferreres[b] and Paula B. Andrade[a]*

[a]REQUIMTE/Serviço de Farmacognosia, Faculdade de Farmácia, Universidade do Porto, R. Aníbal Cunha, 164, 4050-047 Porto, Portugal and [b]Research Group on Quality, Safety and Bioactivity of Plant Foods, Department of Food Science and Technology, CEBAS (CSIC), P.O. Box 164, 30100 Campus University Espinardo, Murcia, Spain

Abstract: The metabolome comprises all metabolites in a biological organism, which constitute the end products of its gene expression. Metabolomics consists on the systematic study of the chemical fingerprints resulting from specific cellular processes or, more particularly, the study of an organism's profile of low molecular weight metabolites. Thus, metabolomics is perhaps the ultimate level of post-genomic analysis as it can reveal changes in metabolite fluxes that are controlled by only minor changes within gene expression. Classical phytochemical approaches often comprised a rather tedious and time consuming process of isolation, dereplication of known substances, followed by structure elucidation and quantification. However, it is important to highlight that, in many situations, the effects of natural products are not due to a single compound, but to a mixture of related and unrelated ones. Thus, metabolomics provides an efficient tool for the quality control and authentication of medicines of natural origin, contributing as well to the characterization of different species. Several combined techniques have been applied in the measurements of intracellular metabolites, whether qualitative or quantitative, which reveal the biochemical status of the organism. This review offers an insight on the methods used in the metabolomics analysis (LC-MS, GC-MS, HPLC-DAD, NMR) of several natural matrices with protective health potential, with special emphasis on the determination of phenolics profiles, once these represent the most abundant and widely spread class of plant natural compounds, additionally exhibiting interesting biological activities.

WHAT IS METABOLOMICS?

A straight-forward definition of metabolomics would be "a science that seeks to identify and quantify the complete set of metabolites in a cell or tissue and to do so as quickly as possible and without bias" [1]. As one can easily guess, such definition cannot still be fulfilled nowadays so that, usually, the analysis is focused on sub-metabolome fractions (for instance, the sub-metabolome extractable with a certain solvent or detectable with a certain analytical technique). The remarkable number of different metabolites is the main contributor to the failure to fulfill the definition. Metabonomics, a word commonly misused as metabolomics, refers to the measurement of metabolite profiles, activities, and reactions toward the environment, medication, or disease, of a given tissue or biological fluid [2].

Metabolomics constitutes the endpoint of the "omics cascade" (Fig. **1**) and is the closest to phenotype. Even so, there is no single-instrument platform that can currently analyze all metabolites [3]. Nevertheless, although the understanding of living organisms at the molecular level is still taking its first steps, it is already evident that all contributors to the "omics cascade" will be important pivots and play a central role in this new science.

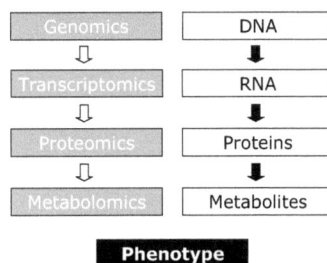

Figure 1: The "omics cascade".

The metabolome (the complete set of an organism's metabolites) represents a vast number of components that belong to a wide variety of classes, such as amino acids, lipids, organic acids, nucleotides, among many others. These compounds are very diverse in their physical and chemical properties and occur in a wide concentration range. For example, within lipids alone, not only high abundance compounds, such as fatty acids, triglycerides, or phospholipids, are encountered, but also trace level components with important regulatory effects, such as eicosanoids derived from arachidonic acid [3].

In addition to the chemical characteristics of the metabolites in study, the idea that metabolite distributions are subjected to high temporal and spatial variability should also be present at all times. These situations include, for instance, circadian fluctuations in hormones in mammalian organisms or diet-dependent biological variability [4]. A careful experimental design is therefore crucial for the success of these studies.

Metabolites are often simply viewed as one of the end-products of gene expression and protein activity. However, it is increasingly understood that metabolites themselves modulate macromolecular processes through, for example, feedback inhibition and as signaling molecules. Metabolomic studies are intended to provide an integrated view of the functional status of an organism. Metabolites represent a diverse range of structures, physicochemical properties, stabilities and abundances. A key consideration in effective metabolomic pursuits is, therefore, the establishment of an optimal balance between quantitative accuracy and the range of metabolites measured [5].

METABOLITE PROFILING

Metabolic profiling stands for the analysis of a group of metabolites either related to a specific metabolic pathway or a class of compounds. The analysis of flavonoids by High Pressure Liquid Chromatography-Mass Spectrometry (HPLC-MS) or of terpenes by Gas Chromatography-Mass Spectrometry (GC-MS) are examples for metabolic profiling.

As it can be easily noticed, by analyzing only a specific set of metabolites on its own, thus ignoring the remaining ones, a truly metabolomic analysis is not accomplished. However, by assembling a whole suit of quantitative methods that analyze key metabolites from different biochemical pathways, metabolite profiling gives rise to metabolomics.

Within metabolite profiling, some metabolites have been so widely studied that some new nomenclatures appeared in the past years. That is the case of lipidomics, which assesses the qualitative and quantitative information about the constitution of the cellular lipidome (sub-compartment of the metabolome, comprising lipid classes, subclasses and lipid signaling molecules) and provides insights into biochemical mechanisms of lipid metabolism, lipid–lipid and lipid–protein interactions [6].

TARGET METABOLITE ANALYSIS

A more directed approach, target metabolite analysis, aims at the measurement of selected analytes, such as biomarkers of disease or toxicant exposure, or substrates and products of enzymatic reactions [7]. Based on the questions asked, metabolites are selected for analysis and specific analytical methods are developed for their determination [3].

In this case, as the metabolite being investigated is already know, techniques that provide less structural information than MS and Nuclear Magnetic Ressonance (NMR) can be applied, such as GC-Flame Ionization Detection (GC-FID) or HPLC-diode array detection (DAD), when authentic standards are applied.

METABOLIC FINGERPRINTING

Fingerprinting techniques involve collecting spectra of unpurified solvent extracts in standardized conditions and ignore, initially, the problem of making individual assignments of peaks, which are frequently overlapping. Multivariate statistical methods such as principal component analysis (PCA) are used to compare sets of spectra to identify clusters of similarity or difference so that conclusions can be drawn about the classification of individual plant samples. The identities of metabolites responsible for differences between classes can be investigated from loading plots generated by PCA and related techniques [8]. As so, only peaks that are different between samples are analyzed, thus avoiding the time-consuming task of identifying all peaks. The most used technique for metabolic fingerprinting is NMR and several types of application can be found in the literature, such as multivariate analysis of unassigned ^{1}H NMR spectra, used to compare the overall metabolic composition of wild-type, mutant, and transgenic plant materials, and to assess the impact of stress conditions on the plant metabolome, among many others.

When working with cell cultures, in addition to metabolic fingerprinting of intracellular metabolites, the analysis of extracellular metabolites excreted into the culture medium or taken up from the medium by cells can provide valuable information on their phenotype and physiological state. Pattern analysis of metabolites in conditioned cell culture media is called metabolic footprinting [9, 10].

Although NMR is the most used approach in fingerprinting studies, mass spectrometry-based investigations are also possible. Here, the metabolite fingerprints are represented by *m/z* values and corresponding intensities of the detected ions. When a separation step takes place prior to the MS analysis, retention times are also used to index metabolites. That can then be used for sample classification using multivariate data analysis techniques.

Regardless of the utility of this kind of studies, using metabolomics exclusively for fingerprinting, without identifying the metabolites that cause clustering of experimental groups, will only deliver a classification tool but not directly contribute to biochemical knowledge and understanding of underlying mechanisms of action [3].

ANALYTICAL TECHNIQUES USED IN METABOLOMICS

Without a doubt, MS and NMR constitute the most widespread and promising techniques for metabolomics, with Liquid Chomatography-Mass Spectrometry (LC-MS), GC-MS, Capillary Electrophoresis- Mass Spectrometryy (CE-MS) and LC-NMR being the most used analytical approaches.

The selection of the most suitable technique is generally a compromise between speed, selectivity, and sensitivity. Data obtained by NMR and MS is often complementary and, preferably, both techniques should be used when possible.

Mass Spectrometry

MS has established itself as the method of choice. MS is often connected to chromatographic separation, like with GC and HPLC.

MS is favored for its high sensitivity and selectivity because it can detect ''NMR-invisible'' moieties, such as sulfates. However, it is not as robust as NMR, with low reproducibility unless standards are employed, it might fail to discriminate between certain classes of compounds due to ionization methods employed [11].

Mass Spectrometry – Gas Chromatography

GC-MS is a relatively low-cost technique that provides high separation efficiencies that can resolve complex biological mixtures. Gas chromatography is a powerful tool to analyse volatile compounds. With the correct derivatization process, even non-volatile compounds may be analyzed. Samples are injected into an inert gas stream and swept into a tube, which is packed with a solid support coated with a resolving liquid phase. Absorptive interactions between the components in the gas stream and the coating lead to a differential separation of the components of the mixture, which are then swept in order through a detector flow cell.

The combination of GC and MS yields an instrument capable of separating mixtures into their individual components, identifying and providing quantitative and qualitative information on the amounts and chemical structure of each compound.

A complete identification and ulterior quantification of volatiles in complex matrices, such as natural products, needs some previous proceedings due to volatiles being present in a wide range of concentrations. Some volatiles can be present in high levels (mg/L or mg/Kg) while others occur in very low ones (ng/L or ng/Kg). When an Ion Trap analyser is used, if a molecule is present in high amounts, the consequence is an "overloading" in the trap, which causes a distortion of mass spectrum of the molecule that may prevent a correct identification. In opposition, when volatile molecules are present in low levels (ng or even pg levels) a very accurate spectrum can be obtained, which is usually coincident with those of most libraries. Consequently, and due to this specificity/limitation, quantification can be analytically difficult. Linearity studies must be performed in certain ranges of concentrations and, when compounds are present in high concentration, dilution of samples is recommended

Mass Spectrometry – Liquid Chromatography

LC coupled to MS is a powerful alternative that offers high selectivity and sensitivity. Nowadays, techniques, such as LC-DAD-MS, and particularly LC-DAD-ESI/MS, are regarded as very useful tools for the analysis of natural matrices [12].

The amount of information obtained by multisignal MS or MS-MS renders two levels for compounds identification: positive and provisional. Positive identification can be achieved when reference compounds are available, thus allowing comparison of both retention time and UV spectra. However, when no standards are available, generally provisional identification takes place. In the case of phenolics, although the identity of subunits, such as aglycones, sugars and acyl moieties, is elucidated, the positions of glycosidic and acyl linkages remain unknown, with the exception of some fragmentation patterns that allow a positive identification of the glycosidic bonds (Fig. **2**) [13]. In many cases, a taxonomic approach may help in the identification; however, NMR analysis is usually necessary. Nevertheless, the level of identification required in food and botanical analysis is a question worth postulating. For example, in the analysis of beverages, such as wines, and oils, most of the times profile analysis yielding provisional identification is enough for quality control and detection of adulteration.

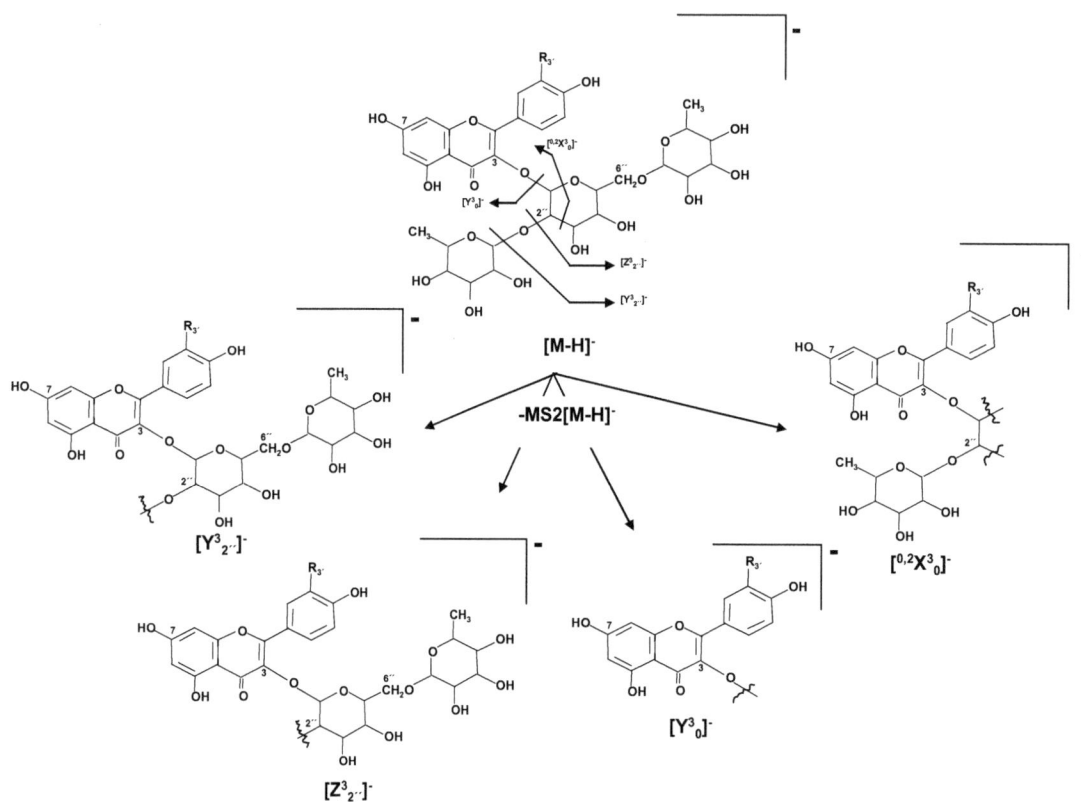

Figure 2: MS2[M-H]- fragmentation of flavonol-3-*O*-(2,6-di-*O*-rhamnosyl-hexoside). Compounds: 7 and 8 (R3′: OH); 9 and 10 (R3′: H); 12 and 14 (R3′: OCH3). From [20] (with kind permission of American Chemical Society).

Spectrometry by UV-Vis and MS, when in total ion count mode, yield detection limits in the range of 10 ng. When single ion monitoring (SIM) mode is used, MS analysis provides better detection limits, usually below 1 ng [12]. SIM mode, however, causes loss of valuable information concerning fragmentation pattern, which is very important for the identification of many compounds.

With the advent of atmospheric pressure ionization (API) sources, LC-MS coupling became more efficient and accessible, causing this technique to be one of the most used nowadays. Comparison of different API sources, such as electrospray ionization (ESI), atmospheric pressure photoionization (APPI) or atmospheric pressure chemical ionization (APCI) is available in literature [13].

When working with phenolics, the highest sensitivity can be obtained by the use of ESI in negative mode, usually involving an eluent consisting of an acidic ammonium acetate buffer. At positive ion mode, the lowest detection limits involves the use of formic acid 0.5%. Also, negative ion mode results in limited fragmentation, which is particularly suited for molecular mass determination, especially when compounds' concentration is low [13].

In the advent of formation of adducts with solvent or acid molecules, or even molecular complexes, the peak at the highest m/z ratio may not be the molecular ion species ($[M+H]^+$ in positive mode and $[M-H]^-$ in negative mode). Instead, $[2M+H]^+$ or $[2M-H]^-$ may be formed. This issue may be corrected by an increase in cone voltage, which diminishes formation of both adducts and complexes [14].

While negative mode ion is very useful to identify known compounds, the first-order mass spectrum yielded by positive mode can provide more structural information. So, the combined use of both ionization modes can give additional accuracy to the molecular mass determination, which is particularly relevant when noise levels are high [12].

Besides spectroscopic data, chromatographic retention times can also add further knowledge on the compounds chemistry. In a general way, for C18- or C8-reversed phase columns, more polar compounds elute first. Thus, increasing glycosylation results in retention time's decrease. Differently, acylation, methylation or prenylation have a distinct effect, rising retention time, although the position where this occurs may play a significant role on chromatographic behavior.

Nuclear Magnetic Ressonance (NMR)

NMR is probably the analytical method providing the most comprehensive structural information, including stereochemical details, which are key attributes in the complete identification and characterization of molecules.

The theory behind NMR is that molecules containing at least one atom with a none zero magnetic moment are potentially detectable by NMR, with the isotopes with a non zero moment including ^1H, ^{13}C, ^{14}N, ^{15}N and ^{31}P. As so, in a biological context, virtually all molecules would generate, at least, one NMR signal. These signals are characterized by their frequency (chemical shift), intensity, fine structure and magnetic relaxation properties, all of which reflect the precise environment of the detected nucleus. [15].

When working in on-line flow NMR, the acquisition time is limited by the short presence of the sample in the detection coil, as a consequence of the flow rates commonly used, thus resulting in poor signal-to-noise ratio values. The on-line mode is the simplest because it does not require any synchronization between the HPLC and the NMR system, maintaining good HPLC resolution, but it has the lowest sensitivity. LC-NMR spectra are acquired continuously during the separation and are stored as a set of scans as discrete increments. The on-line data are processed as for a 2-D NMR experiment: one dimension of this plot represents the NMR ppm scale whilst the other represents the time scale.

Also, when flow rate is reduced by a factor of 3-10, better signal/noise can be registered, followed by an increase in experimental acquisition time, which may lead to diffusion processes that can influence the separation of peaks eluting from the LC column. In order to surpass this problem, accumulation of peaks into storage loops for off-line NMR at a later stage has been proposed [16].

Sensitivity is perhaps the most important requirement for metabolomics. Here, ^1H NMR, with a detection threshold of perhaps 5 nmol, is several orders of magnitude less sensitive than MS, which has a detection threshold of 10^{-12} mols. However, in this field there has been increasing success in improving NMR sensitivity, for instance by the use of cryoprobes. Cryogenic probe heads, in which the sensitivity is increased by cooling the detection system, offers the prospect of a substantial improvement in the detection of signals that are at the limit of detection in conventional probe heads. Cryogenic probe heads are mainly used to record spectra from macromolecules, but they are also suitable for metabolic analysis and the first results with these probe heads confirm that they can deliver substantial gains in sensitivity [15]. The procedure consists on cooling the receiver coil to cryogenic temperatures, while the sample remains at ambient temperature. This limits the noise voltage associated with signal detection and, when compared with regular probes, signal-to-noise ratio is ameliorated by a factor of 3-4.

One unavoidable advantage of NMR is that it is not a destructive technique, meaning that the same sample can be further analyzed with other techniques after NMR analysis. Although 1D NMR studies are extremely useful in classifying similar groups of samples, problems with large numbers of overlapping peaks can make actual identification of large numbers of metabolites difficult. 2D NMR studies can help to overcome these problems. The use of 2D NMR for metabolomics is usually restricted to the characterization of unidentified compounds from the 1D spectrum [8].

SELECTED EXAMPLES FROM LITERATURE

Metabolite Profiling in *C. Roseus*: Phenolics

Catharanthus roseus (L.) G. Don (formerly *Vinca rosea* L., Apocynaceae) is commonly known as the Madagascar periwinkle and was originally an endemic subshrub species of Madagascar. The leaves of *C. roseus* were used in traditional medicine as an oral hypoglycemic agent and the study of this activity led to the discovery of two terpenoid indole alkaloids (TIA), vinblastine and vincristine [17], the first natural anticancer agents to be clinically used and, since they are present in very low levels on *C. roseus* leaves, the TIA pathway has been intensively investigated [18, 19]. The screening of phenolic compounds of seeds, stems, leaves and petals of *C. roseus* (cv. Little Bright Eye) was achieved by HPLC-DAD-ESI-MS/MS [20]. This work is classified as a metabolite profiling rather than targeted metabolite analysis, due to the applied analytical technique, which would allow the detection of several classes of phenolics, should they be present, although only phenolic acids and flavonoids were detected.

The HPLC-DAD-ESI-MS/MS screening of the hydroalcoholic extracts of *C. roseus* material revealed the presence of numerous flavonoids (compounds 2 and 5-18) (Fig. **3**), whose UV spectra were typical of flavonol-3-*O*-glycosyl derivatives [21]. In their MS (MS2 or MS3) fragmentations ions were observed at m/z 285, 300/301 or 315 with high abundance (base peak), corresponding to the deprotonated aglycon ions of kaempferol, quercetin or isorhamnetin, respectively (Fig. **3A** and **3B**, Table **1**). Deprotonated molecular ions at m/z 755 (compound 7) and 739 (compound 9) and MS2 base peak ions at m/z 300 and 285 respectively, pointed to quercetin and kaempferol triglycosides with two rhamnoses and one hexose. This MS2 fragmentation type, in which the base peak corresponds to the deprotonated aglycon ion, indicates that the triglycoside is linked only to one phenolic hydroxyl [22]. Thus, these compounds were identified as quercetin-3-*O*-(2,6-di-*O*-rhamnosyl-galactoside) (7) and kaempferol- 3-*O*-(2,6-di-*O*-rhamnosyl-galactoside) (9), already detected in *C.roseus* leaves [23] and stems [24]. The MS2 fragmentation of these compounds was in accordance with the proposed structures. As discussed above, the base peak corresponds to the deprotonated aglycon ion ($[Y^3_0]^-$), as expected for flavonoid glycosides with one substitution. On the other hand, it can be observed that the fragmentation of the rhamnose in the 2″ position gives rise to the ions $[Y^3_{2''}]$- ($[(M- H) - 146]^-$) and $[Z^3_{2''}]$- ($[Y^3_{2''}-18]^-$, $[(M - H) - 164]^-$) [25] (Fig. **2**). In some cases the $[Z^3_{2''}]^-$ ion exhibits a very high abundance, even being the base peak, but the simultaneous loss of rhamnosyl radical and water indicates an interglycosidic bond, and not a link to phenolic hydroxyl [26, 27]. Other obvious peaks resulted from the internal cleavage of galactose to originate the ion $[^{0,2}X^3_0]$-, a fragment that preserves the rhamnose linked at 2″ position (Fig. **2**), and $[^{0,2}X^3_0-146]$- ion, which lost the rhamnose at the 2″ position. These compounds were also present in seeds and petals: compound 9 was the main compound of the seeds and very abundant in petals, while compound **7** was important in the seeds but vestigial in the petals (Fig. **3**). In addition, compounds 8, 10 and 14, isomers of 7, 9 and 12, respectively, and displaying the same MS2 fragmentation were detected. The longer retention time in reversed phase HPLC of 8, 10 and 14 relative to their isomers (Fig. **3**) indicates that glucose could be the hexose, derivatives of which elute after those of galactose [28]. These compounds were therefore tentatively characterized as quercetin-3-*O*-(2,6-di-*O*rhamnosyl-glucoside) (8), kaempferol-3-*O*-(2,6-di-*O*-rhamnosylglucoside) (10) and isorhamnetin-3-*O*-(2,6-di-*O*-rhamnosylglucoside) (14). Compound 8 was detected solely in the petals and in trace amounts. Six rhamnohexoside derivatives of quercetin, kaempferol and isorhamnetin (compounds 11, 13, 15-18), were also detected. In their MS2 fragmentation mainly the deprotonated aglycon ion was observed, which indicates a 1→6 interglycosidic linkage [26]. These data, in conjunction with the reversed phase HPLC chromatographic behavior, suggest these compounds to be similar to the triglycosides without rhamnose in the 2″ position previously mentioned. Thus, quercetin-3-*O*-(6-*O*-rhamnosyl-galactoside) (11), quercetin-3- *O*-(6-*O*-rhamnosyl-glucoside) (13), kaempferol-3-*O*-(6-*O*-rhamnosyl- galactoside) (15), kaempferol-3-*O*-(6-*O*-rhamnosyl-glucoside) (16), isorhamnetin-3-*O*-(6-*O*-rhamnosyl-galactoside) (17) and isorhamnetin-3-*O*-(6-*O*-rhamnosyl-glucoside) (18) were tentatively identified. Compounds 15 and 18 were found in small amounts in the seeds, while petals contained compounds 11, 13, 15-18, compound 15 being the most abundant in this material (Fig. **3**). Three compounds (2, 5 and 6) exhibited a MS2 fragmentation in which the ion corresponding to the loss of 162 u from the [M - H]⁻ was only seen. In the MS3[(M - H) → (M - H - 162)]⁻ event of compound 2 the fragmentation was similar to that of the triglycosides referred to above, whereas the MS3 of compounds 5 and 6 resembles that of the diglycosides group. These data indicate that these compounds are derivatives of the previous ones, containing an additional glycosylation at the 7 position with hexose [22]. For biosynthetic reasons, kaempferol-3-*O*-(2,6-di-*O*-rhamnosyl-galactoside)-7-*O*-hexoside (2) must proceed from 9, which is the most abundant in seeds. Compounds 5 and 6 can derive from 15 and 16, respectively, the main phenolics in petals, although in *Vinca minor* kaempferol- 3-*O*-(6-*O*-rhamnosyl-glucoside)-7-*O*-glucoside was found [23]. Thus, compounds 5 and 6 are isomers of kaempferol-3-*O*-(6-*O*-rhamnosyl-hexoside)-7-*O*-hexoside, tentatively kaempferol-3-*O*-(6-*O*-rhamnosyl-galactoside)-7-*O*-galactoside   (5)   and   kaempferol-   3-*O*-(6-*O*-rhamnosyl-galactoside)-7-*O*-glucoside (6).

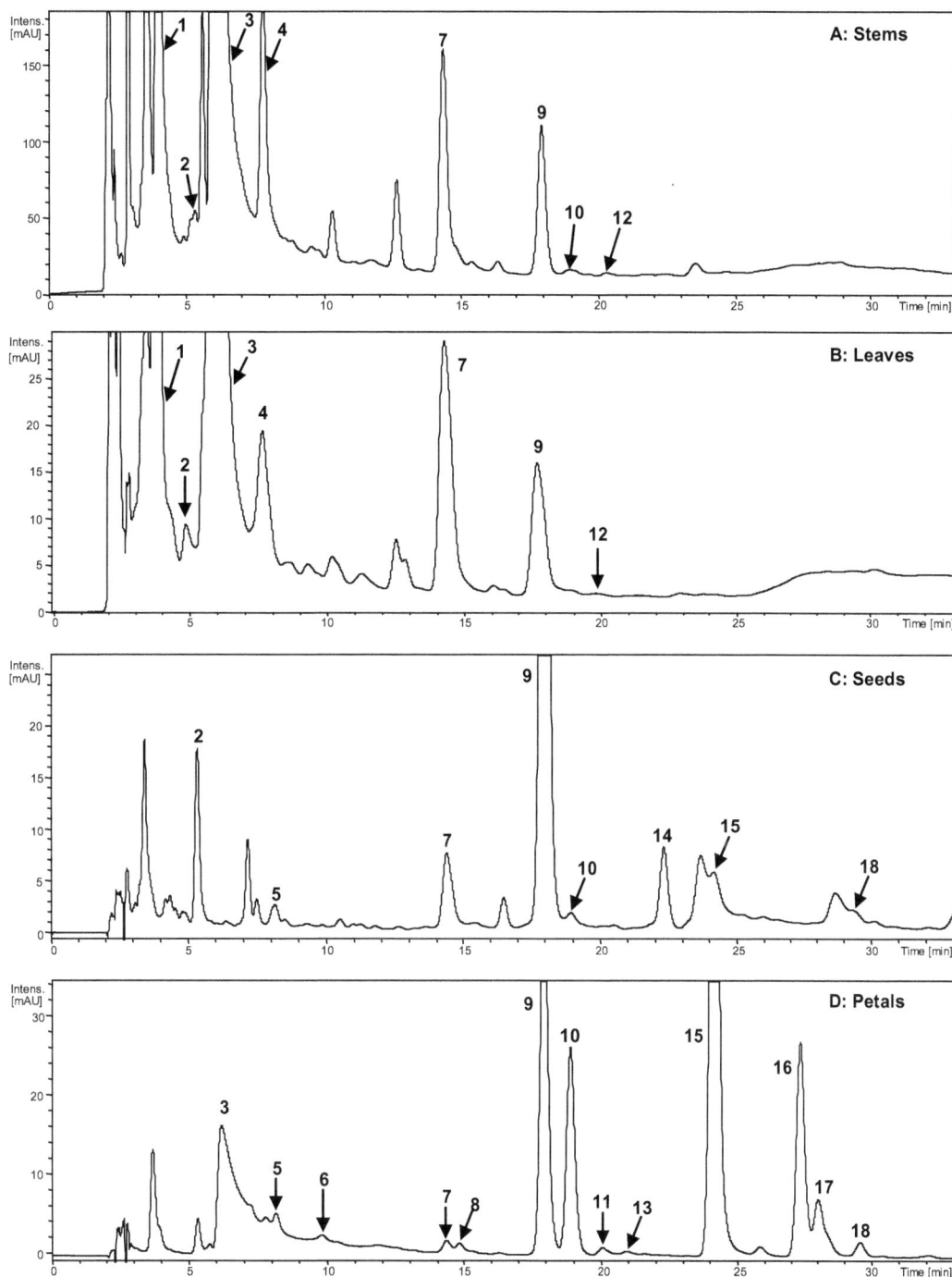

Figure 3: HPLC-DAD chromatogram of *Catharanthus roseus* extracts: (A) Stems; (B) Leaves; (C) Seeds; (D) Petals. Detection at 340 nm. (1) 3-*O*-caffeoylquinic acid; (2) kaempferol-3-*O*-(2,6-di-*O*-rhamnosyl-galactoside)-7-*O*-hexoside; (3) 4-*O*-caffeoylquinic acid; (4) 5-*O*-caffeoylquinic acid; (5) kaempferol-3-*O*-(6-*O*-rhamnosyl-galactoside)-7-*O*-galactoside; (6) kaempferol-3-*O*-(6-*O*-rhamnosyl-galactoside)-7-*O*-glucoside;   (7)   quercetin-3-*O*-(2,6-di-*O*-rhamnosyl-galactoside);   (8) quercetin-3-*O*-(2,6-di-*O*-rhamnosyl-glucoside); (9) kaempferol-3-*O*-(2,6-di-*O*-rhamnosyl-galactoside); (10) kaempferol-3-*O*-(2,6-di-*O*-rhamnosyl-glucoside);   (11)   quercetin-3-*O*-(6-*O*-rhamnosyl-galactoside);   (12)   isorhamnetin-3-*O*-(2,6-di-*O*-rhamnosyl-galactoside);   (13)   quercetin-3-*O*-(6-*O*-rhamnosyl-glucoside);   (14)   isorhamnetin-3-*O*-(2,6-di-*O*-rhamnosyl-glucoside);   (15)   kaempferol-3-*O*-(6-*O*-rhamnosyl-galactoside);   (16)   kaempferol-3-*O*-(6-*O*-rhamnosyl-glucoside);   (17) isorhamnetin-3-*O*-(6-*O*-rhamnosyl-galactoside); (18) isorhamnetin-3-*O*-(6-*O*-rhamnosyl-glucoside). From [20] (with kind permission of American Chemical Society).

Table 1: tR, UV, and -MS data for flavonol glycosides from hydroalcoholic extracts of *Catharanthus roseus*[a]. From [20] (with kind permission of American Chemical Society).

Flavonol-3-O-(2,6-di-O-rhamnosyl-galactoside)-7-O-hexoside

Compounds[b]	R_t (min)	UV (nm)	[M-H]- (m/z)	-MS2[M-H]- (m/z) (%)	-MS3[(M-H)→(M-H-162)]- (m/z) (%)				
				Y^7_0- (-162)	$Y^7_0 Y^3_{2'}$- (-146)	$Y^7_0 Z^3_{2'}$- (-164)	$Y^7_0 \,^{0,2}X^3_0$ (-266)	$[Y^7_0 \,^{0,2}X^3_0$-146]- (-412)	$Y^7_0 Y^3_0$- Aglc-H/2H
2 K-3-(2,6-Rh-Gal)-7-Hx	5.3	265,219sh,347	901	739(100)	593(30)	575(40)	473(20)		285(100)
5 K-3-(6-Rh-Gal)-7-Gal	8.2	265,347	755	593(100)				327(16)	285(100)
6 K-3-(6-Rh-Gal)-7-Glc	9.8	---[c]	755	593(100)					285(100)

Flavonol-3-O-(2,6-di-O-rhamnosyl-hexoside)

Compounds	R_t (min)	UV (nm)	[M-H]- (m/z)	-MS2[M-H]-, (m/z) (%)				
				$Y^3_{2'}$- (-146)	$Z^3_{2'}$- (-164)	$^{0,2}X^3_0$ (-266)	$[^{0,2}X^3_0$-146]- (-412)	Y^3_0- Aglc-H/2H
7 Q-3-(2,6-Rh-Gal)	14.2	255,267sh,299sh,353	755	609(17)	591(36)	489(18)	343(35)	300(100)
8 Q-3-(2,6-Rh-Glc)	14.8	255,267sh,299sh,354	755	609(17)	591(44)	489(14)	343(10)	300(100)
9 K-3-(2,6-Rh-Gal)	17.9	265,296sh,347	739	593(15)	575(44)	473(10)	327(23)	285(100)
10 K-3-(2,6-Rh-Glc)	18.9	265,299sh,348	739	593(20)	575(60)	473(18)	327(10)	284(100)
12 I-3-(2,6-Rh-Gal)	20.2	---[c]	769	623(22)	605(26)	502(5)	357(6)	315(100)
14 I-3-(2,6-Rh-Glc)	22.2	255,267sh,300sh,354	769	623(12)	605(9)	502(1)	357(12)	315(100)

Flavonol-3-O-(6-O-rhamnosyl-hexoside)

Compounds	R_t (min)	UV (nm)	[M-H]- (m/z)	-MS2[M-H]-, (m/z) (%)	
				$^{0,2}X^3_0$ (-266)	Y^3_0- Aglc-H/2H
11 Q-3-(6-Rh-Gal)	20.0	---[c]	609	343(10)	300(100)
13 Q-3-(6-Rh-Glc)	20.9	---[c]	609		301(100)
15 K-3-(6-Rh-Gal)	24.1	265,295sh,347	593	327(10)	285(100)
16 K-3-(6-Rh-Glc)	27.3	265,295sh,347	593		285(100)
17 I-3-(6-Rh-Gal)	27.9	255,266sh,301sh,355	623		315(100)
18 I-3-(6-Rh-Glc)	29.6	255,266sh,301sh,355	623		315(100)

[a] Main observed fragments. Other ions were found but they have not been included. [b] Q: quercetin. K: kaempferol. I: isorhamnetin. Hx: hexoside. Gal: galactoside. Glc: glucoside. Rh: rhamnoside. [c] Compounds hidden by others or in traces. Their UV spectra have not been properly observed.

In addition to the flavonoids described above, phenolic acids were detected: 3-O-caffeoylquinic acid (1) (tR, 3.9 min; UV, 299sh, 325 nm; MS, 353 [M - H]-; MS2[M - H]-, 191 (100), 179 (48)), 4-Ocaffeoylquinic acid (3) (tR, 5.9 min; UV, 299sh, 325 nm; MS, 353 [M - H]-; MS2[M - H]-, 191 (80), 179 (60), 173 (100)) and 5-O-caffeoylquinic acid (4) (tR, 7.7 min; UV, 299sh, 325 nm; MS, 353 [M - H]-; MS2[M - H]-, 191 (100), 179 (2)) in stems and leaves and compound 3 in petals (Fig. **3**), according to the method of Clifford and colleagues [29]. 5-O-Caffeoylquinic acid was detected in *C. roseus* leaves [30]. The results show that leaves and stems are particularly rich in caffeoylquinic acids in comparison with seeds and petals, while these later mainly present a great variety of flavonoids. This may relate with the proposed functions of caffeoylquinic acids as protectors against herbivorism and infection [31], a function particularly relevant for the organs involved in vegetative growth (stems and leaves). On the other hand, the abundance of flavonoids in petals and seeds may be associated with the frequently reported function of this group of compounds in the attraction of pollinators and seed dispersers [32].

Metabolite Profiling in *C. Roseus*: Volatiles

Guedes de Pinho and colleagues [33] performed further metabolomic studies on *C. roseus*, this time focusing on this species' volatile compounds. By using GC-MS and data analysis with PCA, the different plant parts could be distinguished be the means of their volatile composition. Fresh plant and aqueous lyophilized extract were subjected to extraction with organic solvent (dichloromethane) and Head Space – Solid Phase Microexctraction (HS-SPME) for analyzing volatile, semi-volatile and non-volatile compounds in *C. roseus* plant material. After several tests in order to choose the most suitable fibre, Divinylbenzene/Polydimethylsiloxane (DVB/PDMS) one was selected.

SPME allowed the determination of 12 aldehydes, 14 alcohols, 9 esters, 10 nitrogen containing compounds, 17 terpenic compounds (including aliphatic mono and diterpenes), 12 carotenoid derivatives, 6 ketones, 1 hydoxycinnamic acid and 3 phenol compounds. Some of them were present in flowers, leaves and stems, while others existed only in a certain organ of the plant as it is shown in Fig. **4**.

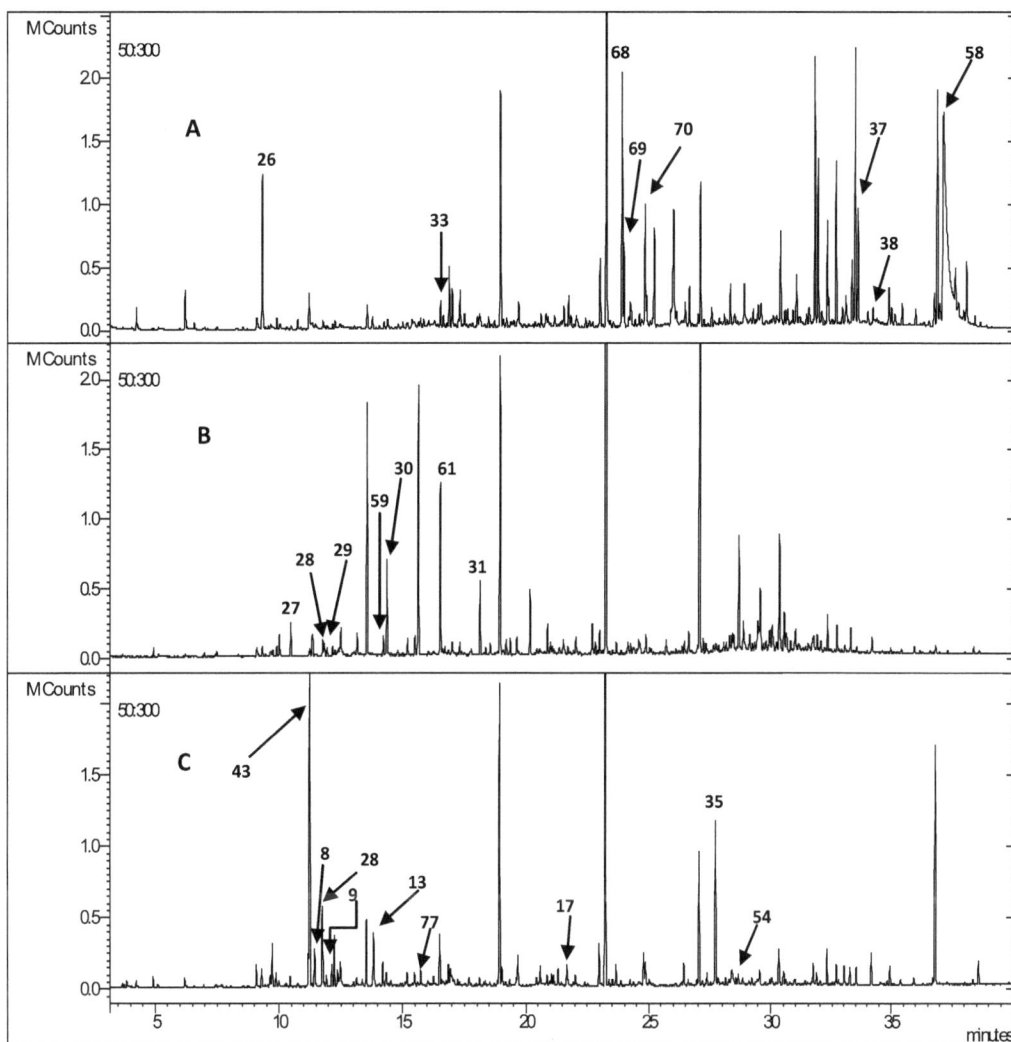

Figure 4: Chromatograms of the SPME using DVB/PDMS fibre analysis in leaves (A), stems (B) and flowers (C) of *C. roseus*. Compound names : 8 – benzyl alcohol, 9 – 1-phenylethanol, 13 – 2-phenylethanol, 17 – *cis*-jasmone, 26 – benzaldehyde, 27 – octanal, 28 – phenylacetaldehyde, 29 – *trans*-octenal, 30 – *cis*-nonenal, 31 – *cis*-2-decenal, 33 – ethylhexanoate, 35 – methyljasmonate, 37 – palmitic acid methyl ester, 28 – palmitic acid ethyl ester, 43 – limonene, 54 – α-bisabolol, 58 – *trans*-phytol, 59 – *cis,trans*-2,6-nonadienal, 61 – *trans,cis*-2,4-decadienal, 68 – β-ionone, 69 – 2,3-epoxy-α-ionone, 70 – dihydroactinolide, 77 – 2-isobutyl-3-methoxypyrazine. From [33] (with kind permission of Elsevier).

Flowers were richer in phenylacetaldehyde and in the correspondent alcohol, 2-phenylethanol, than the other plant organs. 2-Phenylethanol is responsible for the rose-note aroma. These two molecules have an important biological function in plants. The latter has long been known to possess antimicrobial properties [34] and its synthesis by plant reproductive structures may indicate a protective role for flowers and fruits. Both 2-phenylacetaldehyde and 2-phenylethanol are also potent insect attractants [35, 36]. Flowers also exhibited high amounts of mono- and diterpenic compounds. These molecules have been found before only in the essential oil of the leaves of *C. roseus* [37]. Among monoterpenes, it can be highlighted their high limonene amounts. Monoterpenes are, among the most volatile compounds, those with more pleasant aroma descriptors. Diterpenic compounds, such as α-bisabolol and manool and their oxide compounds were found in low levels in the headspace of flowers and leaves. Bisabolol is equally a constituent of the essential oil from German chamomile (*Matricaria recutita*) and *Myoporum grassifolium*. It has a weak sweet floral aroma and is used in various fragrances. Also, it has been used for hundreds of years in cosmetics because of its perceived skin healing properties, also presenting anti-bacterial and anti-fungal activities [38]. *C. roseus* flowers were also rich in methyl jasmonate. Jasmonates are a group of plant stress hormones [39]. Upon exposure to stress (e.g., wounding and pathogens), jasmonates are produced in plants and cause the induction of a proteinase inhibitor [40]. A coordinated activation of programmed cell death and defense mechanisms often accompany the antimicrobial response of plants [41]. In addition, jasmonates can suppress the proliferation of human cancer cells and induce their death. Methyl jasmonate induced death in breast and prostate carcinoma cells, as well as in melanoma, lymphoma, and leukemia cells [42, 43]. It is a chemical inducer of secondary metabolism and it was demonstrated that methyl jasmonate increased the activity of tabersonine epoxidase in hair root cultures of *C. roseus* [44]. It can act as either an attractant or a repellent for various insects [43]. The volatile profile of leaves also comprised different classes of compounds. Among aldehydes and alcohols, high amounts of benzaldehyde and *cis*-hex-3-en-1-ol, could be noted, respectively. Leaves also presented high levels of carotenoid derivatives, such as β-ionone, 2,3-epoxy-β-ionone, β-cyclocitral and dihydroactinilidiolide. All these molecules are degradation products of carotenoids, such as carotene and lutein [45]. β-Ionone and β-cyclocitral are known to be important contributors to the flavor aroma of several fruits and wines [45, 46, 47]. Additionally, leaves contained important amounts of esters compounds (12.8%), including isopropyl and methyl esters of fatty acids, di- and trisaturated.

Among terpenic compounds, leaves were richer in *trans*-geranylcetone and *trans*-phytol. Phytol is a natural linear diterpene alcohol which is involved in the synthesis of vitamins E and K1. It is also a decomposition product of chlorophyll [48]. γ-Decalactone was another compound present in high amounts in the headspace of leaves.

A great variety of volatile nitrogen containing compounds was also found in leaves, namely pyridine and pyrazine, and thiazole compounds. *C. roseus* is a plant known for the presence of important alkaloids with recognized health value, namely anticancer activity [49]. These alkaloids are nitrogen compounds with l-tryptophane as precursor, with a known biosynthetic pathway.

No data could be found in the literature concerning volatiles of *C. roseus* stems. Stems showed higher levels in particular compounds, namely 2-isobutyl-3-methoxypyrazine. Pyrazine compounds are heterocyclic nitrogen containing compounds with unique organoleptic properties. Methoxypyrazines (MP) are very potent odorants and have a distinctive smell, similar to freshly cut green bell pepper or green peas. Human olfactory thresholds for MP are extremely low, in the range of 2 ng/L in water [50], and some recent works attribute some antimicrobial properties to pyrazines [51].

Considering ketones, particular attention may be focused on the 6-methyl-5-hepten-2-one, found in all parts of plant, but with higher contents in stems. This compound has been reported to be an oxidative by-product or degradation product derived fromlicopene, farnesene, citral or conjugated tri-enols [52, 53].

Dichloromethane extraction allowed the identification of 14 additional compounds. Among these compounds some alkaloid-like molecules could be identified, as well as 2,2,7,7-tetramethyltricyclo(6.2.1.0(1,6))undec-4-en-3-one, α-farnesene, and phytol (3,7,11,15-tetramethyl-2-hexadecen-1-ol). 2,2,7,7-Tetramethyltricyclo(6.2.1.0(1,6))undec -4-en-3-one was recently reported as one of the major constituents of the essential oil of *Aristolochia mollissima* [54], which has been proved to have antimicrobial activity and cytotoxicity against four cancer cell lines (ACHN, Bel-7402, Hep G2 and HeLa).

Stems were richer in alkaloid compounds than leaves and flowers. In order to assemble the different identified compounds according to the organ of the plant (leaves, stems and flowers), a principal component analysis (PCA) was performed, using the results obtained from the HS-SPME analysis. Fig. **5** shows the projection of chemical variables (sum of compounds of each chemical family) into the plans F1 and F2. Three distinct groups have been formed. Succinctly, flowers were richer in terpene molecules (including limonene), aldehyde compounds, esters compounds, namely methyljasmonate, and phenols (due to the high amounts of eugenol). Leaves were well correlated to carotenoid derivatives, sum of ketones and ester compounds. Finally, stems were in good correlation with nitrogen containing compounds, alcohol and miscellaneous compounds. In order to select, among all volatiles, those which could be markers of each organ of the plant, an agglomerative hierarchic cluster analysis (HCA) was performed. By this way it was possible to restrict the volatiles to nine compounds. Leaves could be characterized by their levels in hexanol, benzaldehyde, palmitic acid methyl ester and *trans*-phytol, flowers by their contents in 1-phenylethanol, limonene and other terpenes, and, finally, stems by their α-ionone and *trans*-2-decen-1-ol amounts.

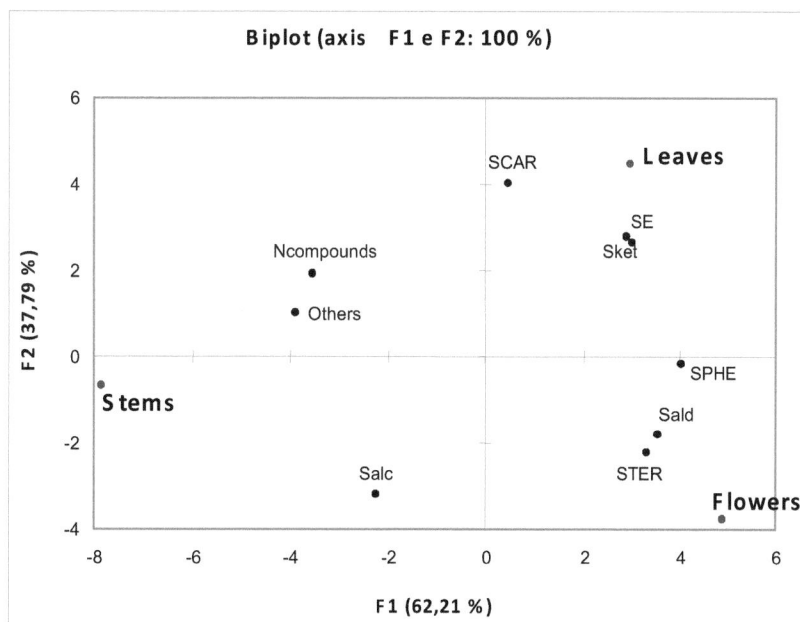

Figure 5: Principal component analysis of all volatiles compounds analyzed by HS-SPME-GC-MS grouped by family classes in flowers, stems and leaves. SCAR – Sum of carotenoid molecules, Sket – Sum of ketones, SPHE – Sum of phenols, SE- Sum of esters compounds, Salc – Sum of alcohols, Sald – Sum of aldehydes, STER – Sum of terpenes, Ncompounds – nitrogen containing compounds. From [33] (with kind permission of Elsevier).

METABOLOMICS IN CHEMICAL ECOLOGY

Phenolics Metabolization by *Pieris Brassicae*

Metabolomics can also be a valuable tool for the study of ecosystems, especially insect plant-interactions. The knowledge of the chemical profile of insects and their host plants may provide important leads towards the understanding of the chemical responses of plants upon predation, or the metabolic fate of the compounds ingested by the insect.

In the work of Ferreres *et al.* [55], the phenolic profiles of aqueous extracts of several materials (larvae, excrements, butterfly) of *Pieris brassicae* insect and one of its host plant, kale (*Brassica oleracea* L. var. *acephala*) leaves, were determined by HPLC/UV–DAD/MSn-ESI.

The principal phenolics in *P. brassicae* excrement (Fig. **6**) were coincident with the main ones in kale leaves [55]. The non-acylated glycosyl flavonols 2, 6, 7, 10, 11, 23, 27, and 30 were detected (Fig. **6**). Notably, the last of these was the most abundant, while in kale leaves it was found in small amounts. Its higher concentration may be due to deglycosylation at C-7 of 7 and of other kaempferol-3-*O*-sophoroside-7-*O*-glycosides, like 10 and 11, as well as from deacylation and deglycosylation of acyl derivatives of the previous two compounds (4, 8, and 16–21). Other non-acylated glycosides present in excrement, and not detected in kale leaves, were kaempferol (38 and 44) and isorhamnetin (40 and 46) derivatives, which should also arise from deglycosylation at C-7 of 6/10, 12/15, 22/23, and 24, respectively. Kaempferol-3-*O*-glucoside (45) was also detected. Regarding glycosylflavonol acylated derivatives common to kale leaves, compounds 16–19 were detected, the most abundant in the native extract of the leaves. Other glycosylflavonol acylated derivatives not found in kale were kaempferol- 3-*O*-(acyl)sophorotrioside with sinapic (37), ferulic (39) and *p*-coumaric acid (41), and acylated derivatives of kaempferol-3-*O*-sophoroside with ferulic and *p*-coumaric acid (47 and 48, respectively). These compounds should also result from deglycosylation at C-7 of kaempferol-3-*O*-(acyl)sophorotrioside/ sophoroside-7-*O*-glycosides. Sulphate derivatives were also noted: kaempferol-3-*O*-sophoroside sulphate (36), quercetin-3-*O*-glucoside sulphate (42) and several kaempferol-3-*O*-glucoside sulphate derivative isomers (Rt 32.5–33.5 min), which were assigned together as 43 (Fig. **6**). The MS of these sulphated derivatives showed a loss of 80 amu to yield the base peak. In general, and as before [56], *P. brassicae* metabolic processes involve deglycosylation at C-7, deacylation and sulphation. Assuming that the majority of sulphated derivatives are monoglucosides, a second deglycosylation step also occurs [56].

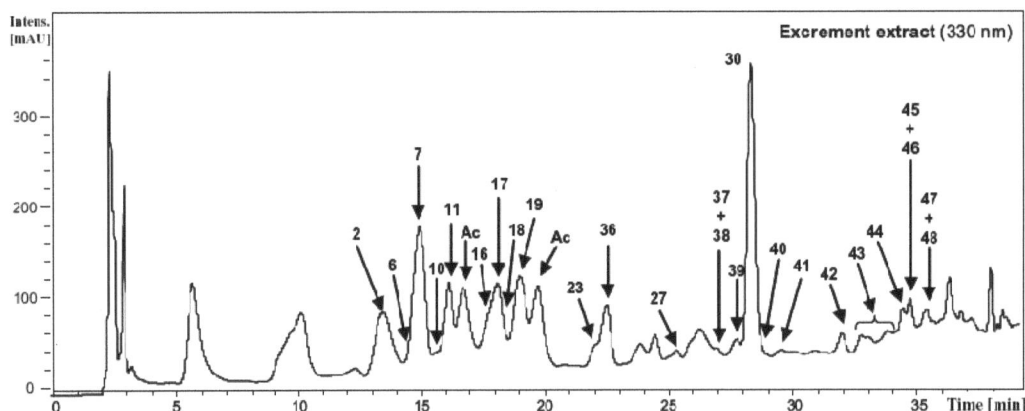

Figure 6: HPLC/UV-DAD phenolic profile of *P. brassicae* excrement aqueous extract. Detection at 330 nm. Peaks: (Ac) acylated derivatives; (2) quercetin-3-*O*-sophoroside-7-*O*-glucoside, (6) kaempferol-3-*O*-sophorotrioside-7-*O*-glucoside, (7) kaempferol-3-*O*-sophoroside-7-*O*-glucoside, (10) kaempferol-3-*O*-sophorotrioside-7-*O*-diglucoside, (11) kaempferol-3-*O*-sophoroside-7-*O*-diglucoside, (16) kaempferol-3-*O*-(sinapoyl)sophoroside-7-*O*-glucoside, (17) kaempferol-3-*O*-(sinapoyl)sophoroside-7-*O*- diglucoside, (18) kaempferol-3-*O*-(feruloyl)sophoroside-7-*O*-glucoside, (19) kaempferol-3-*O*-(feruloyl)sophoroside-7-*O*- diglucoside, (23) kaempferol-3-*O*-gentiobioside-7-*O*-diglucoside, (27) quercetin-3-*O*-sophoroside, (30) kaempferol-3-*O*-sophoroside (36) kaempferol-3-*O*-sophoroside sulphate, (37) kaempferol-3-*O*-(sinapoyl)sophorotrioside, (38) kaempferol-3-*O*-sophorotrioside, (39) kaempferol-3-*O*-(feruloyl)sophorotrioside, (40) isorhamnetin-3-*O*-sophoroside, (41) kaempferol-3-*O*-(p-coumaroyl)sophorotrioside, (42) quercetin-3-*O*-glucoside sulphate, (43) kaempferol-3-*O*-glucoside sulphate, (44) kaempferol-3-*O*-gentiobioside, (45) kaempferol-3-*O*-glucoside, (46) isorhamnetin-3-*O*-gentiobioside, (47) kaempferol-3-*O*-(feruloyl)sophoroside (isomer), and (48) kaempferol-3-*O*-(p-coumaroyl)sophoroside. From [55] (with kind permission of Elsevier).

The HPLC/UV–DAD/MSn-ESI analysis of both *P. brassicae* larvae and butterflies revealed phenolics in very low amounts. To check for the occurrence of flavonoid derivatives, HPLC-MSn was used by extracting the MSn ions at m/z 284–285, 300–301 and 314–315 (extracted ion chromatogram, EIC), whose presence would lead to kaempferol, quercetin and isorhamnetin glycosides, respectively. The ions presenting a loss of 80 amu during their fragmentation were also extracted, since this may indicate the presence of sulphate derivatives (Constant Neutral Loss Chromatogram). Compounds 27, 30, 36, and 43 were detected in the larvae, while in the butterfly, no flavonoid derivative was found. This study provided evidence that the larvae sequesters and metabolizes kale's phenolic compounds, namely through deacylation, deglycosylation and sulphating reactions.

Changes in *B. Oleracea.* Var *Costata* After *P. Brassicae* Predation

The role of volatile compounds in shaping insect-plants relations is a relatively new area of research which has known a great impulse over the last few years.

In plants, there is a constitutive emission of volatile compounds that are released from the surface of the leaf and/or accumulated in storage sites. Terpenoids constitute the most important group of volatiles that are emitted by plants, consisting predominantly of monoterpenes, sesquiterpenes and their derivatives, homoterpenes. These volatiles play different roles in herbivore elimination, either by attraction of parasitoids that increase herbivore mortality (indirect defense) or by directly reducing herbivores. However, some environment stimuli, such as feeding [57] or oviposition [58], can change both qualitatively and/or quantitatively the blend of volatile constituents [59].

The influence of *P. brassicae* feeding on kale was monitored by Fernandes and colleagues [60], by evaluating its effect in the volatiles released by the plant through time. It should be highlighted that the profile of the insect itself was analysed, and for that purpose, HS-SPME was performed directly into the alive specimen, as showed in Fig. **7**.

Several chemical classes of compounds could be found in kale, prior and after the insect's attack. The same occurred between the insect and the plant (Fig. **8**).

With the exception of aldehydes, all classes of compounds were found before and after insect's attack; however, some compounds of each class could be detected only after insect feeding, mainly terpenes. Compounds such as α-thujene, sabinene, β-pinene, psi-cumene, *m*-cymene, *o*-cymene, *p*-cymene, *l*-camphor, longifolene and geranylcetone are examples of this situation.

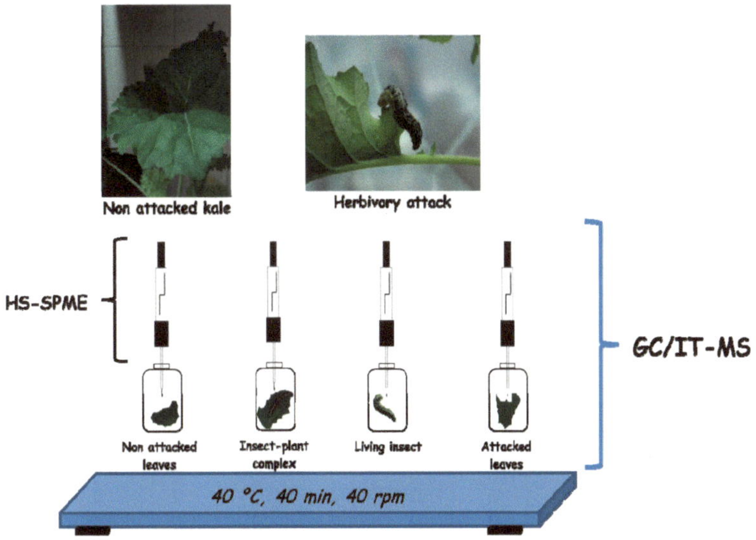

Figure 7: Experimental design of *P. brassicae*/*B. oleracea* var. *acephala* analysis.

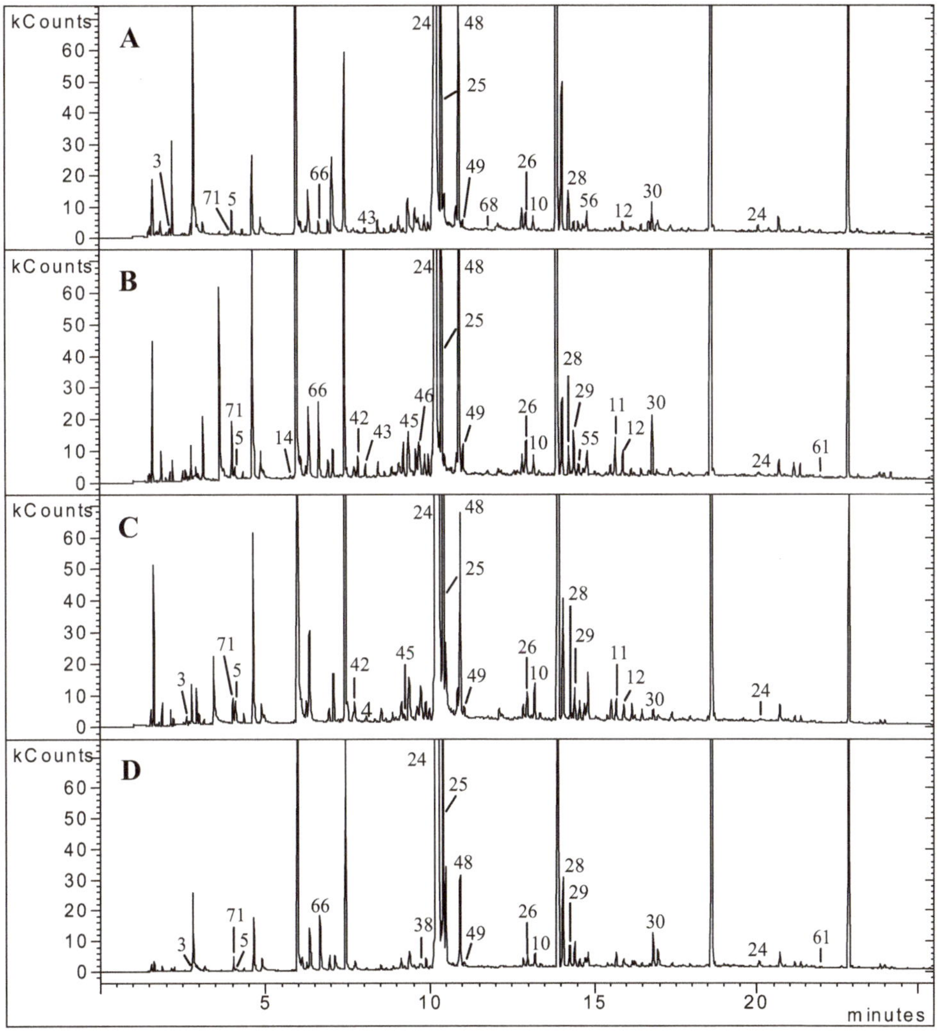

Figure 8: Chromatographic profile of HS-SPME combined with GC/IT-MS using Divinylbenzene/PDMS fibre. Non attacked kale (A), kale after 4 hours (B) and after 24 hours of insect's attack (C) and kale after mechanical damage (D). Identity of compounds: (3) 1-Penten-3-ol; (5) (*Z*)-2-Penten-1-ol; (10) (*E*)-2-Nonen-1-ol; (11) Undecanole; (12) (E)-2-Decen-

1-ol; (14) (*E*)-2-Hexenal;24(*Z*)-3-Hexenylacetate; (25) Acetic acid, hexyl ester; (26) Propanoic acid, 4-hexen-1-yl estere; (28) Butanoic acid,4-hexen-1-yl ester; (29) 2-Ethylhexyl acetate; (30) Pentanoic acid, 4-hexen-1-yl ester; (38) 6-Methyl-5-heptene-2-one; (42) α-Thujene; (43) α-Pinene; (45) Sabinene; (46) β-Pinene, (48) Limonene; (49) Eucalyptol; (55) l-camphore; (56) ρ-Menthone; (61) *β*-Caryophyllene; (66) Allyl Isothiocyanate; (68) 2-Methylbutyl isothiocyanate; (71) Toluene. From [60] (with kind permission of Elsevier).

Terpenes were the class more affected by predation. After 1 hour of insect attack, kale terpenes' amount had increased by over 315%. Although there was a tendency for this quantity to decrease through time (Fig. **9**), after 24 hours their amounts were still ca. 90 % higher than those prior to the attack. After 1 hour, alcohols had decreased by about 30% and aldehydes, that were absent in non attacked leaves, appeared, with hexanal, (*E*)-2-hexenal and heptanal being detected. Hexanal was the only aldehyde that could be detected in kale after 24 hours predation.

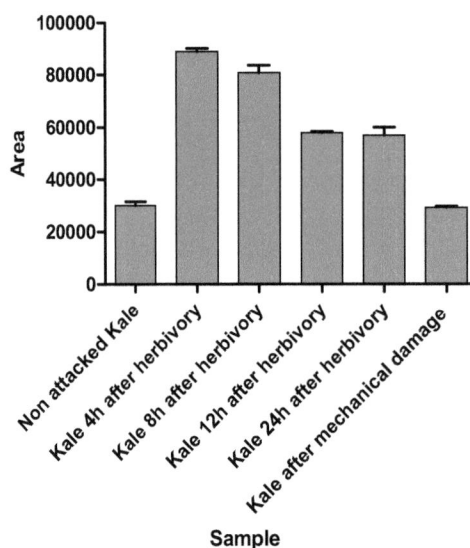

Figure 9: Variation in total terpenes content in non attacked kale, kale after insect's attack and after mechanical damage. Values show areas mean ± SE of 3 experiments. From [60] (with kind permission of Elsevier).

A pattern in the time of appearance of the compounds could be noticed. Among the seven terpenoids that could be found exclusively after insect predation, five were detected only 4 hours after the attack, being absent 1 hour after herbivory. On the other hand, three ester compounds (acetic acid, butyl ester; acetic acid, heptyl ester; 2-ethylhexyl ester) were absent before the attack, but were found 1 hour immediately after predation. These results strongly suggest that the synthesis of terpenes occured *de novo*, while the referred esters are probably accumulated in the leaves and released after predation.

In all experiments, esters were the main class of compounds, always accounting for more than 90 % of the volatiles. In fact, this value results from the contribution of one single compound, (*Z*)-3- hexenyl acetate, which, alone, accounted for 70% to 92% of the identified compounds in the different experiments. Although high amounts of this compound have been reported [61], to the best of our knowledge this was the first time that such high proportion of 3-hexenyl acetate is found. This compound has been extensively described in literature as being crucial in shaping insect-plant interactions [62, 63]. In this study, the amounts of (*Z*)-3-hexenyl acetate suffered an increase of ca. 10% 1 hour after insect's attack (Table 1). Analysis after 12 hours past the attack revealed an increase by some 15% (Fig. **4**). Interestingly, if the damage to the leaf was caused mechanically, instead of an increase in (*Z*)-3-hexenyl acetate its quantities, compared with basal emissions, would diminish by ca. 25% (Fig. **4**). This result, as well as the absence of the alcohol (*Z*)-3-hexen-1-ol, had already been described in a similar study involving one plant from the same species, although from a different cultivar, *B. oleracea* var. *gemnifera* [62].

Regardless of the differences between volatiles emitted after predation and those that result from mechanical damage, it can be said that in both situations the chemicals released are similar, albeit different from a quantitative point of view.

The vegetable species used in these experiments, kale, contains glucosinolates, which can be found through most cruciferous. These secondary metabolites are the precursors of the volatile isothiocyanates and are involved in defense against predation, as isothiocyanates are toxic upon ingestion, contact, or when present in the gas phase [64]. his defense system consists of glucosinolates and myrosinases, which are thioglucoside glucohydrolases that hydrolyze the thioglucosidic bond of glucosinolates, yielding glucose and an unstable aglycone. Spontaneous rearrangement of the aglycone then leads to the formation of an isothiocyanate [65].

In this study, allylisothiocyanate was detected in small amounts in plants that had not been attacked by insects or mechanically damaged. Given the fact that, in intact plant tissue, glucosinolate hydrolysis is prevented by spatial separation of myrosinases and glucosinolates by storage in different cells [66], the presence of allylisothiocyanate in non attacked leaves must mean that some kind of damage has been delivered to the leaf. In fact, the volatile analysis was not performed in the plant as a whole, as leaves were gently removed from the plant and analyzed. Nevertheless, as expected, the amounts of this compound rose as a consequence of insect predation, having increased by over 70%. As expected, the amounts of allylisothiocyanate in the leaves of the mechanically damaged plant were far higher than the basal values, over 96%. Overall, increase in allylisothiocyanate was much higher in mechanically damaged leaves than those that suffered insect's attack. This result was explained by authors as a consequence of a higher degree delivered by manual damage than that of insect's chewing.

Thus, compounds emitted by insect damaged leaves shared remarkable chemical similarities, as was carefully described by Mattiaci and colleagues [62].

Such structural resemblance displayed by a wide range of species could indicate that a common group of biosynthetic pathways are triggered [63].

With this work, further knowledge concerning the specialist *P. brassicae* and its interactions with one of its host plants was provided, which can be important in pest management, chemical ecology and entomology.

Metabolomics Applied to Macroalgae

Marine products represent an exciting area for the application of metabolomics science. It has been known for some years now that marine organisms often produce a set of metabolites with very particular chemistry, that sometimes cannot be found in land organism.

Valentão and colleagues, 2009 [67] conducted a study in two macroalgae species, green *Codium tomentosum* Stackhouse and red *Plocamium cartilagineum* (Linnaeus) P. S. Dixon from the Atlantic Ocean surrounding Portugal.

In both species, the absence of phenolic compounds was confirmed by HPLC-DAD and the attention was centered in other metabolites, such as organic acids and volatile compounds.

The two species showed different organic acids profile: *C. tomentosum* was characterized by the presence of oxalic, aconitic, ketoglutaric, pyruvic, malic, malonic and fumaric acids, while *P. cartilagineum* presented oxalic, ketoglutaric, pyruvic and acetic acids. In *P. cartilagineum* these compounds were present in vestigial amounts, while *C. tomentosum* exhibited a higher content, being oxalic acid the main compound.

GC-MS analysis of volatiles yielded the identification of forty-one compounds, which included alcohols, aldehydes, esters, halogenated compounds, ketones, monoterpenes, norisoprenoid derivatives, among others. Norisoprenoid derivatives and aldehydes were predominant. The main volatiles in green and red seaweeds were limonene and benzophenone, respectively.

Regarding terpenes, the authors present an interesting result, as five terpenes usually found in terrestrial plants [68], were identified and no halogenated terpene was detected. In *P. cartilagineum* only menthone was identified, while the terpenoid composition of *C. tomentosum* was more complex, with limonene being the major identified compound. This compound revealed to be chemopreventive and chemotherapeutic against many rodent solid tumour types [69].

These results draw a new attention to the biosynthetic pathways of macroalgae and terrestrial plants, as it would seem that they may not be as different as initially thought.

Metabolomics in Cell Culture

Metabolomic studies in cell cultures represent an interesting new area and helps elucidating the role of environmental stimuli in shaping the biochemical pathways that conditionate the phenotype.

A metabolomic approach followed by principal components and partial least square analysis was used to investigate the effect of environmental factors on two *Daucus carota* L. cv. Flakkese cell lines (R3M and R4G), selected for their ability to produce anthocyanins in the light and the dark, respectively [70]. LC–MS analysis was chosen, and methanolic extracts were used.

The HPLC–DAD–MS analysis of methanolic extracts of R3M and R4G cell lines allowed the detection of 122 molecules, with half of them being putatively identified. In addition to HPLC-MS analysis, HPLC-DAD was equally used. Comparison between both techniques showed that, although in a general way HPLC-MS was more sensible, it had reduced sensitivity specifically for coumaric acid derivatives. Nevertheless, HPLC–MS allowed the detection of molecules not detectable with HPLC–DAD, including several hydroxybenzoic acids, more abundant in R4G, and sinapic acid derivatives. Using HPLC-DAD, only one hydroxybenzoic acid derivative and one sinapic acid derivative had been found.

Overall, anthocyanins, hydroxycinnamic and hydroxybenzoic acids and their derivatives accounted for some 70% of the detected signal.

In both cell lines, cyanidin-(sinapoyl)-pentose-hexose-hexose was the most abundant anthocyanin. The two cell lines showed quantitative differences in: individual anthocyanin content, with cyanidin-(feruloyl)-pentose-hexose-hexose particularly abundant in R4G and cyanidin-pentose-hexose-hexose more abundant in R3M; hydroxycinnamic acid derivatives, more abundant in R3M; hydroxybenzoic acid derivatives, mainly detected in R4G.

The authors then studied the role and the interaction between anthocyanins, hydroxycinnamic and hydroxybenzoic acids in the two cell lines, and under application of two types of stress: enhancement of agitation and reduction of the batch volume, in the presence or absence of light.

In R3M, the two stressing conditions (batch volume of 20 ml, and enhancement of agitation) had a strong effect in the light (about 70% average increase), but no effect in the dark. In R4G, the two stresses induced only a slight increase in anthocyanin content in the dark (about 25% on average) but a much higher increase (average 90%), when combined with the light treatment. The authors were also able to draw many other conclusions (please refer to the original work).

Metabolomics in Quality Control

There is no doubt that the robustness of metabolomics offers a range of possibilities for its application. One of these applications is quality control of either foodstuffs or medicinal plants. With the increasing number of commercial plant extracts, their quality control as a tool for preventing adulteration or wrongful designations is a requirement.

Mattoli and colleagues [71] conducted a study in which several medicinal plant extracts (artichoke (*Cynara scolymus* L.), black cohosh (*Cimicifuga racemosa* L. Nutt.), dropwort (*Filipendula vulgaris* Moench or *Filipendula hexapetala* Gilib), everlasting (*Helichrysum italicum* Don), meadowsweet (*Spiraea ulmaria* L. or *Filipendula ulmaria* L. Maxim), sage (*Salvia officinalis* L.), sunflower (*Helianthus annuus* L.) and yarrow (*Achillea millefolium L.*)) were simultaneously analyzed by MS/ESI (positive and negative mode) and ¹H-NMR without previous chromatographic separation.

The positive ion ESI spectra of the extracts investigated were highly complicated in all cases, with a high level of chemical background, as it would be expected given the complexity of the extracts. In contrast, the negative ion ESI mass spectra were much better defined and several compounds could be identified, including acids, flavonols and flavonoids. The analysis of authentic standards confirmed the identification and, with this, the authors demonstrated the usefulness of MS/ESI in negative mode as a tool for phytochemical analysis and quality control in this species, without previous separation of the constituents.

The ¹H-NMR spectra of all the extracts exhibit a quite complicated aliphatic region and a reasonably resolved aromatic region. Because of the better resolution and the analytical value of the aromatic region, this region was chosen for use in the possible identification of the fingerprint of extracts

The positive and negative ion ESI-MS and ^{1}H-NMR analysis of the extracts under investigation led to a large amount of data, which was then submitted to a chemometric evaluation. Overall, it was possible to classify each species based on their ^{1}H NMR aromatic region and their most abundant ions of the MS analysis.

CONCLUSION

There is increasing evidence of the usefulness of metabolomics towards the chemical valorization of matrices from natural origin, including both terrestrial and aquatic species. It has been shown to be a most promising tool in conjunction with information provided by other approaches, like biological screening for active compounds, or in understanding bio and ecological interactions.

REFERENCES

[1]   Sumner LW, Mendes P, Dixon RA. Plant metabolomics: large-scale phytochemistry in the functional genomics era. Phytochemistry 2003; 62: 817-36.
[2]   Shyur L-F, Yang N-S. Metabolomics for phytomedicine research and drug development. Curr Opin Chem Biol 2008; 12: 66-71.
[3]   Dettmer K, Aronov PA, Hammock BD. Mass spectrometry-based metabolomics. Mass Spectrom Rev 2007; 26: 51–78.
[4]   Vigneau-Callahan KE, Shestopalov AI, Milbury PE, Matson WR, Kristal BS. Characterization of dietdependent metabolic serotypes: Analytical and biological variability issues in rats. J Nutr 2001; 131: 924S-32S.
[5]   Dixon RA, Gang DR, Charlton AJ, Fiehn O, Kuiper HA, Reynolds TL, Tjeerdema RS, Jeffery EH, German JB, Ridley WP, Seiber JN. Applications of metabolomics in agriculture. J Agric Food Chem 2006; 54: 8984-94.
[6]   German JB, Roberts MA, Watkins SM. Genomics and metabolomics as markers for the interaction of diet and health: Lessons from lipids. J Nutr 2003; 133: 2078S–83S.
[7]   Fiehn O. Metabolomics - The link between genotypes and phenotypes. Plant Mol Biol 2002; 48: 155-71.
[8]   Ward JL, Baker JM, Beale MH. Recent applications of NMR spectrometry in plant metabolomics. FEBS J 2007; 274: 1126-31.
[9]   Allen J, Davey HM, Broadhurst D, Heald JK, Rowland JJ, Oliver SG, Kell DB. High-throughput classification of yeast mutants for functional genomics using metabolic footprinting. Nat Biotechnol 2003; 21: 692–96.
[10]   Allen J, Davey HM, Broadhurst D, Rowland JJ, Oliver SG, Kell DB. Discrimination of modes of action of antifungal substances by use of metabolic footprinting. Appl Environ Microbiol 2004; 70: 6157–65.
[11]   Xiayan L, Legido-Quigley C. Advances in separation science applied to metabonomics. Electrophoresis 2008; 29: 3724-36.
[12]   Cuyckens F, Claeys M. Mass spectrometry in the structural analysis of flavonoids. J. Mass Spectrom 2004; 39: 1-15.
[13]   Rauha JP, Vuorela H, Kostiainen R. Effect of eluent on the ionization efficiency of flavonoids by ion spray, atmospheric pressure chemical ionization, and atmospheric pressure photoionization mass spectrometry. J Mass Spectrom 2001; 36: 1269-80.
[14]   de Rijke E, Zappey H, Ariese F, Gooijer C, Brinkman UATh. Liquid chromatography with atmospheric pressure chemical ionization and electrospray ionization mass spectrometry of flavonoids with triple-quadrupole and ion-trap instruments. J Chromatogr A 2003; 984: 45-58.
[15]   Krishnan P, Kruger NJ, Ratcliffe RG. Metabolite fingerprinting and profiling in plants using NMR. J Exp Bot 2005; 56: 255-65.
[16]   Wolfender J-L, Ndjoko K, Hostettmann K. Liquid chromatography with ultraviolet absorbance–mass spectrometric detection and with nuclear magnetic resonance spectrometry: a powerful combination for the on-line structural investigation of plant metabolites. J Chromatogr A 2003; 1000: 437-55.
[17]   Sottomayor M, Ros Barceló A. In: Atta-ur-Rahman, Ed. Studies in natural products chemistry (Bioactive natural products). Amsterdam, Elsevier Science Publishers 2005; pp. 813-57.
[18]   Sottomayor M, Lopes Cardoso I, Pereira LG, Ros Barceló A. Peroxidase and the biosynthesis of terpenoid indole alkaloids in the medicinal plant *Catharanthus roseus* (L.) G. Don. Phytochem Rev 2004; 3: 159-71.
[19]   van der Heijden R, Jacobs DI, Snoeijer W, Hallard D, Verpoorte R. The *Catharanthus* alkaloids: pharmacognosy and biotechnology. Curr Med Chem 2004; 11: 607-28.
[20]   Ferreres F, Pereira DM, Valentão P, Andrade PB, Seabra RM, Sottomayor M. New phenolic compounds and antioxidant potential of *Catharanthus roseus*. J Agric Food Chem 2008; 56: 9967-74.
[21]   Mabry TJ, Markham KR, Thomas MB. The systematic identification of flavonoids. New York: Springer 1970.
[22]   Ferreres F, Llorach R, Gil-Izquierdo A. Characterization of the interglycosidic linkage in di-, tri-, tetra- and pentaglycosylated flavonoids and differentiation of positional isomers by liquid chromatography/electrospray ionization tandem mass spectrometry. J Mass Spectrom 2004; 39: 312-21.

[23]   Nishibe S, Takenaka T, Fujikawa T, Yasukawa K, Takido M, Morimitsu Y, Hirota A, Kawamura T, Noro Y. Bioactive phenolic compounds from *Catharanthus roseus* and *Vinca minor*. J Nat Med 1996; 50: 378-83.

[24]   Brun G, Dijoux MG, David B, Mariotte AMA. New flavonol glycoside from *Catharanthus roseus*. Phytochemistry 1999; 50: 167-69.

[25]   Cuyckens F, Rozenberg R, Hoffmann E, Claeys M. Structure characterization of flavonoid *O*-diglucosides by positive and negative nano-electrospray ionization ion trap mass spectrometry. J Mass Spectrom 2001; 36: 1203-10.

[26]   Cuyckens F, Ma YL, Pocsfalvi G, Claeys M. Tandem mass spectral strategies for the structural characterization of flavonoid glycosides. Analusis 2000; 28: 888-95.

[27]   Cavaliere C, Foglia P, Pastorini E, Samperi R, Laganà A. Identification and mass spectrometric characterization of glycosylated flavonoids in *Triticum durum* plants by high-performance liquid chromatography with tandem mass spectrometry. Rapid Commun Mass Spectrom 2005; 19: 3143-58.

[28]   Tsao R, Yang R, Young JC, Zhu H. Polyphenolic profiles in eight apple cultivars using High-Performance Liquid Chromatography (HPLC). J Agric Food Chem 2003, 51: 6347-53.

[29]   Clifford MN, Johnston KL, Knight S, Kuhnert,, N. Hierarchical scheme for LC-MSn identification of chlorogenic acids. J Agric Food Chem 2003, 51: 2900-11.

[30]   Choi YH, Tapias EC, Kim HK, Lefeber AWM, Erkelens C, Verhoeven JThJ, Brzin J, Zel J, Verpoorte R. Metabolic discrimination of *Catharanthus roseus* leaves infected by phytoplasma using ^{1}H-NMR spectrometry and multivariate data analysis. Plant Physiol 2004; 135: 2398-410.

[31]   Harrison HFJr, Mitchell TR, Peterson JK, Wechter WP, Majetich GF, Snook ME. Contents of caffeoylquinic acid compounds in the storage roots of sixteen sweetpotato genotypes and their potential biological activity. J Am Soc Hortic Sci 2008; 133: 492-500.

[32]   Croteau R, Kutchan TM, Lewis NG. In: Buchanan BB, Gruissem W, Jones RL, Eds. Biochemistry and molecular biology of plants. Rockville, MD, American Society of Plant Physiologists 2000; pp. 1250-318.

[33]   Guedes de Pinho P, Gonçalves RF, Valentão P, Pereira DM, Seabra RM, Andrade PB, Sottomayor M. Volatile composition of *Catharanthus roseus* (L.) G. Don using solid-phase microextraction and gas chromatography/mass spectrometry. J Pharm Biomed Anal 2009; 49: 674-85.

[34]   Baldwin EA, Scott JW, Shewmaker CK, Schuch W. Flavor trivia and tomato aroma: Biochemistry and possible mechanisms for control of important aroma components. Hort Sci 2000; 35: 1013-22.

[35]   Berrah G, Konetzka WA. Selective and reversible inhibition of the synthesis of bacterial deoxyribonucleic acid by phenethyl alcohol. J Bacteriol 1962; 83: 738-44.

[36]   Pichersky E, Gershenzon J. The formation and function of plant volatiles: perfumes for pollinator attraction and defense. Curr Opin Plant Biol 2002; 5: 237-43.

[37]   Brun G, Bessière JM, Dijoux-Franca MJ, David B, Mariotte AM. Volatile components of *Catharanthus roseus* (L.) G. Don (Apocynaceae). Flavour Fragr J 2001; 16: 116-9.

[38]   Demetzos C, Kolocouris A, Anastasaki T. A simple and rapid method for the differentiation of C-13 manoyl oxide epimers in biologically important samples using GC–MS analysis supported with NMR spectrometry and computational chemistry results. Bioorg Med Chem Lett 2002; 12: 3605-9.

[39]   Ziegenbein FC, Hanssen HP, Konig HA. Secondary metabolites from *Ganoderma lucidum* and *Spongiporus leucomallellu*. Phytochemistry 2006; 67: 202-11.

[40]   Sembdner G, Parthier B. The biochemistry and the physiological and molecular actions of jasmonates. Ann Rev Plant Physiol Plant Mol Biol 1993; 44: 569-89.

[41]   Farmer EE, Ryan CA. Interplant communication: airborne methyl jasmonate induces synthesis of proteinase inhibitors in plant leaves. Proc Natl Acad Sci 1990; 87: 7713-6.

[42]   Mitler R, Lam E. Sacrifice in the face of foes: Pathogen-induced programmed cell death in plants. Trends Microbiol 1996; 4: 10-5.

[43]   Fingrut O, Flescher E. Plant stress hormones suppress the proliferation and induce apoptosis in human cancer cells. Leukemia 2002; 16: 608-16.

[44]   Rodriguez S, Compagnon V, Crouch NP, St-Pierre B, De Luca V. Jasmonate-induced epoxidation of tabersonine by a cytochrome P-450 in hairy root cultures of *Catharanthus roseus*. Phytochemistry 2003; 64: 401-9.

[45]   Silva Ferreira AC, Monteiro J, Oliveira C, Guedes de Pinho P. Study of major aromatic compounds in port wines from carotenoid degradation. Food Chem 2008; 110: 83-7.

[46]   Kotseridis Y, Baumes RL, Bertrand A, Skouroumounis GK. Quantitative determination of 2-methoxy-3-isobutylpyrazine in red wines and grapes of Bordeaux using a stable isotope dilution assay. J Chromatogr A 1999; 848: 317-25.

[47]   Silva Ferreira AC, Guedes de Pinho P. Nor-isoprenoids profile during port wine ageing - influence of some technological parameters. Anal Chim Acta 2004; 2004: 169-76.

[48]   Schwartz SJ, von Elbe JH. Kinetics of chlorophyll degradation to pyropheophytin in vegetables. J Food Sci 1983; 48: 1303-6.

[49]   Nobili S, Lippi D, Witort W, Donnini M, Bausi L, Mini E, Capaccioli S. Natural compounds for cancer treatment and prevention. Pharmacol Res 2009; 59: 365-78.

[50]   Cai L, Koziel JA, O'Neal ME. Determination of characteristic odorants from *Harmonia axyridis* beetles using *in vivo* solid-phase microextraction and multidimensional gas chromatography-mass spectrometry-olfactometry. J Chromatogr A 2007; 1147: 66-78.

[51]   Premkumar T, Govindarajan S. Antimicrobial study of pyrazine, pyrazole and imidazole carboxylic acids and their hydrazinium salts. World J Microbiol Biotechnol 2005; 21: 479-80.

[52]   Lewinsohn E, Sitrit Y, Bar E, Azulay Y, Meir A, Zamir D, Tadmor Y. Carotenoid pigmentation affects the volatile composition of tomato and watermelon fruits, as revealed by comparative genetic analyses. J Agric Food Chem 2005; 53: 3142-8.

[53]   Cole ER, Kapur NS. The stability of lycopene I - Degradation by oxygen. J Food Sci 1957; 8: 360-5.

[54]   Yu JQ, Liao ZX, Cai XQ, Lei JC, Zou GL. Composition, antimicrobial activity and cytotoxicity of essential oils from *Aristolochia mollissima*. Environ Toxicol Pharmacol 2007; 23: 162-7.

[55]   Ferreres F, Fernandes F, Oliveira JMA, Valentão P, Pereira JA, Andrade PB. Metabolic profiling and biological capacity of *Pieris brassicae* fed with kale (*Brassica oleracea* L. var. *acephala*). Food Chem Toxicol 2009; 47: 1209-20.

[56]   Ferreres F, Valentão P, Pereira JA, Bento A, Noites A, Seabra RM, Andrade PB. HPLC-DAD–MS/MS-ESI screening of phenolic compounds in *Pieris brassicae* L. reared on *Brassica rapa* var. *rapa* L. J Agric Food Chem 2008; 56: 844-53.

[57]   Howe GA, Jander G. Plant immunity to insect herbivores. Annu Rev Plant Biol 2008; 59: 41-66.

[58]   Meiners T, Hilker M. Induction of plant synomones by oviposition of a phytophagous insect. J Chem Ecol 2000; 26: 221-32.

[59]   Bukovinszky T, Gols R, Posthumus MA, Vet LEM, Van Lenteren JC. Variation in plant volatiles and attraction of the parasitoid *Diadegma semiclausum* (Hellén). J Chem Ecol 2005; 31: 461-80.

[60]   Fernandes F, Pereira DM, Guedes de Pinho P, Valentão P, Pereira JA, Bento A, Andrade PB. HS-SPME and GC/IT-MS applied to an alive system: *Pieris brassicae* fed with kale. Food Chem 2009; (submitted).

[61]   Geervliet JBF, Posthumus MA, Vet LEM, Dicke M. Comparative analysis of headspace volatiles from different caterpillar-infested or uninfested food plants of *Pieris* species. J Chem Ecol 1997; 23: 2935-54.

[62]   Mattiacci L, Rocca, BA, Scascighini N, D'Alessandro M, Hern A, Dorn S. Systemically induced plant volatiles emitted at the time of danger. J Chem Ecol 2001; 27: 2233-52.

[63]   Paré PW, Tumlinson JH. Plant volatiles as a defense against insect herbivores. Plant Physiol 1999; 121: 325-31.

[64]   Agrawal AA, Kurashige NS. A Role for isothiocyanates in plant resistance against the specialist herbivore *Pieris rapae*. J Chem Ecol 2003; 29: 1403-15.

[65]   Mumm R, Posthumus MA, Dicke M. Significance of terpenoids in induced indirect plant defence against herbivorous arthropods. Plant Cell Environ 2008; 31: 575-85.

[66]   Andréasson E, Jørgensen LB. In: Romeo JT, Ed. Integrative phytochemistry: from ethnobotany to molecular ecology. Amsterdam, Pergamon 2003; pp. 79-99.

[67]   Valentão P, Trindade P, Gomes D, Guedes de Pinho P, Mouga T, Andrade PB. *Codium tomentosum* and *Plocamium cartilagineum*: chemical approach and antioxidant potential. Food Chem 2009; (submitted).

[68]   Grayson DH. Monoterpenoids. Nat Prod Rep 1997; 14: 477-522.

[69]   Crowell PL, Gould MN. Chemoprevention and therapy of cancer by d-limonene. Crit Rev Oncog 1994; 5: 1-22.

[70]   Ceoldo S, Toffali K, Mantovani S, Baldan G, Levi M, Guzzo F. Metabolomics of *Daucus carota* cultured cell lines under stressing conditions reveals interactions between phenolic compounds. Plant Sci 2009; 176: 553-65.

[71]   Mattoli L, Cangi F, Maidecchi A, Ghiara C, Ragazzi E, Tubaro M, Stella L, Tisato F, Traldi P. Metabolomic fingerprinting of plant extracts J Mass Spectrom 2006; 41: 1534-45.

CHAPTER 2

Optimizing the Generalization Ability of Artificial Neural Networks in ELISA Protocols by Employing Different Topologies and GENETIC Operators

Constantinos Kousoulos, Yannis Dotsikas and Yannis L. Loukas*

Division of Pharmaceutical Chemistry, Department of Pharmacy, University of Athens, Panepistimioupoli Zografou GR - 157 71, Athens, Greece

Abstract: The aim of the present work was to present the ability of Artificial Neural Networks (ANN) in successfully predicting the response of an Enzyme-linked Immunosorbent assay (ELISA) from the relative input parameters and to further enhance its performance by use of different network architectures, learning algorithms and genetic operators. Representatives of three major categories of ANN topologies were investigated, namely Multilayer Feed-Forward (MLF), Generalized Feed-Forward (GenFF) and Radial Basis Function (RBF) which were trained with back-propagation with momentum and a scaled conjugate gradient learning algorithm. Tuning of the input and hidden layer size was performed by use of a genetic algorithm, while different combinations of genetic operators were used for the optimal GenFF network in order to increase its predictive ability. The major advantage of this approach was the simultaneous data-driven, modeling and optimization process which demands no *a priori* knowledge of variable correlations and can be employed in setting up new assays.

INTRODUCTION

Artificial Neural Networks (ANN) have recently received a great deal of attention in many fields of study. Their remarkable capabilities in modeling complex real-world problems are the main reason for their widespread popularity amongst scientists from different areas and fields. Despite the absence of a universally accepted definition, most would agree that ANN are structures comprised of densely interconnected adaptive simple processing elements (known as artificial neurons or nodes) that are capable of performing massively parallel computations for data processing and knowledge representation [1-2]. Rosenblatt [3] developed the first ANN on computer, namely the 'perceptron', which was capable of dealing with linearly separable problems. The incorporation of an additional layer of neurons between the input and the output layer, known as multilayer feed-forward (MLF) or multilayer perceptron (MLP), signaled the beginning of a new era for ANN. This second breakthrough enabled ANN to overcome the problems of the perceptron-based networks, stemming from their inadequacy in dealing with complex, non-linear problems.

Neural networks have been used for a wide variety of applications, especially in the field of chemistry, where statistical methods are traditionally employed [5-11], since they have proved to be robust and easy to implement for various problems. ANN offer a number of advantages over statistical techniques including the requirement for less formal statistical training, the ability to implicitly detect complex, non-linear relationships between dependent and independent variables, the ability to detect all possible interactions between predictor variables, as well as the availability of multiple training algorithms [12-14]. There are however many pitfalls in the use of MLF networks, such as the inability to extract in a direct way theoretical information on the model (black box nature) [15], their proneness to overfitting, as well as the empirical nature of model development. They constitute a cumbersome task, since they involve selecting heuristically an appropriate non-linear model from numerous alternates [16].

During the last decade, there has been increased use and exploitation of the synergism between artificial neural networks and genetic algorithms (GA) [17-20]. Genetic algorithms are the implementation of various search paradigms inspired by natural evolution. They follow the 'survival of the fittest' and 'genetic propagation of characteristics' principles of biological evolution for searching the solution space of an optimization problem. Instead of continuously restructuring the ANN architecture, this 'trial and error' procedure can be circumvented by the use of a genetic algorithm, which will automatically generate and evaluate a number of possible solutions to the above problem in order to provide the optimum configuration for the network.

Immunoassays constitute a powerful tool in the area of analytical chemistry that has been extensively employed in research and routine applications [21]. The development and optimization of a novel assay requires careful and systematic investigation of many factors. Problems arise, due to the complex nature of immunoassays that involve multiple assay steps. The traditional method of optimizing an immunoassay is to identify the key variables through preliminary studies, during the assay development phase. In order to optimize many variables that influence a response, one should study each factor independently varying it, while holding all others constant. Since most immunoassay systems exhibit interactions between two or more factors, this method can often lead to incomplete understanding of the behavior of the assay and limited predictive ability. An alternative approach is to employ statistical experimental design techniques that allow for the investigation of many factors simultaneously. A few attempts have been appeared in the literature [22-24] and it is profound that have helped a lot in optimizing immunoassay procedures with much fewer experiments. The non-linear nature and the multivariate character of an immunoassay system render neural networks as the ideal tool not only to optimize, but to predict the behavior for an extended range of values of all factors, with subsequent savings of time and materials. So far, one study has been found in the literature [25] employing a simple neural network (three-layered feed-forward) for the prediction of the response as a function of four input variables in separate cross validation sets.

In the present study, an opioid analgesic, fentanyl, was used as a model-antigen in order to examine the suitability of ANN in predicting the response of an ELISA, as a function of antigen concentration and a set of relative parameters. To this end, three types of neural networks were applied, namely MLF, Generalized Feed-Forward (GenFF) and Radial Basis Function (RBF), to a data set of 120 observations which were obtained by measuring the absorbance given by random combinations of 5 input parameters. For the training of the above networks two algorithms were used, gradient descent with momentum and scaled conjugate gradient, representing 1st and 2nd order methods respectively. Structure optimization was achieved by use of a GA resulting in a hybrid model. ANN formalism was integrated with GA for the determination of the optimum number of input (independent) variables and hidden nodes as well as the appropriate values for the momentum and learning rate parameters of the back-propagation algorithm. To our knowledge, this kind of synergism is implemented for the first time in ELISAs. Finally, in an attempt to gain further improvement in the performance of the network that provided the lowest prediction error, additional training took place using several combinations of genetic operators.

THEORY

MLF and Generalized Feed-forward

ANN topologies or architectures [26] are formed by organized nodes (or neurons) into layers and linking these layers of neurons with modifiable weighted interconnections, known as synaptic weights. A multilayer feed-forward neural network (MLF) or a multilayer perceptron (MLP) is the most widely used type of network and consists of an input layer, with nodes representing input variables to the problem, an output layer, with nodes representing the dependent variables, as well as one or more hidden layers containing nodes to help capture the nonlinearity in the data. All nodes from one layer are connected to all nodes of the following layer. There may be zero, one or more hidden layers. Networks with one hidden layer are considered as universal approximators and make up the vast majority of the ANN architectures.

The input into a node is a weighted sum of the outputs from nodes connected to it. Each unit takes its net input and applies an activation function to it, known as the transfer function. Common choices for transfer functions are non-linear ones, such as the sigmoid or the tangent hyperbolic as well as the linear function.

A GenFF network [27] can be visualized as a MLF that contains additional connections bypassing one or more layers. In this special case of MLF, a direct feed through of signals from the input layer is allowed, resulting in more efficient training of the layers, which are thus closer to the input.

RBF Neural Networks

Radial Basis Function networks (RBF) are a variant of three-layer feed-forward networks. They differ strongly from them in the activation (transfer) functions and how they are used. They are extremely powerful tools and have been implemented for various applications in the field of chemometrics [8]. In a RBF network units respond (nonlinearly) to the distance of points from the center represented by the radial unit. The response surface of a single radial unit is, therefore, a Gaussian (bell-shaped) function, peaked at the center, and descending outwards. A radial unit is defined by its center point and a radius. A point in N-dimensional space is defined using N numbers, so the center of a radial unit is stored as weights and the radius (or deviation) value is stored as the threshold.

A RBF network, therefore, has a hidden layer of radial units, each actually modeling a Gaussian response surface, followed by an output layer containing linear units with linear transfer function.

Learning Algorithms

In MLF networks, given a set of input patterns with associated known outputs, the objective is to train the network, using supervised learning to estimate the functional relationship between the inputs and outputs. The network can then be used to model or predict a response corresponding to a new input pattern. To accomplish learning, some form of an objective function is required. The goal is to use the objective function to optimize the weights. The most common performance metric is the sum of squared errors defined as:

$$E = \frac{1}{2} \sum_{p=1}^{n} \sum_{k=1}^{o} (y_{pk} - \hat{y}_{pk})$$

where the subscript p refers to the patterns with a total of n patterns, the subscript k, to the output unit with a total of o output units, y is the observed response and $\hat{y}$ is the predicted response.

The most common learning algorithm is the back-propagation learning rule (BP). BP is based on searching an error surface (error as a function of ANN weights) using gradient descent for point(s) with minimum error. In other words, the main objective of the algorithm is to find the set of weights that minimize the objective function. Unfortunately, BP algorithm is usually characterized by a poor convergence rate and depends on parameters, which have to be specified by the user, as no theoretical basis for choosing them exists. The values of these parameters, namely learning rate and momentum constant, are often crucial for the success of the algorithm.

On the other hand, second order learning methods use not only the slope of the performance surface but also the curvature in order to adjust the weights. However, such methods require the inversion of a curvature matrix (Levenberg-Marquardt) or the storage of an iterative approximation of that inverse (quasi-Newton methods such as BFGS), which makes them computationally expensive. Conjugate gradient techniques are capable of exactly minimizing a d-dimensional unconstrained quadratic problem in d iterations without requiring explicit knowledge of the curvature matrix. These techniques are intrinsically "offline" [28] and require a line search that involves several calculations of either the error function or its derivative, which result in increased complexity. A variation of the conjugate gradient method, the Scaled Conjugate Gradient (SCG), introduced by Møller [29], avoids the line search by scaling the step size. A significant advantage of scaled conjugate gradient learning is that it is parameterless, as the user doesn't need to set learning rates or momentum terms.

Genetic Algorithms (GAs)

Genetic algorithms [30] are general-purpose search algorithms based upon the principles of evolution observed in nature. GAs combine selection, crossover and mutation operators with the goal of finding the best solution to a problem. Each solution is called a chromosome. A chromosome is made up of a collection of genes, which are simply the parameters to be optimized. For the ANN problem each chromosome contains information about the construction of the neural network. There are many possibilities to represent the candidate solutions. The parameter values may be scaled and can be encoded as integer values or real values. An often used representation is the binary encoding of the parameter values. In this particular encoding method the characters in the bit string are zero and one.

GA starts with the random generation of an initial set of individuals (chromosomes), the initial population. The individuals are evaluated and ranked. Since the number of individuals in each population is kept constant, for each new chromosome an old one has to be discarded, in general the one with the worst fitness value. There are two basic operators to generate a new individual: *mutation* and *crossover*. During mutation, a couple of bits are flipped at random. This can result in entirely new gene values being added to the gene pool. With these new values the genetic algorithm may be able to arrive to a better solution than was previously possible. Mutation is an important part of the genetic search as it helps to prevent the population from stagnating at any local optima. On the other hand, crossover simulates the sexual generation of a child or offspring from two parents. This is performed by taking parts of the bit-string of one of the parents-chromosomes and the other parts from the other parent and combining both in the child. There are three basic kinds of crossover: one-point, two-point and uniform. The idea behind crossover is that the new chromosome may be better than both of the parents if it takes the best characteristics from each one of them.

Selection is a genetic operator that chooses a chromosome from the current generation's population for inclusion in the next generation's population. There are many methods for selecting the best chromosomes. The most common are: *Roulette selection, Tournament selection* and *Elitist selection.*

EXPERIMENTAL

Reagents

Anti-fentanyl rabbit polyclonal antibody (9.2 μmol L^{-1}) and fentanyl-BSA conjugate were obtained from Biostride Inc. (USA). Sulfo-NHS-LC-Biotin was obtained from Molecular Probes (Eugene, OR) and BSA (RIA and ELISA grade) was brought from Calbiochem (Germany). TMB peroxide substrate solution was obtained from Pierce (Rockford, IL). Tween 20 was purchased from ICN Biomedicals (Germany). Streptavidin labeled with type VI peroxidase, and all other reagents were from Sigma (Greece). All aqueous solutions and buffers were prepared using water de-ionized and doubly distilled (Resistivity > 18 MΩ cm).

The washing solution used in this protocol was a PBS buffer (pH = 7.40) containing 0.05% (v/v) Tween 20 and the assay buffer was the washing solution containing 0.1% BSA (w/v). Dilutions of conjugates and standards were made using the assay buffer. The coating buffer consisted of a 0.100 M carbonate/bicarbonate buffer (pH = 9.60) and the blocking solution was a PBS buffer containing 1% BSA (w/v). Fentanyl solutions were kept in polypropylene tubes, due to its adsorption onto glass surface.

Instrumentation

All measurements were performed with a Fluostar Galaxy (BMG LabTechnologies GmbH, Germany) multifunctional microplate reader. Absorbance optics was installed for the experiments and the absorbance was measured at 450 nm with background subtraction at 620 nm. The 96-well microtiter plates (transparent polystyrene plates with a Maxisorp surface) were obtained from Nunc (Nalge Nunc, UK). All plates were washed with a fully automated Tecan Columbus (Tecan, Austria) 96-well microplate washer.

ELISA Protocol

The competitive ELISA protocol [21, 22] is schematically presented in Fig. **1**. In brief, the wells of 96-well plates were filled with 100 μL of fentanyl antibody dilution (1:500, 1:1000, 1:1500, 1:1750, 1:2000, 1:2500) in coating buffer and incubated for 14 h at 4 °C. Next, the antibody solution was removed and the plates were post-coated with 200 μL of blocking solution for 1 h at room temperature. The plate was aspirated and washed four times with 300 μL of washing solution and then the microwells were filled with 50 μl of fentanyl standard solutions (0, 0.100, 0.200, 0.400, 0.500, 0.800, 1.000, 1.500, 2.000 ng mL^{-1}) and 50 μL of the diluted (1:200, 1:400, 1:500, 1:600, 1:700, 1:800, 1:900, 1:1000) biotinylated fentanyl-BSA conjugate in assay buffer and incubated for (0.5, 1.0, 2.0, 3.0, 14.0 h) at room temperature. The wells were aspirated and rewashed six times by means of the same washing solutions and 100 μL of streptavidin-HRP dilution (1:500, 1:1000, 1:1250, 1:1500, 1:2000, 1:2500) from a stock solution 125 mg L^{-1} in assay buffer were dispensed and incubated for 30 min at room temperature. After washing once again six times, TMB peroxide substrate solution was added and the plate was kept in dark. 30 min later the enzymatic reaction was stopped with 100 μL H$_2$SO$_4$ (2 M), resulting in a yellow colored solution. The absorbance was measured at 450 nm with background subtraction at 620 nm. The preparation of the biotinylated fentanyl-BSA has been previously described [21, 22].

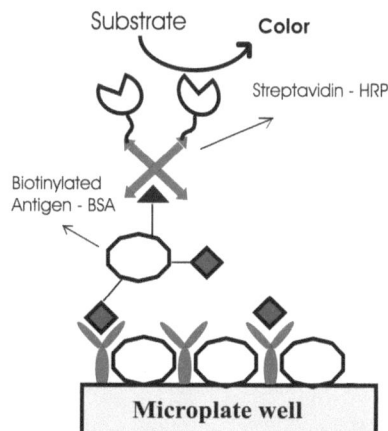

Figure 1: Schematic illustration of the concept of the developed ELISA protocol.

ANN Architectures

The input pattern consisted of five variables: Antibody dilution (Ab), fentanyl concentration (C), Biotinylated fentanyl-BSA conjugate dilution (BSA), Incubation time for the competition between antigen and tracer (t), and Streptavidin-HRP conjugate dilution (Str-HRP). The ranges of these variables were previously mentioned. Output consisted of a single element, Absorbance (A). A set of 120 random combinations of the five input elements was run in the laboratory and the absorbance values were determined.

Once the input variables and the dataset were selected the next step was the division of the dataset in three subsets, namely the training, validation and test subsets. The main requirement during training is the data representativity, meaning that the samples in the dataset should be (evenly) spread over the expected range of data variability. To avoid the risk of not selecting representative samples during training, the Kohonen self-organizing map approach was adopted. The technique's main objective is to map objects from n-dimensional into two-dimensional space. The samples for the validation and test subsets were selected in the same unbiased way. Next, normalization of the three subsets took place to the range [-0.9, 0.9] in order to prevent variables with large values from dominating the model as well as premature saturation of hidden nodes. It should also be mentioned that from a selection of different activation functions the best results were obtained using a tangent hyperbolic (tanh) transfer function in the hidden layer and a linear function in the output layer of the MLF and GenFF networks. The first is necessary for modeling the non-linearities in the data while the second is generally preferred in function approximation tasks. In the case of the RBF topology, the hidden layer consisted of nodes with Gaussian activation functions followed by an output layer, which performed a linear combination of the radial units outputs.

For the training two different learning algorithms were used, namely gradient descent with momentum and conjugate gradient. The momentum term is commonly used in weight updating to help the search escape local minima and functions, as a low pass filter smoothing out progress over small bumps in the error surface, by remembering the previous weight change. It also accelerates the weight updates when there is a need to reduce the learning rate in order to avoid oscillation.

As in any prediction-function approximation model, the selection of appropriate model inputs is extremely important. It is well-known that ANN are data driven approaches capable of determining the critical model inputs. However, presenting a large number of inputs usually increases network size leading to a considerable reduction of processing speed and to a substantial increase in the amount of data required to estimate the connection weights efficiently. Another critical aspect in ANN model development is the choice of the number of nodes in the hidden layer and hence the number of connection weights as this will determine, to a great extent, the network's performance. When insufficient hidden units are selected to model nonlinear relationships within the data, the prediction error will generally be large as a result of a lack of fit. On the contrary, an abundant number of hidden units will result in an overfit and the prediction error will increase proportionally. Moreover, as it has already been mentioned, the BP algorithm depends on two user-defined parameters. Employing a high learning rate will accelerate training but it will also increase the danger of system oscillation while a small one will result in slow convergence of weights to an optimum and a high probability of getting stuck at a local optimum.

Determining the topology of a neural network, as well as the correct values for the above-mentioned parameters is a cumbersome problem-specific task. GAs eventually free the designer of this drudgery of trial-and-error, which is inevitable in conventional designing. To this end a GA was implemented to optimize network architectures and to configure the network in terms of the number of inputs, hidden layer nodes, momentum and learning rate values. The initial network architectures comprised of five input and one output nodes. The upper and lower bounds for the remaining values to be optimized are presented in Table **1**.

Table 1: Upper and lower bounds for the parameters to be optimized by the GA

	MLF	GenFF	RBF
Hidden layer nodes / cluster centers	1-29	1-29	1-50
Momentum	0-1	0-1	0-1
Learning rate	0-1	0-1	0-1

Initially, the candidate solutions were coded in the form of binary strings (chromosomes). GA then began by randomly creating an initial population (a collection of chromosomes), evaluated this population through multiple generations (using the genetic operators selection, crossover, mutation) in the search for a good solution for the specific problem. In the present study, the size of the population was 50, the probability of crossover was 0.9, the probability of mutation was 0.01 and the number of evolution generations was 100. The resulting networks were trained for a maximum of 5000 epochs.

RESULTS AND DISCUSSION

ANN systems were simulated using Matlab Neural Network Toolbox running on a Pentium IV platform. Training continued until there was no further decrease in validation error. The quality of the resulting models was assessed by the term MSE:

$$MSE = \sum_{i=1}^{N} \sum_{j=1}^{g} \frac{\left(y_{ij} - out_{ij}\right)^2}{N_g}$$

where N is the number of objects in the examined data set (train, validate or test), g is the number of output variables, y_{ij} is the element of target matrix y $(N \times g)$ for the data considered (i.e training, validate or test set) and out_{ij} is the element of the output matrix out $(N \times g)$ of the neural network.

The data set was divided in the unbiased way described above, into a training set of 80 data, a validation set and a test set of 20 data each; the training data set was used for model building, while the validation data set was utilized for model validation. The resultant model was not biased towards the training data set and thus it was likely to have a better generalization capability for unseen data. Model validation is the process by which the input vectors from input/output data sets, on which the network was not trained, are presented to the trained neural network model to see how well this model predicts the corresponding data set output values. When validation data is presented to the network as well as training data, the neural network model is selected to have parameters associated with the minimum validation data model error. The basic idea behind using a validation data set for model validation is that after a certain point in the training, the model begins overfitting the training data set. In principle, the model error for the validation data set tends to decrease as the training takes place up to the point that overfitting begins, and then the model error for the validation data suddenly increases.

Gradient-based methods cannot be efficiently used for optimizing the input space of an ANN model as opposed to a member of the group of stochastic optimization formalisms, known as GAs. In the present study each solution (set of parameter values) of the GA population was used to construct a network, which is then trained on the training set of observations. These coded strings of information were then evaluated and selected (proportionally to their response), in order to undergo reproduction, by means of the cross-over operator (one-point) and, to a lesser extent, of the mutation operator. The selection of individuals for cross-over and mutation was biased towards good individuals. The operator used for the selection procedure was "roulette" with the selection chance being proportional to the (scaled) fitness value. GA was designed to select variables and values for the above-mentioned parameters, with the aim of minimizing the MSE in the validation set, which thus served as the fitness criterion.

Genetic optimization was applied to three kinds of ANN, namely, MLF, GenFF and RBF, which were trained by means of gradient descent with momentum and scaled conjugate gradient. In Table **2** an example of the genetically optimized number of input and hidden layer units, as well as the optimal values for the momentum and learning rate that produced the lowest MSE for each type of network are presented. It is worth noting that in all cases, four input variables, namely **antibody dilution**, **fentanyl concentration**, **biotinylated fentanyl-BSA conjugate dilution** and **streptavidin-HRP conjugate dilution** were selected for the construction of the optimal ANN models. The variable of incubation time was discarded as it proved to have insignificant contribution to the model. This becomes evident from Table **3** where correlations coefficients between input and output variables are computed for the optimal MLF model. Similar results were obtained for GenFF and RBF models.

Table 2: Number of input units and hidden layer units for each of the genetically optimized networks.

	MLF[a]	MLF[b]	GenFF[a]	GenFF[b]	RBF[a]	RBF[b]
Input values	4	4	4	4	4	4
Hidden nodes / cluster centres	8	4	3	4	26	15
Momentum	0.899 0.394	-	0.849 0.390	-	0.849 0.388	-
Learning rate	0.728 0.597	-	0.692 0.470	-	0.693 0.468	-

[a] ANN trained by gradient descent with momentum. The first value in momentum and learning rate rows corresponds to the hidden layer while the second to the output layer.
[b] ANN trained by the conjugate gradient algorithm.

Table 3: Partial correlations between all variables (input and output) involved in the optimal MLF model.

	1 (C)	2(Antibody)	3 (biot)	4 (str-HRP)	5 (t)	A (output)
1 (C, ng mL^{-1})	1					
2 (Antibody dilution)	-0.0818	1				
3 (biotinylated tracer dilution)	0.2575	-0.3205	1			
4 (streptavidin-HRP conj. dilution)	0.2862	0.1357	0	1		
5 (t, h)	-0.0553	0.1903	1.11×10^{-17}	-0.1323	1	
Absorbance (output)	-0.6723	-0.2365	-0.6272	-0.4104	-0.1394	1

The optimal model for each kind of network was tested for its ability to generalize in the same testing set of observations in order to compare their predictive ability by means of RMSEP (Root Mean Square Error of Predictions):

$$RMSEP = \sqrt{\frac{\sum_{i}^{n} |y_{obs} - y_{pred}|^2}{n}}$$

where y_{obs} is the experimental (observed) output value, while y_{pred} is the predicted output value.

The results obtained in Table 4 suggest clearly that the optimal kind of network was the Generalized Feed Forward. Both MLP and GenFF networks demonstrated equivalent predictive ability and can be successfully implemented for non-linear approximation problems such as the present one. Nevertheless, GenFF network had the advantage of increased flexibility and freedom in data flow, due to the fact that the input and output layers were directly connected to each other. Moreover, the conjugate gradient learning algorithm proved to be more effective than the classic approach of back propagation with momentum and learning rate both in the MLP and the GenFF networks. This did not apply also in the RBF network which was the one with the lowest predictive ability in general. It is worth noting that even with the use of genetic optimization for the determination of the optimal momentum and learning rate values, the RMSEP value of the corresponding networks wasn't smaller than the respective RMSEP value of the networks employing conjugate gradient algorithm.

Table 4: RMSEP values obtained for each optimal model during testing

Training algorithm	MLP	GenFF	RBF
Conjugate grad. (RMSEP)	0.104	0.102	0.187
Momentum (RMSEP)	0.112	0.118	0.168

As already mentioned, in all of the above networks GA operated under roulette selection based on rank and with one-point crossover. In an attempt to further improve network performance, the optimal GenFF network topology was submitted to a set of training procedures each time with a different set of genetic operators. Selection of the best set of solutions (chromosome) was based either on roulette or tournament approach with rank or fitness value. Other genetic operators employed were *top percent* and *best*. The selected individuals were then submitted to one-point, two-point or uniform crossover, while the mutation probability remained stable at 0.01. The algorithm would proceed for 100 generations with each generation consisting of 50 chromosomes. The resulting RMSEP value for each of the combinations tested can be seen in Table **5**.

Table 5: RMSEP values obtained for each combination of genetic operators during testing of the resultant GenFF ANN.

Combinations of genetic operators	RMSEP
Roulette – Rank – One-point	0.103
Roulette – Rank – Two-point	0.171
Roulette – Rank – Uniform	0.0857
Roulette – Fitness – One-point	0.0970
Roulette – Fitness – Two-point	0.0970
Roulette – Fitness – Uniform	0.106
Top percent (10%) – One-point	0.187
Top percent (10%) – Two-point	0.109
Top percent (10%) – Uniform	0.117
Tournament (N=2) – One-point	0.148
Tournament (N=2) – Two-point	0.143
Tournament (N=2) – Uniform	0.117
Best – One-point	0.0903
Best – Two-point	0.0987
Best – Uniform	0.130

Based on the results of Table 5 one can easily observe the great variance in RMSEP values even in the same category of selection method. This is obvious in the case of roulette selection based on rank of individuals. When uniform crossover was applied, the lowest value of error overall was achieved as opposed to one of the highest values with two-point crossover. This demonstrates and also highlights the great importance of the crossover operator. Fig. **2** shows the linear regression ($r^2 = 0.988$) of the measured vs predicted absorbances for the optimal GenFF network obtained with the use of roulette selection based on rank with uniform crossover while Fig. **3** shows the proximity and relative agreement between the predicted and measured curves.

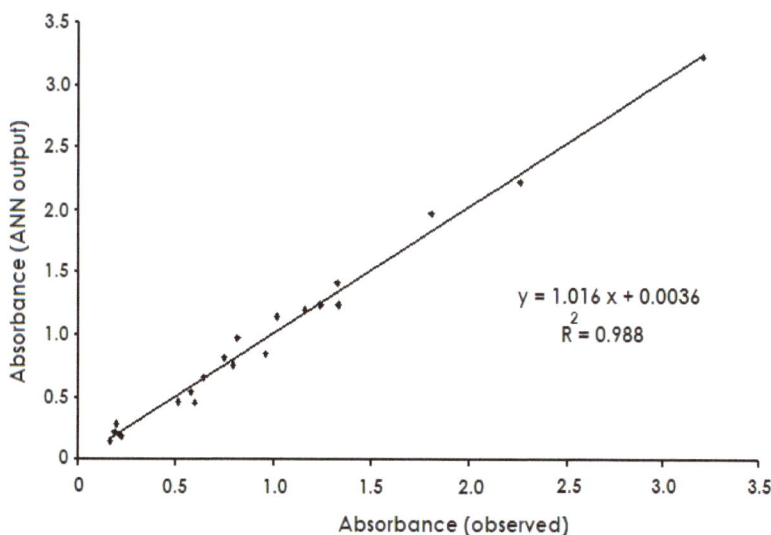

$$y = 1.016\,x + 0.0036$$
$$R^2 = 0.988$$

Figure 2: Correlation between obtained and predicted absorbance values for the test set by applying the network with the highest generalization and predictive ability (GenFF).

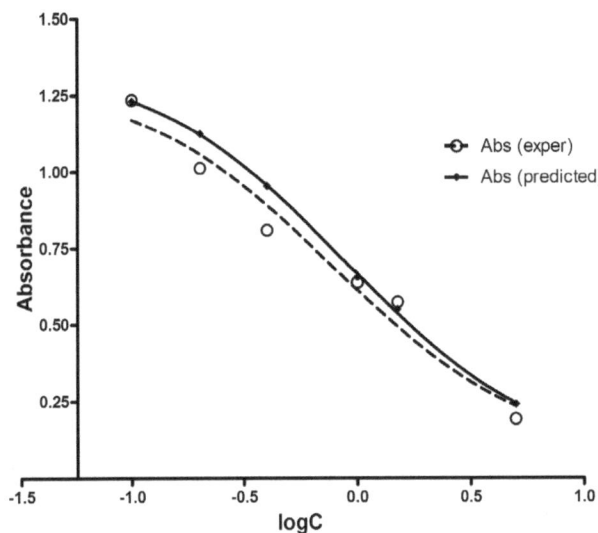

Figure 3: Correlation of the curves obtained from experimental data (intermittent line) and from predictions by ANN (continuous line) for the model antigen-fentanyl.

CONCLUSIONS

In the present study a hybrid process modeling and optimization methodology has been employed. The synergism between ANN and GA has proven to be both suitable as well as reliable in predicting the response of an ELISA, as a function of antigen concentration and a set of relative parameters. Different ANN architectures and training algorithms were used in order to achieve the highest predictive ability. Structure optimization was performed by use of a GA which enabled both the rapid identification of the optimal number of input variables and hidden layer nodes, as well as the determination of the correct values for learning rate and momentum. Furthermore, the combination of different genetic operators for more improvement of the GenFF network predictive ability led to a solution close to the best (perhaps global) with a correlation coefficient of 0.988. On the basis of the obtained results, the previous methodology could be successfully implemented for setting up new assays which is both an expensive and cumbersome task demanding a great number of experiments and the consumption of costly reagents.

REFERENCES

[1]   Basheer IA, Hajmeer M. Artificial neural networks: Fundamentals, computing, design, and application. J Microbiol Methods 2000; 43: 3-31.

[2]   McCulloh WS, Pitts W. A logical calculus of ideas immanent in Nervous activity. Bull Math Biophy 1943; 5: 115-133.

[3]   Rosenblatt F. The perceptron: A probabilistic model for information storage and organization in the brain. Psycholog Rev 1958; 65: 386-408.

[4]   Rumelhart DE, Hinton GE, Williams RJ, in: Rumelhart DE, McClelland JL, (Eds.), Learning internal representation by error propagation. Parallel Distributed Processing: Exploration in the Microstructure of Cognition, Vol. 1, MIT Press, Cambridge, 1986, Ch. 8.

[5]   Massart DL, Handbook of Chemometrics and Qualimetrics, Elsevier, Amsterdam, 1997, Ch. 44.

[6]   Agatonovich-Kustrin S, Beresford R. Basic concepts of artificial neural network (ANN) modeling and its application in pharmaceutical research. J Pharm Biomed Anal 2000; 22: 717-727.

[7]   Manallack DT, Livingstone DJ. Neural networks in drug discovery: Have they lived up to their promise?. Eur J Med Chem 1999; 34: 195-208.

[8]   Loukas YL. Radial basis function networks in host-guest interactions: Instant and accurate formation constant calculations. Anal Chim Acta 2000; 417: 221-229.

[9]   Loukas YL. Artificial neural networks in liquid chromatography: Efficient and improved quantitative structure-retention relationship models. J Chromatogr A 2000; 904: 119-129.

[10]   Loukas YL. Quantitative structure-binding relationships (QSBR) and artificial neural networks: Improved predictions in drug: Cyclodextrin inclusion complexes. Int J Pharm 2001; 226: 207-211.

[11] Loukas YL. Adaptive neuro-fuzzy inference system: An instant and architecture-free predictor for improved QSAR studies. J Med Chem 2001; 44: 2772-2783.

[12] Cheng B, Titterington DM. Neural networks: a review from a statistical perspective (with discussion). Stat Sci 1994; 9: 2-54.

[13] Warner B, Misra M. Understanding Neural Networks as Statistical Tools. Am Stat 1996: 50: 284-293.

[14] Tu J., Advantages and disadvantages of using artificial neural networks versus logistic regression for predicting medical outcomes. J Clin Epidemiol 1996; 49: 1225-1231.

[15] Mjalli FS, Al-Asheh1 S, Alfadalaa HE. Use of artificial neural network black-box modeling for the prediction of wastewater treatment plants performance. J Environ Manage 2007; 83: 329-338.

[16] Hornik K, Stinchcombe M, White H. Multilayer feedforward networks are universal approximators. Neural Networks 1989; 2: 359-366.

[17] Jouan-Rimbaud D., Massart DL, de Noord OE. Random correlation in variable selection for multivariate calibration with a genetic algorithm. Chemometr Intell Lab 1996; 35: 213-220.

[18] Jouan-Rimbaud D, Massart DL, Leardi R, de Noord OE. Genetic Algorithms as a Tool for Wavelength Selection in Multivariate Calibration. Anal Chem 1995; 67: 4295-4301.

[19] Sexton RS, Dorsey RE, Johnson JD. Toward global optimization of neural networks: A comparison of the genetic algorithm and backpropagation. Decis Support Syst 1998; 22: 171-185.

[20] Niculescu SP. Artificial neural networks and genetic algorithms in QSAR, J Mol Struc-Theochem 2003; 622: 71-83.

[21] Gosling JP. A decade of development in immunoassay methodology. Clin Chem 1990; 36; 1408-1427.

[22] Bunch DA, Rocke DM, Harrison RO. Statistical design of ELISA protocols. J Immunol Methods 1990; 132: 247-254.

[23] Sittampalam GS, Smith WC, Miyakawa TW, Smith DR, McMorris C. Application of experimental design techniques to optimize a competitive ELISA. J Immunol Methods 1996; 190; 151-161.

[24] Jeney C, Dobay O, Lengyel A, Ádam É, Nász I. Taguchi optimisation of ELISA procedures. J Immunol Methods 1999; 223: 137-146.

[25] Vertosick FT, Rehn T. Predicting behavior of an enzyme-linked immunoassay model by using commercially available neural network software. Clin Chem 1993; 39: 2478-2482.

[26] Lawrence J, Introduction to Neural Networks, Design, Theory and Applications, California Scientific Software Press, Nevada City, 1994, 10.

[27] Arulampalam G, Bouzerdoum A. A generalized feedforward neural network architecture for classification and regression. Neural Networks 2003; 16: 561-568.

[28] Bishop CM. Neural Networks for Pattern Recognition, Oxford Univ. Press, Oxford, 1995; 9.

[29] Møller MF. A scaled conjugate gradient algorithm for fast supervised learning. Neural Networks, 1993; 6: 525-533.

[30] Shapiro AF. The merging of neural networks, fuzzy logic, and genetic algorithms. Insur Math Econ 2002; 31: 115-131.

30   *Reviews in Pharmaceutical and Biomedical Analysis*, 2010, 30-51

CHAPTER 3

Bioinformatics Tools for Mass Spectrometry-based Proteomics Analysis

D.Di Silvestre[1], S. Daminelli[1], P. Brunetti[1] and PL. Mauri[1*]

[1]Institute for Biomedical Technologies Proteomics and Metabolomics Unit - CNR Via Fratelli Cervi 93 20090 Segrate (Milan) Italy

Abstract: In recent years mass spectrometry-based proteomics became very important and now it is the leading approach employed in high-throughput analysis. Its relevance increased thanks to availability of genome-sequence database and the development of high sensitivity instruments allows a rapid and automated proteins profiling. The need to analyze complex biological samples at a large-scale level required the development of computational tools to analyze and statistically evaluate data generated from mass spectrometry (MS) experiments. These aspects have stimulated the young emerging field of bioinformatics in proteomics to introduce new software and algorithms to handle large and heterogeneous data sets and to improve the knowledge of discovery process. This review discusses of the most recent progresses in bioinformatics tools useful in mass spectrometry-based proteomics. In particular we will be focusing on software applications applied to proteomics profiling biomarker discovery and cluster analysis. Finally since most known mechanisms leading to biological processes involve different molecules here are reported the most recent methodologies to investigate biological systems through their underlying interactions with particular attention to protein-protein interaction.

INTRODUCTION

After "genomics" a variety of "-omics" sub-disciplines such as proteomics metabolomics or lipidomics have begun to emerge each one with their own set of instruments techniques reagents and software. The technologies that have driven these new areas of research consist of DNA and protein microarrays mass spectrometry and a number of other methodologies that enable high-throughput analyses; under this aspect the field of bioinformatics has grown in parallel giving a great contribution due to the large scale of data produced.

Concerning proteomics analysis during the last few years the related technologies have become more and more important showing enormous data-gathering capabilities to discover specific biomarkers by analyzing tissues and bio-fluids such as serum plasma and urine that may reflect the disease status. In this scenario clinical proteomics defined as a subset of proteomics activities in the field of medicine aim to provide clinicians with new tools to accurately diagnose disease and treat patients in an individualized manner [1, 2].

The availability of genome-sequence databases and developments in liquid chromatography (LC) and mass spectrometry (MS) have had a big impact on proteomics. In fact in addition to the classical methods such as gel electrophoresis [3] mass spectrometry-based proteomics are increasing their importance. Moreover recent technological evolutions have led to the development of instruments with high sensitivity and specificity that have permitted the detailed study of complex biological fluids.

In order for MS-based proteomics to be successful clinically effective novel biomarkers must have high specificity and must be sufficiently robust [4]. This demands rigorous quantification strategies and the need for well-designed large-scale clinical trials to validate the use of novel proteomic signatures [2].

In the last few years several methods for protein quantitation and specific bioinformatics tools have been developed but quantification of complex biological samples remains a challenge [5]. In addition to problems presented by the sample itself many other factors such as specimen collection pre-fractionation methodology instrumentation set-up handling and processing database mining statistical analysis and data storage may affect the ultimate success of proteomic analysis [6].

Just like data generation data processing is also a fundamental aspect in proteomics [7]. The availability of mass spectrometry methods to analyze complex biological samples at large-scale level created the need for computational tools to analyze and statistically evaluate data generated from LC-MS experiments [8]. This reason gave impulse to the young emerging field of bioinformatics in proteomics which is introducing new softwares and algorithms to handle large and heterogeneous data sets and to improve the knowledge of discovery process.

C.K. Zacharis and P.D. Tzanavaras (Eds)

Actually one of the major challenges for biologist is understanding the complex interplay between genes proteins and small molecules into the cells. "-Omics" instruments and specially proteomics ones produce enormous amounts of data and standardized design of data management and reprocessing system are two key points for successful research [7]. As for these aspects bioinformatics is beginning to provide both conceptual bases and practical methods for detecting systemic functional behaviours of the cell [9]. Simultaneously computational tools have become critical for the integration representation and visualization of heterogeneous biological data.

In this review after a rapid overview about the most important high-throughput proteomics technologies and quantification methods we explore progress made in bioinformatics applied to proteomics. Specifically we will focus on quantitative methods for biomarker discovery cluster analysis and future challenges faced in this cutting edge area of research. In particular the usefulness of proteotypic peptides is emerging [10] as it can represent an important step to improve quantification investigation and diagnosis strategies.

Finally since most known mechanisms leading to disease involve different molecules this review focuses on the recent bioinformatics tools to investigate biological systems through their underlying interactions with particular attention to protein-protein interaction.

PROTEOMIC TECHNOLOGIES

The selection of the appropriate technology or combination of technologies to answer biological questions is essential for maximum coverage of the selected sub-proteome and to ensure the full interpretation and downstream utility of the data [11].

The extreme complexity of the proteome and the high dynamic range of proteins abundance required the development of protein fractionation techniques in order to identify low-abundance proteins in complex protein mixtures [12]. In this context it has been developed the strategy that combines isoelectric focusing in immobilized pH gradient strips (in-gel IEF) mass spectrometry (MS) and bioinformatics (IEF-LC-MS/MS) [13, 14].

The improvement of this approach has enabled complex proteomes such as plasma to be profiled with considerably greater dynamic range of coverage allowing confident identification of many proteins at low ng/mL concentrations [15]. Despite that none of the current proteomics technologies whether gel- or MS-based are able to identify the whole proteome by themselves [16].

The traditional way to separate proteins in complex mixtures has been 2DE and protein identification using 2DE usually relies on MS and the application of peptide mass fingerprinting [17, 18]. The 2DE technology remains widely used for studying protein profiles likely due to the relatively low-cost required to set up a laboratory compared with the much larger capital investment required for advanced technologie. Nevertheless 2DE is considered to be lacking in proteome coverage particularly for proteins having extreme Isoelectric point (pI) Molecular Weight (MW) and for membrane proteins.

In the last few years more stringent requirements for publishing identifications based on simple peptide mass fingerprinting have been put into effect. In fact additional information like interpreted peptide fragmentation spectra using tandem mass spectrometry (MS/MS) is now necessary to accept identification [19]. This has driven "shotgun proteomics" to be the most widely used method for protein identification [20].

Similarly to the shotgun sequencing approach in genomics the term "shotgun proteomics" describes a method for systematic protein identification using a combination of liquid chromatography (LC) and tandem mass spectrometry (MS/MS) [21]. In this context the MudPIT (Multidimensional Protein Identification Technology) approach has been developed [22]. In detail it involves the production of peptides from enzymatic digestion of a complex protein mixture then the initial separation first by means of strong cation exchange (SCX) using steps of increasing salt concentration followed then by C_{18}-based reverse phase (RP) chromatography using an acetonitrile gradient. Eluted peptides are directly analyzed by detecting MS and MS/MS spectra so the whole experiment can be fully automated. More importantly idle time on the most expensive part of the system the mass spectrometry is kept to a minimum (Fig. **1**).

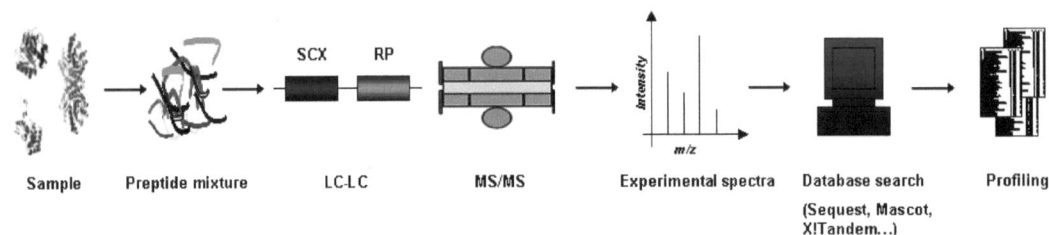

Figure 1: MudPIT (Multidimensional Protein Identification Technology) approach.

A different and interesting high-throughput proteomics method that permits rapid profiling of many samples is surface-enhanced laser desorption ionization time-of-flight (SELDI-TOF). It is based on retention of proteins due to different solid-phase chromatographic surfaces (called a Protein Chip Array) including ion exchange immobilized metals or antibodies combined with a time-of-flight mass analyzer (TOF) MS analysis allowing peak profiles of proteins similar to a second dimension of separation based on m/z values [23]. Although results from SELDI-based diagnostic studies have generated both excitement and scepticism and did not allow direct identification of proteins they remain an interesting proposition for clinical investigations in combination with other approaches for protein identification and characterization.

Among other techniques that can be used as part of a multidimensional separation set-up coupled directly to a MS there is capillary electrophoresis (CE). Advantages related to this in the context of shotgun MS-based proteomics MudPIT approach have been developed. The workflow for MudPIT scheduled different steps and the whole experiment can be automated. Protein sample is enzimatically digested (usually with trypsin) and the resulting peptide mixture is fractionated by means of liquid chromatography (usually via strong cation exchange (SCX) and reverse phase (RP)) and analyzed via tandem mass spectrometry (MS/MS). The MS precursor ion intensities can be used for peptide quantification while tandem mass (MS/MS) spectra contain sequence information. In order to perform the protein identification experimental mass spectra are correlated to peptide sequences by matching with the theoretical mass spectra produced *in silico* from referential databases.

Separation technology include high resolution power high concentration and narrow analyte bands but still now major hurdles remain about its direct interface with MS and other separation systems such as nano-liquid chromatography [24]. In fact at opposed to ion source of MS CE runs at very high-voltage (thousands of kV) and standard CE protocols use buffers that are not compatible with MS analysis. However some newer developments trying to eliminate these drawbacks [25].

Algorithms and Softwares for Proteomics

In proteomics analysis by means of gel electrophoresis several software for the comprehensive visualization exploration and analysis of 2DE gel data have been developed. Software packages include Delta2D, ImageMaster, Melanie, PDQuest, Progenesis and REDFIN - among others; in particular DeCyder, Delta2D, Progenesis and REDFIN can handle analysis of DIGE (Difference gel electrophoresis) experiments also.

Although this technology is widely utilized at the present time is affected by some limitations. In fact for example while PDQuest and Progenesis tend to agree on the quantification and analysis of well-defined well-separated protein spots they deliver different results and analysis tendencies with less-defined less-separated spots [3].

Systematic identification of proteins by means of "shotgun proteomics" involves production of thousands of spectra (10^4 for each sample) that must be interpreted. Due to the amount of data produced elaboration of mass spectrometry data is strictly related to bioinformatics application. In this direction in order to process data efficiently new software and algorithms are continuously being developed to improve protein identification and characterization in terms of high-throughput and statistical accuracy.

Tandem mass spectra (MS/MS) interpretation is based on the comparison of acquired experimental spectra versus a database of theoretical peptide fragments [26]. To perform this specific job several software and algorithms commercial or available for free have been developed in the last few years. The oldest and best-recognized algorithm that executes this function has been developed by Yates *et al.* and called SEQUEST [27]. It allows to define the peptides sequence and can automatically search post-translational modifications. To assess the quality of the match between experimental and virtual tandem mass spectra SEQUEST uses a cross correlation (Xcorr) function. Xcorr is not a probabilistic score and therefore further validation by various statistical approaches was needed. In particular DTASelect [28] Peptide Prophet and Protein Prophet [29] are widely used. At the same time the last Bioworks versions and the new software Discoverer (Thermo Fisher Scientific inc.) that are based on SEQUEST algorithm have been updated with new tools that also allow a probabilistic evaluation.

The same fame of the SEQUEST algorithm belongs to MASCOT as well probably the most widely used software for mass spectra interpretation [30]. Based on the MOWSE algorithm [31] it is probability-based and available both as a limited web service and as a commercial standalone platform.

A free available alternative to MASCOT and SEQUEST is X!TANDEM [32]. Compared to other algorithms its major advantage is the automatic search of modified peptides with a relatively low computational time.

A lot of other software and algorithms are available. Similar to X!Tandem there is OMSSA [33] an open source software based on a BLAST-like statistical model. Comparable to SEQUEST are SONAR [34] and SALSA [35] while ProbID [36] and PROFOUND [37] use a Bayesian model to calculate the probability that the peptide identification is true. Another interesting statistical approach to validate protein identification is found in Probity [38]. It allows the calculation of the statistical significance of each peptide identification reporting the risk that identification is a false positive in relation to database size.

Finally the most recent of commercial search engines are Phenyx that incorporates the true probabilistic scoring system OLAV [39] SCOPE [40] that uses a stochastic model and Spectrum Mill that is inserted in MassHunter workstation. It's supplied by Agilent with its mass spectrometers allows the identification of the proteins via fast database searches with automatic or manual match validation and unique algorithms that minimize false positives; further it also offers de novo spectral interpretation and the identification of the abundance differences of two-fold or greater without complicated isotope labeling [41].

As described above several computational solutions are now available to correlate experimental MS/MS spectra to peptide sequences generated in silico using database search engines such as SEQUEST MASCOT and others. Efforts made in this direction required computing clusters or parallel virtual machines comprised of clusters of a number of CPUs to remove the bottleneck of data handling. In this manner it's possible to increase database search speed by dividing the job into separate tasks and performing them in parallel [42]. In fact the amount of spectra collected in a typical MudPIT experiment can range from tens to hundreds of thousands foreach analyzed sample and such a high demand on computer processing and power is fundamental.

QUANTITATIVE ANALYSIS

Proteomic analyses have recently emerged as valuable strategies to identify molecular alterations in a variety of disease states. In this context quantification of proteome differences between two or more biological systems remains a challenging technical task in proteomics and it is a crucial aspect in understanding the regulation of cellular mechanisms.

In addition to the classical methods of differential protein gel or blot staining [3] mass spectrometry-based quantitation is rapidly increasing its importance. In this field quantification techniques are based upon two distinct methods one that incorporate labels into peptides or proteins prior to MS analysis and the other label-free methods [43-45].

Label Quantitation

In "label-methods" different stable isotopes are introduced into proteins or peptides to create a specific mass tag that can be recognized by a mass spectrometer to quantify proteins differences. Labeled peptides are chemically identical to their native counterparts and their behaviour during chromatographic separation or mass spectrometric analysis is identical but distinguishable owing only to a mass difference. The ratio of signal intensities among different labeled peptide pairs gives the measure of relative abundance of peptides/proteins among two or more different biological states.

In literature many ways permitting incorporation of a label into a protein or a peptide have been reported. An exhaustive and detailed description of these approaches lies outside the scope of this review so here we will only shortly describe the main aspects of major techniques.

Isotopic tags can be incorporated *in vivo* or *in vitro* during sample preparation [4546]. Initially *in vivo* labeling was described for total labeling of bacteria using 15N-enriched cell culture medium [47]. A shortcoming of this approach is the number of labeled nitrogen atoms that can vary from peptide to peptide. What overcomes this problem is the 'stable-isotope labeling by aminoacid in cell culture (SILAC) approach introduced by Mann and co-workers in 2002 [48]. In this methodology in fact the medium contains $^{13}C_6$-arginine and $^{13}C_6$-lysine which ensures that all tryptic cleavage products of a protein (except for the C-terminal peptide) carry at least one labeled amino acid resulting in a different mass shift over the non-labeled counterpart. The main advantages of SILAC are correlated to the differently treated samples that can be combined at the level of intact cells prior to sample preparation; this minimizes the potential biases error introduced by biochemical and mass spectrometric procedures. At the same time a possible limitation could be low cellular growth in adapted media.

To overcome this and other problems inherent to metabolic labelling of proteins and peptides a post-biosynthetic labelling by chemical or enzymatic derivatization *in vitro* has been developed. In their pioneering work Gygi and colleagues introduced *in vitro* strategy named "isotope-coded affinity tags" (ICAT) [43] in which cysteine residues are tagged with a reagent containing either eight or zero deuterium atoms as well as a biotin group that can be exploited for affinity purification of labelled peptides and subsequent MS analysis. Although the use of cysteine significantly reduces the complexity of the peptide mixture its use excludes from the analysis those proteins that do not contain this aminoacid. This limitation has driven the development of new strategies which employ reactive residues that occur more frequently in proteins. Among these techniques we list the isotope coded protein label (ICPL) [49] the isobaric tags for relative and absolute quantification (iTRAQ) [50] and tandem mass tags (TMT) [51].

The iTRAQ approach is widely used and it's based on the covalent labeling of the N-terminus and side chain amines of peptides with tags of varying mass. Samples tagged with different tags are pooled and usually fractionated by nano-liquid chromatography and analyzed by tandem mass spectrometry (MS/MS). Potential problem of iTRAQ experiments is that coelution in LC separation of peptides with similar mass that contribute to the same reporter ions could interfere with the quantification.

Stable-isotope labeling provides a clear benefit as the comparison between two or more samples is performed in the same analysis. However there are some inherent drawbacks to this approach due to the additional steps required for protein labeling efficiency and high costs of reagents/media.

Label Free Quantitation

As previously reported label-based approaches for protein quantitation are not always practical or feasible for different reasons such as high cost. There are simpler alternative methods called "label-free quantitation" and they are based on non-labeled samples [52]. These approaches correlate the mass spectrometric signal the number of identified peptides or other statistical parameters with the relative or absolute protein quantity. For instance as for the strategy that correlates the mass spectrometric signal with protein quantity the ion chromatograms for every peptide are extracted from an LC-MS/MS run and their mass spectrometric peak areas are integrated over the chromatographic time scale. The intensity value for each peptide in one experiment can be compared to the respective signals in one or more other experiments to yield relative quantitative information [5354].

A wide variety of other quantitative proteomic methods has been described in literature [55] and several studies revealed a relationship between protein abundance and various sampling statistics. In this scenario Gao *et al.* suggested that peptide hits are related to protein abundance [56] Florens *et al.* used protein coverage [57] Wang *et al.* normalized peptide signal intensities [44] while others used peptide SEQUEST Xcorr sum [58] or SEQUEST Score evaluation [5960].

One of the most frequently used "label-free quantitation" approaches is based on the evaluation of spectral counting (SpC) [2261]. Introduced by Yates and Washburn it is founded on the empirical observation that the more a particular protein is present in a sample the more tandem MS spectra will be collected for peptides of that protein. Further they assume that the linearity of response is the same for every protein. From these assumptions it's possible to obtain relative quantification by comparing the number of these spectra from a set of experiments. The advantage of the spectral counting approach is related to the benefits obtained from extensive MS/MS data acquisition across the chromatographic time scale; however although it is very intuitive and simply spectral counting approach is still controversial because it does not measure any direct physical property of a peptide.

Related to the spectral count there is the normalized spectral abundance factor (NSAF) and its natural log transformation that has been used for quantitative proteomics analysis by evaluation with statistical t-test [62].

By means of label-free approaches some researchers attempted to estimate the absolute protein expression levels. In particular Rappsilber *et al.* computed a protein abundance index (PAI or emPAI) dividing the number of observed peptides by the number of all possible detectable tryptic peptides for each protein [6364]. Regarding this approach important advances have been made by using computational models that predict which peptides of a given protein are likely to be detected by the mass spectrometer and thus would form a better basis for quantification. Despite this consideration estimation of the absolute protein content in complex mixtures is an open challenge and for reliable absolute quantitation the use of internal standard is fundamental [65].

Among proteomic quantification techniques by means mass spectrometry the disadvantage of using label-free approaches is that they are the least accurate because all the systematic and non-systematic variations between experiments can be reflected in the data obtained; therefore every effort should be made to control reproducibility at each step and this is possible by performing replicates analyses [66]. At the same time there are many reasons why label-free quantification should be considered. In fact it is more straightforward and less expensive than isotopically coded reagents and it seems to be particularly relevant for the direct comparison of samples in a fully automated high-throughput setting with a low time-consuming. Moreover in terms of analytical strategy there are no limitations to the number of experiments that can be compared.

Bioinformatics Tools for Quantitative Analysis

A current bottleneck in the rapid advance of proteomic technologies is the closed nature and slow development cycle of vendor-supplied software solutions [67]. This limitation has driven many laboratories to develop algorithms and in-house software for identification visualization and also quantitation using mass spectrometry data.

In relation to this aspect Mortensen *et al* created an open source software called MSQuant. It permits analysis directly on the mass spectrometric data and supports relative protein quantitation based both label and free-label ion intensity approaches. Moreover it iteratively calibrates MS data improving mass accuracy and reduces false positive identification [67].

Another interesting software recently described by Park *et al.* is Census [68]. It is available for free compatible with labeling approaches as well as with label-free analysis accepts single-stage mass spectrometry (MS) or tandem mass spectrometry (MS/MS) and high- and low-resolution mass spectrometry data. Furthermore it contains RelEx [69] and improves quantitation of low-abundance proteins.

An important aspect underlined by Carvalho *et al* is represented by normalization methods used in quantitative analysis by SpC [70]. Proteomics data normalization in fact is a fundamental aspect necessary to compare samples with the purpose of identifying differences and minimizing false positive results. To address this open issue they developed PatternLab a software that as well as different normalization strategies implements ACFold and nSVM (natural support vector machine) methods to identify protein expression differences [71].

ProteinQuant Suite is a new useful software developed by Mann and co-worker to evaluate protein quantitation [20]. It comprises three standalone complementary computer utilities namely ProtParser ProteinQuant and Turbo RAW2MGF. In particular Turbo RAW2MGF is very attractive because it allows software application to data collected from different types of mass spectrometers.

A similar feature is contained in ProtQuant software developed by Bridges *et al* [58]. It is a Java-based tool for label-free protein quantification and through its graphical interface accepts multiple file formats without size limitations.

The utility previously described focuses attention on an important question regarding common file format for mass spectrometric proteomics data. Over the years different manufacturers of mass spectrometers have developed various proprietary data formats for handling such data which makes it difficult for academic scientists to directly manipulate their data. To address this limitation several open XML-based data formats have been developed by the Trans-Proteomic Pipeline at the Institute for Systems Biology to facilitate data manipulation [7273]. In particular the Human Proteome Organization (HUPO) has developed a common file format called mzData which offers similar functionality to mzXML. The existence of two standard formats for proteomics data is an undesirable state thus mzData and mzXML developers are currently developing a unique format called mzML [74]. Of course this has determined the development of software viewers for mzXML and mzData such as MZmine [75] TOPPview [76] or many others.

The list of computational tools developed for quantitative analysis is very long. Among others the Corra software developed by Brusniak *et al.* [77] is particularly interesting because it enables appropriate statistical analysis false discovery rate and ultimately informs subsequent targeted identification of differentially abundant peptides by MS/MS. The APEX quantitative tool on the other hand is a free open source Java application which improves basic spectral count methods and aims to calculate absolute protein expression [78] while i-Tracker represent a free available software specific for iTRAQ protocol [79].

Finally MAProMA (Multidimensional Algorithm Protein Map) is a simple in-house software that allows evaluation of differentially expressed proteins by means of Sequest Score and two different algorithms called Dave (Differential average) and DCI (Differential Confidence index) [59]. It allows the comparison of up to 125 protein lists and their visualization in a format more comprehensible to biologists [80].

SRM AND PROTEOTYPIC PEPTIDES

Systems biology requires the detection and quantification of large numbers of analytes and biologically relevant molecules are often below the detection limits of shotgun MS-based proteomics. To overcome this shortcoming targeted proteomics workflows have recently been introduced [81-83]. This new approach is based on "Selected reaction monitoring" (SRM) or "Multiple reaction monitoring" (MRM) when parallel acquisition of SRM transition occurs. These two terms are equivalent but MRM has been deprecated by IUPAC nomenclature [83].

SRM is generally performed by means of triple-quadrupole (QQQ) instruments but recently a new type of MS instrument i.e. the linear ion trap has been used [84-86]. Each experiment is performed by selecting representative peptides of a protein with known *m/z* values (precursor ions) fragmenting them through collision induced dissociation (CID) and monitoring only specific pre-selected daughter fragments (product ions) that are characteristic to each precursor [838687]. The combination of a precursor-product m/z values is known as a 'transition' and is highly specific for a given peptide aminoacid sequence.

The aim of SRM is quantification of predetermined set of peptides previously identified with known MS/MS fragmentation pattern and specific for each targeted protein. This set of peptides has been called "proteotypic peptides". Initially they were defined as peptides in a protein sequence that is most likely to be confidently observed by current MS-based proteomics methods [8188]; while recently the definition has been refined as peptides that uniquely identify the targeted proteins [10].

The usefulness of proteotypic peptides required the development of MS data repositories and computational tools to predict their sequences. Peptide Atlas [89] Human proteinpedia [90] GPM Proteomics database [88] PRIDE [91] and ISPIDER central [92] are the more important specific databases used to collect spectra and sequences of these peptides.

These data resources have also been used from bioinformaticians to attempt to design algorithms for prediction of the most likely MS-observable peptides [1093]. In relation to this aim Webb-Robertson *et al.* [94] applied a support vector machine (SVM) model that uses a simple descriptor space based on 35 properties of aminoacid content charge hydrophilicity and polarity for the quantitative prediction of proteotypic peptides. Similarly Mallick *et al.* [10] developed a computational tool Pepdite Sieve studying physicochemical characteristic properties of about 600000 peptides identified in yeast analysis while Sanders *et al.* [95] used a methodology for constructing artificial neural networks to predict which peptides are potentially observable for a given set of experimental instrumental and analytical conditions for 2DC-MS/MS datasets.

To point targeted MS experiments towards a specific biological question the increasing amounts of information on pathways protein interactions gene expression changes and gene ontologies can be accessed using public databases to establish a list of target proteins.

A software system to support the set up of SRM experiments would guide the user through the critical sequential steps just discussed and represented in (Fig. **2**).

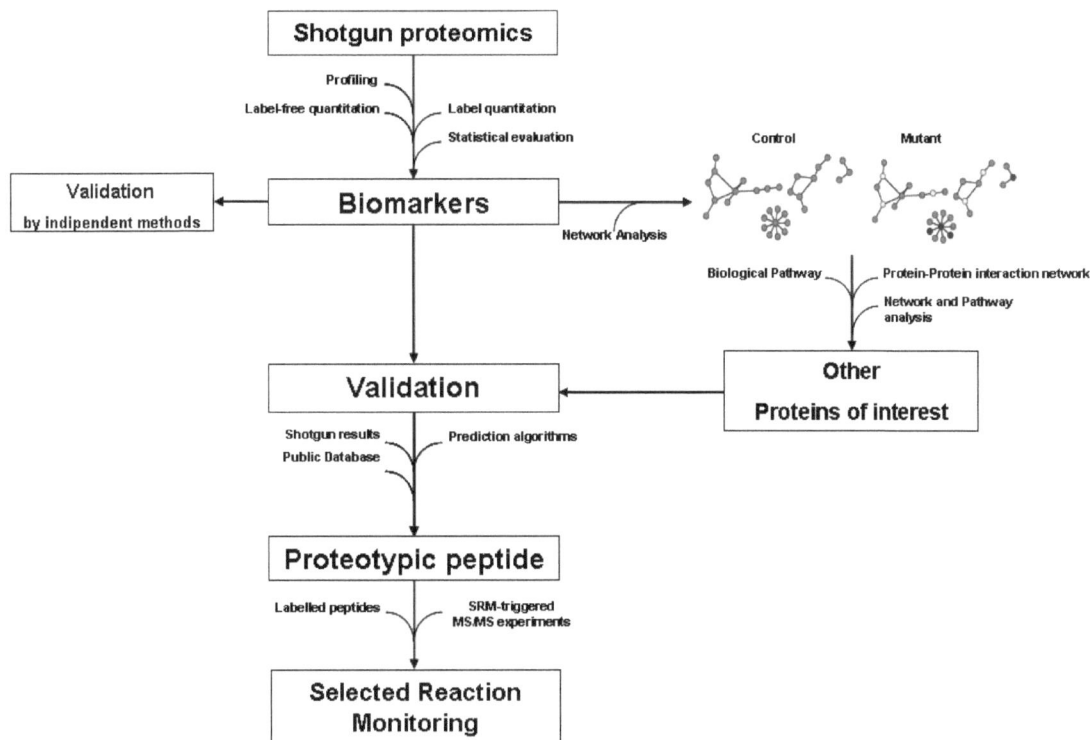

Figure 2: Strategies for biomarker discovery by means of MS-based proteomics.

Until now the majority of the developed software to support the setup of SRM assays for targeted proteomics is made up of commercial solutions. These platform-specific tools include for example MRMPilot (Applied Biosystems) SRM Workflow Software (Thermo Scientific) VerifyE (Waters) and Optimizer (Agilent Technologies). However a free available software TIQAM (Targeted Identification for Quantitative Analysis by multiple reaction monitoring (MRM)) has recently been released [83].

CLUSTERING

The clustering problem has been addressed in many contexts and in many disciplines by researchers; this reflects its broad appeal and usefulness as one important step in exploratory data analysis. However clustering is a difficult combinatorial problem and differences in assumptions and contexts in the scientific communities have made the transfer of useful generic concepts and methodologies slow to occur.

Cluster analysis encompasses different methods and algorithms for grouping objects of similar kinds into respective categories. The best known technique is hierarchical clustering

The figure shows a workflow where different approaches are combined for measuring the relative and/or absolute changes in protein expression. A primary evaluation of biomarker discovery is perfomed using shotgun proteomics and label- or free-label quantitative methods. Biomarkers are used to identify sub-network or specific pathway potentially modulated by different conditions. Identification of potential biomarkers is confirmed via orthogonal methods such as immunoassay or validated with Selected Reaction Monitoring (SRM). Implication of bioinformatics support is crucial in each represented step. Protein profiling protein quantitation validation by means of SRM or network analysis are strictly depended by bionformatics applications.

Especially the agglomerative one in which a graph called dendrogram is produced to represent the distribution of the data considered [96]. It belongs to "unsupervised learning" a technique of statistical data analysis employed in many fields (Fig. **3**).

With respect to the measurement of similarity and to the linking method chosen objects are clustered in nested groups from the unique root till all the leaves like in tree-structure

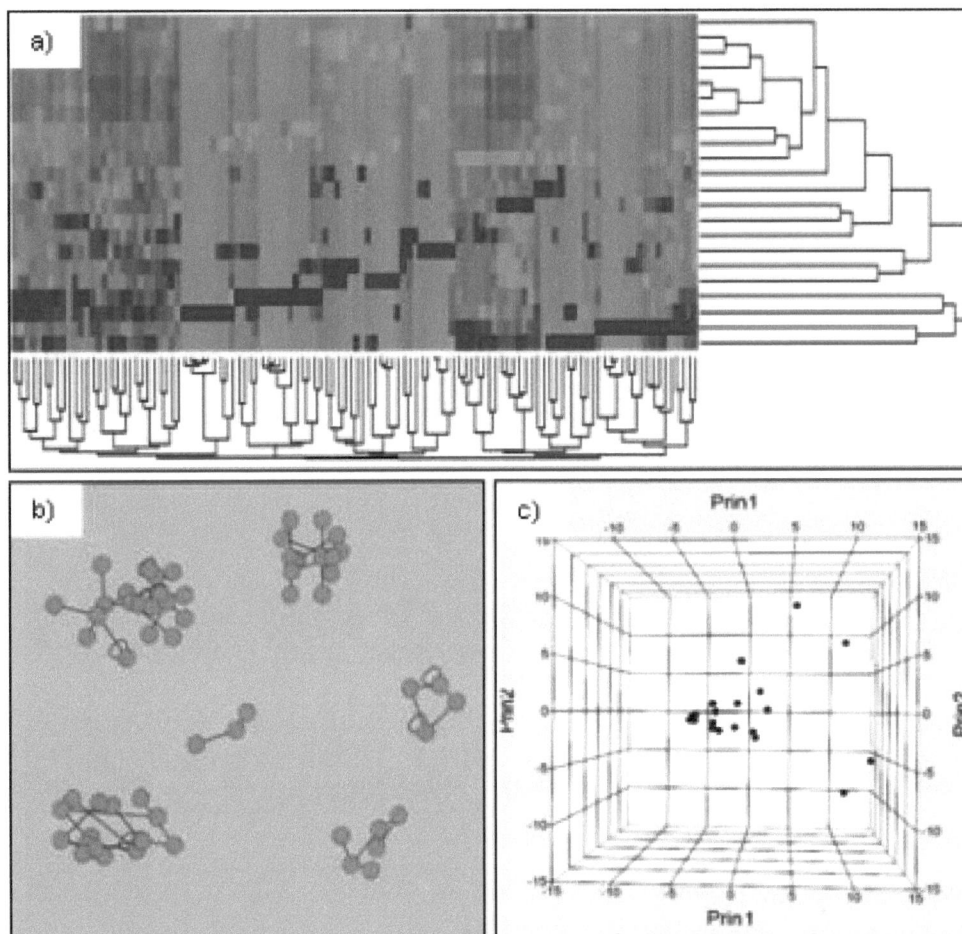

Figure 3: Cluster analysis applications.

Cluster analysis is the unsupervised classification of patterns (observations data items or feature vectors) into groups (clusters). Hierarchical clustering algorithms produce a nested series of partitions based on a criterion for merging or splitting clusters based on similarity. Heat map analysis of proteins or gene expression allows a rapid identification of common traits that significantly change between two o more groups (a). Clustering algorithms are widely used also in network analysis. Several Cytoscape plugins such as ClusterViz [http://clusterviz.sourceforge.net/] or ClusterMaker [http://www.cgl.ucsf.edu/cytoscape/ cluster/clusterMaker.html] include Hierarchical or k-Means for clustering expression or genetic data and MCL (Markov Cluster Algorithm) or FORCE for clustering similarity networks (b). Similarly to hierachical clustering Principal component analysis (PCA) is widely used in many disciplines. It involves a mathematical procedure that transforms a number of possibly correlated variables into a smaller number of uncorrelated variables called principal components (c) representation. The distance measure defines how the similarity of two elements is calculated and the most famous and used metric is the Euclidean distance. Another important parameter is the linkage method. This is a function of the pairwise distances between observations that define which clusters should be merged and which ones should be splitted [97]. In this category there is a large variety of linkage criteria such as single-linkage average-linkage UPGMA or Ward's method.

Since clustering is basically a mathematical approach on which is possible to apply a variety of distance measurements and linkage rules speaking univocally about "cluster analysis" is incorrect. Same procedures that diverge in few features can produce overturned results so it is necessary to specify which methodology has been followed to provide better comprehensible results.

Despite these considerations with the sufficient forethought clustering techniques were employed in a wide variety of biological studies during the last few years in particular they were applied on data derived from high-throughput methodologies such as DNA microarrays [98] or for measuring gene or protein expression in a biological system and for grouping those with similar expression patterns and possibly share common biological pathways [98-102].

Meunier *et al.* published an exhaustive technical analysis regarding hierarchical clustering methodology as a powerful data mining approach for a first exploration of proteomic data [103]. In this context cluster analysis has been applied on data obtained by 2DE to map proteins expression patterns in breast cancer [104] and for the classification of breast tumor tissues [105]. Unsupervised cluster analysis used recent high-throughput proteomic approaches such as MudPIT to partition the cell lines in a manner that reflected their motile/invasive capacity [106].

Clustering has been applied to several proteomic datasets to reduce complexity grouping identified spectra thus reducing potential false positive identifications [107]. Hierarchical cluster analyses and principle component analyses (PCA) were also used to evaluate the prediction capability of extracted data obtained from the analyses with two-dimensional liquid chromatography-tandem mass spectrometry [108]; the results suggested that those methods may be useful for large-scale clinical proteomic profiling.

By means of MudPIT approach the analysis of proteomes from human tissues showed great reproducibility (approximately 80%) and showed potential applications in disease diagnosis and classification [109]. As for biomarker discovery unsupervised cluster analyses have been used in high throughput proteomics profiling of secretomes making it easier to evaluate the process of data reproducibility [110].

However there is still a number of unclear aspects when cluster analysis is applied to proteomics in clinical fields. Considerations regarding precision and sensitivity of the method are linked to quality of data and to clustering parameters such as distance measure or grouping methods and prior to data normalization and reduction. Due to the enormous variety of high-throughput technologies algorithms softwares and data formats the optimization of the clustering methodologies should be done very carefully. No optimal and univoquous settings for clustering parameters have been obtained up to now in proteomics. This surely is the main bottleneck considering the employment of clustering techniques in MS/MS proteomics and will be one of the nodal points in the development of bioinformatics platforms for clinical applications.

The selection of clustering method to be used is therefore a daunting task for the researcher conducting the experiment. An additional related problem is the determination of the number of clusters that is most appropriate for the data considered. Ideally the resulting clusters should not only have good statistical properties (compact well-separated connected and stable) but also give results that are biologically relevant. It has been proposed a variety of measures to validate the results of cluster analysis and to determine which clustering algorithm presents the best performance for specific experiment [111-113].

Many of the clustering methods are in use today and result from a cross-fertilization between several disciplines such as biology mathematics and statistics. The result is a considerable amount of "ad hoc" developments and the 'reinvention' of new and existing algorithms.

Many algorithms for cluster analysis have been developed and implemented by means of different programming languages and environments for statistical computing and graphics such as R a free statistical software (www.r-project.org) and MATLAB (www.mathworks.com) a commercial software; these are solutions widely used by engineers and scientists in industry government and education.

NETWORK ANALYSIS

Reductionism which has dominated biological research for over a century has provided a wealth of knowledge about individual cellular components and their functions. Despite its enormous success it is increasingly clear that a discrete biological function can only rarely be attributed to an individual molecule [114]. Thus one of the aims of post-genomic biomedical research is to systematically catalogue all molecules and their interactions within a living cell.

Understanding the roles and consequences of these interactions is fundamental for the development of systems biology as well as for the development of novel therapies [115]. The post-genomic era is characterized by vast amounts of data from various sources creating a need for new tools to extract biologically meaningful information. A major challenge for biologists and bioinformaticians is to gain understanding of cell functions by integrating these available data into an accurate cellular model that can be used to generate hypotheses for testing.

In recent years systems biology approaches have evolved in two distinct directions namely "computational systems biology" that uses modeling and simulation tools and "data-derived systems biology" that relies on "-omics" datasets [116].

Network Analysis Representation

From these new areas of research some topics such as dynamic network [115] graph inference [117] and graph analysis [118] are in continuous growth. Each one uses systems biology analysis as a method for rationalizing biological knowledge that attempts to go far beyond heat maps and/or gene ontology classification [119].

Biomolecular interactions play a role in the majority of cellular processes that are regulated connecting numerous constituents such as DNA RNA proteins and small molecules. Each of these types of interactions can be interpreted as a network and can be divided into two major categories: pathways and interaction networks.

Biological pathways include metabolic regulatory and signaling networks; while interactions networks make up the second category in which the nodes represent biological entities and edges represent some forms of interaction or relationships [114]. As for protein–protein interactions the networks can be represented as undirected graphs in which nodes represent the proteins and edges represent direct physical interactions (Fig. **4**).

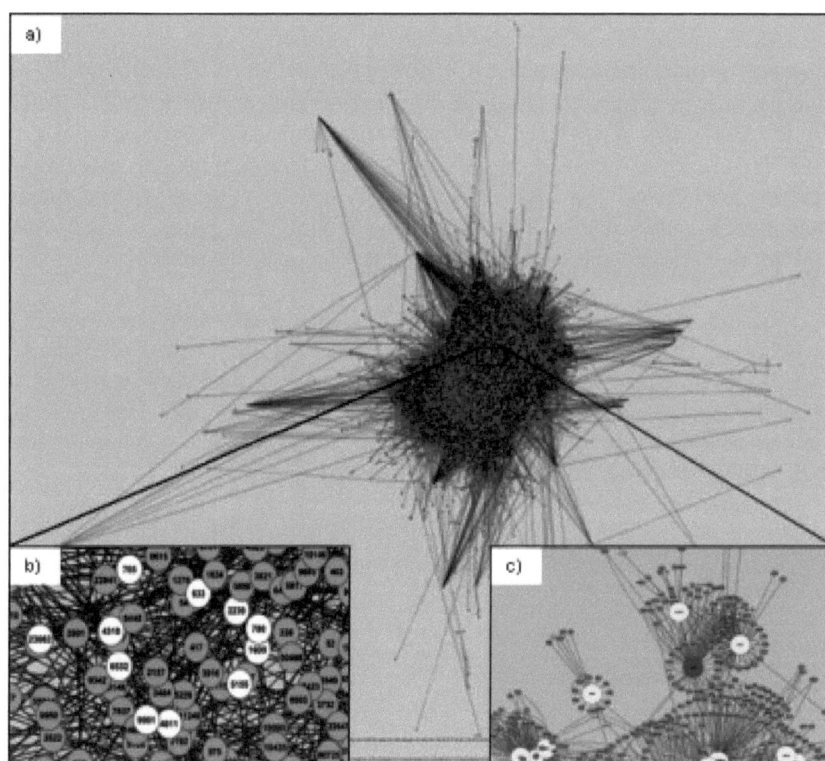

Figure 4: Protein-protein interaction network.

Human protein-protein interaction network created by means of Cytoscape software (a). The network has been downloaded from HPRD [http://www.hprd.org/] database and is composed by literature curated interactions. Protein-protein interaction networks are usually abstracted as undirected graphs where the nodes represent the proteins and edges direct physical interactions; each edge is associated to a probabilistic score (b). Proteomics data obtained by means of shotgun proteomics represent an ideal input to analyze protein-protein interactions networks. Protein profiling and the differentially expressed proteins are mapped on the network to identify branches (sub-networks) which change consistently between investigated conditions (c).

On the other hand gene-regulatory networks can be abstracted to directed graphs where nodes symbolize the genes encoding transcriptional factors (or other types of proteins) and links correspond to the transcriptional regulation. Finally metabolic networks can be represented as bipartite graphs in which nodes are separated into two sets: enzymes and substrates [120].

The distinction between different types of pathways or networks is related to human representation; so it is purely virtual and it is not related to any intrinsic structure in the cell or organism. This abstraction is a natural result of the desire to rationalize complex systems and to facilitate data integration but it must be kept in mind that these networks are neither static nor well defined concepts [121].

While at the beginning networks have been used to represent important biological processes and routinely to show relationships between biologically relevant molecules more recently the use of networks in biology has changed from purely illustrative and didactic to a more analytic purpose and the shift was partially the result of the confluence of advances in computer science and high-throughput techniques in systems biology.

From a visualization standpoint the real power is the ability to map expression mutations or post translational modifications onto pathways to reveal or suggest how the pathway and its components are modulated under different sets conditions included disease states. Thus the ability to analyze a variety of data sources and to map those data onto pathways is crucial [87 106 122].

Regarding proteomics data an ideal input to analyze protein-protein interactions networks would be the protein profiling and the lists of modulated proteins between two or more biological states. The aim is to identify branches (sub-networks) which change consistently between investigated conditions (Fig. **4**). However only a minority among all potential down- and up-regulated proteins is detected in a typical proteomics experiment and therefore measurements alone cover a small fraction of the network only. For this reason any meaningful sub-network will involve many unquantified proteins. Furthermore our current knowledge of bio-molecular interactions in terms of cataloguing interactions and understanding their biophysical properties is still limited to few organisms and is hindered by the limitations of existing technologies primarily in throughput and reproducibility [123].

Database for Network Analysis

In spite of these bottlenecks specific databases and bioinformatics tools are continuously under development and are increasing their importance. The exponential accumulation of molecular-biological intracellular data required the availability of several repositories to collect store and share experimental data.

An exhaustive overview of existing databases is available through the Pathguide website (http://www.pathguide.org/). It is a useful web resource where about 300 biological pathways and interaction networks resources are stored and classified. Among these HPRD [124] MINT [125] IntAct [126] Reactome [127] DIP [128] and BioGrid [129] are some of more important databases where it is possible find information about mammalian protein-protein interaction. Other similar databases such as OPHID [130] HPID [131] IntNetDB [132] STRING [133] and POINT [134] on the other hand infer mammalian protein–protein interactions using orthologs.

Finally systems such as Atlas [135] BIOZON [136] INTEGRATOR [137] or Gaggle [138] provide integration and querying capabilities with heterogeneous biological data.

Software

The main goal of the analytical branch of systems biology is to develop computational tools for which the input are the '-omics' data and the output is a list of activated sub-networks or pathways in the studied biological system [106]. At the same time to help the scientists apprehending their networks of interest it is fundamental to visualize and analyze them under several aspects.

The trend in the development of specific tools is to go beyond 'static' representations of cellular state towards a more dynamic model of cellular processes through the incorporation of expression data sub-cellular localization information and time-dependent behaviour [139]. Today in order to visualize and explore biological networks there are many tools available including well-known examples such as Cytoscape VisANT Pathway Studio or PATIKA [139].

Among these probably Cytoscape is the most famous open source software [140]. It is a Java application whose source code is released under the Lesser General Public License (LGPL). The major strength of Cytoscape is the large user and developer base that continuously develops new plugins for the system. In fact more than 60 plugins are currently available for tasks such as importing and visualizing networks from various data formats generating networks from literature searches or for analyzing them.

PathSys [141] is a software platform for biological pathways analysis querying and visualization that similarly to Cytoscape and its Bionetbuilder plugin includes a general-purpose scalable warehouse of biological information which integrates over 20 curated and publicly contributed data sources biological experimental and PubMed data for the 8 representative genomes (*S. cerevisiae D. melanogaster* etc.). Like PATIKA [142] PathSys supports SQL-like queries that can explore network properties such as connectivity and node degree. Moreover after importing expression data users can apply sorting normalization and clustering algorithms on the data and then create various tables heat maps and network views of the data.

Table 1: Summary of bioinformatics tools resources and databases for proteomics analysis

Data analysis softwares for protein identification in mass spectrometry-based proteomics			
Software	**License**	**WebSite**	**Reference**
SEQUEST	Proprietary	http://fields.scripps.edu/sequest/index.html	[27]
MASCOT	Proprietary	http://www.matrixscience.com/	[30]
X!Tandem	Open Source	http://www.thegpm.org/	[32]
OMSSA	Open Source	http://pubchem.ncbi.nlm.nih.gov/omssa/	[33]
SONAR	Web-based	http://65.219.84.5/service/prowl/sonar.html	[34]
SALSA	Proprietary	http://www.mc.vanderbilt.edu/lieblerlab/salsa_overview.php	[35]
ProbID	Proprietary	https://products.appliedbiosystems.com/	[36]
PROFOUND	Web-based	http://prowl.rockefeller.edu/prowl-cgi/profound.exe	[37]
Phenyx (OLAV)	Proprietary	http://www.genebio.com/products/phenyx/	[39]
Probity (algorithm)	-	-	[38]
SCOPE (algorithm)	-	-	[40]
Spectrum Mill	Proprietary	http://www.chem.agilent.com/	[41]
Scaffold	Proprietary	http://www.proteomesoftware.com/	-
Prospector	Proprietary	http://prospector.ucsf.edu/prospector/mshome.htm	[145]
Sherpa	Proprietary Freeware Old version MacOS	http://www.hairyfatguy.com/Sherpa/	[146]
MyriMatch	Open Source	http://fenchurch.mc.vanderbilt.edu/lab/software.php	[147]
Greylag	Open Source	http://greylag.org/	-
ByOnic	Web-based	http://bio.parc.xerox.com/	[148]
InsPecT	Open Source	http://proteomics.ucsd.edu/Software/Inspect.html	[149]
SIMS	Open Source	http://emililab.med.utoronto.ca/	[150]
Tools for quantitative proteomics			
Software	**License**	**WebSite**	**Reference**
ProteinQuant Suite	-	-	[20]
ProtQuant	Freeware	http://www.agbase.msstate.edu/tools.html	[58]
MSQuant	Open Source	http://msquant.sourceforge.net/	[67]
Census	Proprietary free for academic use	http://fields.scripps.edu/census/index.php	[68]
PatternLab	Proprietary	http://pcarvalho.com/patternlab/downloads/windows/patternlab/	[71]
CORRA	Open Source	http://sourceforge.net/projects/corra/	[77]
APEX	Open Source	http://pfgrc.jcvi.org/index.php/bioinformatics/apex.html	[78]
I-Tracker	Open Source	http://www.cranfield.ac.uk/health/researchareas/bioinformatics/page6801.jsp	[79]
MaProMa	Proprietary	-	[80]
Tools for SRM			
Software	**License**	**WebSite**	**Reference**
MRMPilot (Applied Biosystem)	Proprietary	https://products.appliedbiosystems.com/ab/en/US/adirect/ab?cmd=catNavigate2&catID=605354	-
SRM Workflow (Thermo)	Proprietary	http://www.thermo.com/	-
Verifye (Waters)	Proprietary	http://www.waters.com/	-
Optimizer (Agilent)	Proprietary	http://www.chem.agilent.com/en-US/Support/Downloads/Patches/MHDataAcq/optimizer/Pages/mh_opt_ssb.aspx	-
TIQAM	Open Source	http://tools.proteomecenter.org/TIQAM/TIQAM.html	[83]
Proteotypic peptides repositories			
Software	**WebSite**		**Reference**
GPM Proteomic Database	http://www.thegpm.org/		[88]
Peptide Atlas	http://www.peptideatlas.org/		[89]
Human Proteinpedia	http://www.humanproteinpedia.org/		[90]
PRIDE	http://www.ebi.ac.uk/pride/		[91]
Ispider central	http://www.ispider.manchester.ac.uk/cgi-bin/ispider.pl		[92]
NCBI peptidome	http://www.ncbi.nlm.nih.gov/peptidome/		[151]

Table 1: cont....

Resources for cluster analysis

Software	License	WebSite	Reference
mclust (R package)	Open Source	http://cran.r-project.org/web/packages/mclust/index.html	[112]
cluster (R package)	Open Source	http://cran.r-project.org/web/packages/cluster/	-
fpc (R package)	Open Source	http://cran.r-project.org/web/packages/fpc/index.html	-
pvclust (R package)	Open Source	http://cran.r-project.org/web/packages/pvclust/index.html	[152]
clValid (R package)	Open Source	http://cran.r-project.org/web/packages/clValid/index.html	[153]
SciPy-Cluster (Python Module)	Open Source	http://scipy-cluster.googlecode.com	-
MATLAB Clustering toolbox	No BSB License	http://www.mathworks.com/matlabcentral/fileexchange/7486	-

Tools for Network Analysis

Software	License	WebSite	Reference
Osprey	Proprietary free for academic use	http://biodata.mshri.on.ca/osprey/servlet/Index	[129]
Pathway Studio	Proprietary	http://www.ariadnegenomics.com/products/pathway-studio/	[139]
Cytoscape	Open Source	http://www.cytoscape.org/	[140]
PathSys	Proprietary	http://biologicalnetworks.net/PathSys/index.php	[141]
PATIKA	Proprietary/non-profit use only	http://www.patika.org/	[142]
VisANT	Proprietary free for all no-profit users	http://visant.bu.edu/	[143]
Proviz	Open Source	http://cbi.labri.fr/eng/proviz.htm	[144]
Cell Illustrator	Proprietary player version is free	http://www.cellillustrator.com/home	[154155]
Ingenuity	Proprietary	http://www.ingenuity.com/	-
Cell Designer	Proprietary it is free to use	http://www.celldesigner.org/	[156]

Protein-protein interactions Database

Database	WebSite	Reference
HPRD	http://www.hprd.org/	[124]
MINT	http://mint.bio.uniroma2.it/mint/Welcome.do	[125]
IntAct	http://www.ebi.ac.uk/intact/main.xhtml	[126]
Reactome	http://www.reactome.org/	[127]
DIP	http://dip.doe-mbi.ucla.edu/dip/Main.cgi	[128]
Biogrid	http://www.thebiogrid.org/	[129]
OPHID	http://ophid.utoronto.ca/ophidv2.201/	[130]
HPID	http://wilab.inha.ac.kr/hpid/	[131]
InNetDB		[132]
STRING	http://string.embl.de/	[133]
POINT		[134]
BIOZON	http://www.biozon.org/	[136]
INTEGRATOR		[137]
KEGG	http://www.genome.jp/kegg/pathway.html	[157]
Panther	http://www.pantherdb.org/	[158]
Human Protein Atlas	http://www.proteinatlas.org/	[159]
BIND	www.bind.ca/	[160]
ProNav	http://mysql5.mbi.ucla.edu/	[161]

General proteomics Database

Database	WebSite	Reference
David	http://david.abcc.ncifcrf.gov/	[162]
Uniprot	http://www.uniprot.org/	[163]
Swiss-Prot	http://www.expasy.org/sprot/	[164]
EBI	http://www.ebi.ac.uk/Databases/protein.html	[165]
NCBI	http://www.ncbi.nlm.nih.gov/guide/proteins/	[166]

The "PATIKA Project" is a web-based visual editor that allows accessing to several biological databases that contain pathway information. It has been implemented using "Java Server Pages" and is publicly available for non-profit use. Its main characteristics are the high-quality visualizations using the "Tom Sawyer Visualization" software and the compatibility with SQL-like queries on node and edge properties.

Like Cytoscape VisANT is a Java application that can be extended using plugins and is freely available [143]. It supports creation visualization and analysis of mixed networks i.e. networks containing both directed and undirected links. In particular VisANT implements algorithms for analyzing node degrees clusters path lengths network motifs and network randomizations.

Finally Osprey a Java application that can be used free of charge is one of the first tools specifically designed to visualize and analyze large networks [129] while ProViz [144] is a software implemented in C++ and released under the GNU General Public License GPL.

In contrast to the majority of softwares described above few commercial solutions have been developed. These softwares such as Cell Illustrator (www.cellillustrator.com) or Ingenuity (http://www.ingenuity.com/) are not available free of charge and represent a valid choice for simulating and representing biological systems.

CONCLUSION

Post-genomics research is characterized by a very relevant amount of data which require powerful informatics supports for their organization and interpretation. This necessity has driven bioinformaticians to develop new modelling methods at various levels of sophistication and has opened new frontiers for "specialized" bioinformatics in several fields of pure and applied research.

Mass spectrometry-based proteomics approaches have emerged as powerful technologies suitable to investigate biological problems and their relevance has increased in parallel with a rapid evolution of bioinformatics science in this area of research. As widely reported in this review bioinformatics tools have a primary importance in the support of proteomics research and in improving the development of the technologies themselves.

Proteomics experiments are always accompanied by issues related to analysis and interpretation which require new methodological approaches that can be found in the field of applied mathematics and informatics. Several topics such as data storage protein profiling and biomarker discovery have been widely faced in the last few years. On the other hand other important research areas such as protein-protein interaction or proteotypic peptide investigation are at an initial phase of exploration. However all of these issues are still strictly dependent on bioinformatics applications (Fig. **5**).

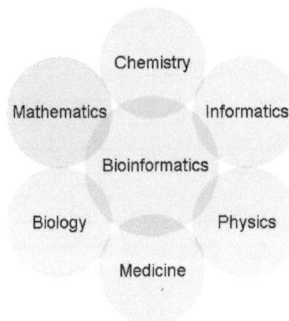

Figure 5: Multidisciplinarity of bioinformatics applications.

The flow of information produced by high-throughput analytical techniques coupled with advances in computing power has enabled scientists to analyze biological systems in novel ways. This aspect has contemporary driven biomedical researchers and bioinformaticians to gain understanding of the workings of the cell by integrating heterogeneous data into an accurate model that can be used to generate hypothesis for testing. For obtaining this goal leading the way to personalized medicine further advances in "-omics" technologies and in bioinformatics science will still be necessary. In this context bioinformatics is surely a key ingredient for the biomedical science and making today's biological and medical sciences a field rich in opportunities.

ACKNOWLEDGEMENTS

Support by joint grants from Fondazione CARIPLO (Proteomic platform Operational Network for Biomedicine Excellence in Lombardy project) is gratefully acknowledged.

REFERENCES

[1]   Vitzthum F, Behrens F, Anderson NL, Shaw JH. Proteomics: from basic research to diagnostic application. A review of requirements and needs. J Proteome Res 2005; 4: 1086-1097.

[2]   Beretta L. Proteomics from the clinical perspective: many hopes and much debate. Nat Methods 2007; 4: 785-786.

[3]   Alban A, David SO, Bjorkesten L, Andersson C, Sloge E, Lewis S, Currie I. A novel experimental design for comparative two-dimensional gel analysis: two-dimensional difference gel electrophoresis incorporating a pooled internal standard. Proteomics 2003; 3: 36-44.

[4]   Sawyers CL. The cancer biomarker problem. Nature 2008; 452: 548-552.

[5]   Zhao Y, Lee WNP, Xiao GG. Quantitative proteomics and biomarker discovery in human cancer. Expert Rev Proteomics 2009; 6: 115-118.

[6]   Marrer E, Dieterle F. Promises of biomarkers in drug development-a reality check. Chem Biol Drug Des 2007; 69: 381-394.

[7]   Hamacher M, Stephan C, Meyer HE, Eisenacher M. Data handling and processing in proteomics. Expert Rev Proteomics 2009; 6: 217- 219.

[8]   Mueller LN, Brusniak MY, Mani DR, Aebersold R. An assessment of software solutions for the analysis of mass spectrometry based quantitative proteomics data. J Proteome Res 2008; 7: 51-61.

[9]   Kanehisa M, Bork P. Bioinformatics in the post-sequence era. Nat Genet 2003; 33: 305-310.

[10] Mallick P, Schirle M, Chen SS, Flory MR, Lee H, Martin D, Ranish J, Raught B, Schmitt R, Werner T, Kuster B, Aebersold R. Computational prediction of proteotypic peptides for quantitative proteomics. Nat Biotechnol 2007; 25: 125-131.

[11] Matt P, Fu Z, Fu Q, Eyk JEV. Biomarker discovery: proteome fractionation and separation in biological samples. Physiol Genomics 2008; 33: 12-17.

[12] Anderson NL, Anderson NG. The human plasma proteome: history character and diagnostic prospects. Mol Cell Proteomics 2002; 1: 845-867.

[13] Giorgianni F, Desiderio DM, Beranova-Giorgiann S. Proteome analysis using isoelectric focusing in immobilized ph gradient gels followed by mass spectrometry. Electrophoresis 2003; 24: 253-259.

[14] Hubner NC, Ren S, Mann M. Peptide separation with immobilized pi strips is an attractive alternative to in-gel protein digestion for proteome analysis. Proteomics 2008; 8: 4862-4872.

[15] Qian WJ, Jacobs JM, Liu T, Camp DG, Smith RD. Advances and challenges in liquid chromatography-mass spectrometry-based proteomics profiling for clinical applications. Mol Cell Proteomics 2006; 5: 1727-1744.

[16] Conrads TP, Hood BL, Petricoin EF, Liotta LA, Veenstra TD. Cancer proteomics: many technologies one goal. Expert Rev Proteomics 2005; 2: 693-703.

[17] Mann M, Hjrup P, Roepstor P. Use of mass spectrometric molecular weight information to identify proteins in sequence databases. Biol Mass Spectrom 1993; 22: 338-345.

[18] Horgan GW. Sample size and replication in 2D gel electrophoresis studies. J. Proteome Res 2007; 6: 2884-2887.

[19] Bradshaw RA, Burlingame AL, Carr S, Aebersold R. Reporting protein identification data: the next generation of guidelines. Mol Cell Proteomics 2006; 5: 787-788.

[20] Mann B, Madera M, Sheng Q, Tang H, Mechref Y, Novotny MV. Proteinquant suite: a bundle of automated software tools for label-free quantitative proteomics. Rapid Commun Mass Spectrom 2008; 22: 3823-3834.

[21] Wolters DA, Washburn MP, Yates JR. An automated multidimensional protein identification technology for shotgun proteomics. Anal Chem 2001; 73: 5683-5690.

[22] Washburn MP, Wolters D, Yates JR. Large-scale analysis of the yeast proteome by multidimensional protein identification technology. Nat Biotechnol 2001; 19: 242-247.

[23] Wulfkuhle JD, Paweletz CP, Steeg PS, Petricoin EF, Liotta L. Proteomic approaches to the diagnosis treatment and monitoring of cancer. Adv Exp Med Biol 2003; 532: 59-68.

[24] Wang Y, Rudnick PA, Evans EL, Li J, Zhuang Z, Devoe DL, Lee CS, Balgley BM. Proteome analysis of microdissected tumor tissue using a capillary isoelectric focusing-based multidimensional separation platform coupled with ESI-tandem MS. Anal Chem 2005; 77: 6549-6556.

[25] Kraly J, Fazal MA, Schoenherr RM, Bonn R, Harwood MM, Turner E, Jones M, Dovichi NJ. Bioanalytical applications of capillary electrophoresis. Anal Chem 2006; 78: 4097-4110.

[26] Sadygov RG, Cociorva D, Yates JR. Large-scale database searching using tandem mass spectra: looking up the answer in the back of the book. Nat Methods 2004; 1: 195-202.

[27] Eng JK, McCormack AL, Yates JRI. An approach to correlate tandem mass spectral data of peptides with amino acid sequences in a protein database. J Am Soc Mass Spectrom 1994; 5: 976-989.

[28] Tabb DL, McDonald WH, Yates JR. Dataselect and contrast: tools for assembling and comparing protein identifications from shotgun proteomics. J Proteome Res 2002; 1: 21-26.

[29] Nesvizhskii AI, Keller A, Kolker E, Aebersold R. A statistical model for identifying proteins by tandem mass spectrometry. Anal Chem 2003; 75: 4646-4658.

[30] Perkins DN, Pappin DJ, Creasy DM, Cottrell JS. Probability-based protein identification by searching sequence databases using mass spectrometry data. Electrophoresis 1999; 20: 3551-3567.

[31] Pappin DJ, Hojrup P, Bleasby AJ. Rapid identification of proteins by peptide-mass fingerprinting. Curr Biol; 1993; 3: 327-332.

[32] Craig R, Beavis RC. Tandem: matching proteins with tandem mass spectra. Bioinformatics 2004; 20: 1466-1467.

[33] Geer LY, Markey SP, Kowalak JA, Wagner L, Xu M, Maynard DM, Yang X, Shi W, Bryant SH. Open mass spectrometry search algorithm. J Proteome Res 2004; 3: 958-964.

[34] Field HI, Fenyo D, Beavis RC. Radars a bioinformatics solution that automates proteome mass spectral analysis optimises protein identification and archives data in a relational database. Proteomics 2002; 2: 36-47.

[35] Liebler DC, Hansen BT, Jones JA, Badghisi H, Mason DE. Mapping protein modifications with liquid chromatography-mass spectrometry and the salsa algorithm. Adv Protein Chem 2003; 65: 195-216.

[36] Zhang N, Aebersold R, Schwikowski B. ProbID: a probabilistic algorithm to identify peptides through sequence database searching using tandem mass spectral data. Proteomics 2002; 2: 1406-1412.

[37] Zhang W, Chait BT. Profound: an expert system for protein identification using mass spectrometric peptide mapping information. Anal Chem 2000; 72: 2482-2489.

[38] Eriksson J, Fenyö D. Probity: a protein identification algorithm with accurate assignment of the statistical significance of the results. J Proteome Res 2004; 3: 32-36.

[39] Colinge J, Masselot A, Giron M, Dessingy T, Magnin J. Olav: towards high-throughput tandem mass spectrometry data identification. Proteomics 2003; 3: 1454-1463.

[40] Bafna V, Edwards N. Scope: a probabilistic model for scoring tandem mass spectra against a peptide database. Bioinformatics 2001; 17: Suppl 1 S13-S21.

[41] Brambilla F, Resta D, Isak I, Zanotti M, Arnoldi A. A label-free internal standard method for the differential analysis of bioactive lupin proteins using nano HPLC-chip coupled with ion trap mass spectrometry. Proteomics 2009; 9: 272-286.

[42] Sadygov RG, Eng J, Durr E, Saraf A, McDonald H, MacCoss MJ, Yates JR. Code developments to improve the efficiency of automated ms/ms spectra interpretation. J Proteome Res 2002; 1: 211- 215.

[43] Gygi SP, Rist B, Gerber SA, Turecek F, Gelb MH, Aebersold R. Quantitative analysis of complex protein mixtures using isotope-coded affinity tags. Nat Biotechnol 1999; 17: 994-999.

[44] Wang W, Zhou H, Lin H, Roy S, Shaler TA, Hill LR, Norton S, Kumar P, Anderle M, Becker CH. Quantification of proteins and metabolites by mass spectrometry without isotopic labeling or spiked standards. Anal Chem 2003; 75: 4818-4826.

[45] Bantscheff M, Schirle M, Sweetman G, Rick J, Kuster B. Quantitative mass spectrometry in proteomics: a critical review. Anal Bioanal Chem 2007; 389: 1017-1031.

[46] Ong SE, Mann M. Mass spectrometry-based proteomics turns quantitative. Nat Chem Biol 2005; 1: 252-262.

[47] Oda Y, Huang K, Cross FR, Cowburn D, Chait BT. Accurate quantitation of protein expression and site-specific phosphorylation. Proc Natl Acad Sci U S A 1999; 96: 6591-6596.

[48] Ong SE, Blagoev B, Kratchmarova I, Kristensen DB, Steen H, Pandey A, Mann M. Stable isotope labeling by amino acids in cell culture SILAC as a simple and accurate approach to expression proteomics. Mol Cell Proteomics 2002; 1: 376-386.

[49] Schmidt A, Kellermann J, Lottspeich F. A novel strategy for quantitative proteomics using isotope-coded protein labels. Proteomics 2005; 5: 4-15.

[50] Ross PL, Huang YN, Marchese JN, Williamson B, Parker K, Hattan S, Khainovski N, Pillai S, Dey S, Daniels S, Purkayastha S, Juhasz P, Martin S, Bartlet-Jones M, He F, Jacobson A, Pappin D.J. Multiplexed protein quantitation in Saccharomyces cerevisiae using amine-reactive isobaric tagging reagents. Mol Cell Proteomics 2004; 3: 1154-1169.

[51] Thompson A, Schäfer J, Kuhn K, Kienle S, Schwarz J, Schmidt G, Neumann T, Johnstone R, Mohammed A.K.A, Hamon C. Tandem mass tags: a novel quantification strategy for comparative analysis of complex protein mixtures by MS/MS. Anal Chem 2003; 75: 1895-1904.

[52] Haqqani AS, Kelly JF, Stanimirovic DB. Quantitative protein profiling by mass spectrometry using label-free proteomics. Methods Mol Biol 2008; 439: 241-256.

[53] Bondarenko PV, Chelius D, Shaler TA. Identification and relative quantitation of protein mixtures by enzymatic digestion followed by capillary reversed-phase liquid chromatography-tandem mass spectrometry. Anal Chem 2002; 74: 4741-4749.

[54] Wang G, Wu WW, Zeng W, Chou CL, Shen RF. Label-free protein quantification using LC-coupled ion trap or ft mass spectrometry: Reproducibility linearity and application with complex proteomes. J. Proteome Res. 2006; 5: 1214-1223.

[55] Hamdan M, Righetti PG. Modern strategies for protein quantification in proteome analysis: advantages and limitations. Mass Spectrom Rev 2002; 21: 287-302.

[56] Gao J, Opiteck GJ, Friedrichs MS, Dongre AR, Hefta SA. Changes in the protein expression of yeast as a function of carbon source. J Proteome Res 2003; 2: 643-649.

[57] Florens L, Washburn MP, Raine JD, Anthony RM, Grainger M, Haynes JD, Moch JK, Muster N, Sacci JB, Tabb DL, Witney AA, Wolters D, Wu Y, Gardner MJ, Holder AA, Sinden RE, Yates JR, Carucci DJ. A proteomic view of the Plasmodium falciparum life cycle. Nature 2002; 419: 520-526.

[58] Bridges SM, Magee GB, Wang N, Williams WP, Burgess SC, Nanduri B. ProtQuant: a tool for the label-free quantification of MudPIT proteomics data. BMC Bioinformatics 2007; 8: Suppl 7 S24.

[59] Mauri P, Scarpa A, Nascimbeni AC, Benazzi L, Parmagnani E, Mafficini A, Peruta MD, Bassi C, Miyazaki K, Sorio C. Identification of proteins released by pancreatic cancer cells by multidimensional protein identification technology: a strategy for identification of novel cancer markers. FASEB J 2005; 19: 1125-1127.

[60] Briani F, Curti S, Rossi F, Carzaniga T, Mauri P, Deho G. Polynucleotide phosphorylase hinders mRNA degradation upon ribosomal protein s1 overexpression in Escherichia coli. RNA 2008; 14: 2417- 2429.

[61] Liu H, Sadygov RG, Yates JR. A model for random sampling and estimation of relative protein abundance in shotgun proteomics. Anal Chem 2004; 76: 4193-4201.

[62] Zybailov B, Mosley AL, Sardiu ME, Coleman MK, Florens L, Washburn MP. Statistical analysis of membrane proteome expression changes in Saccharomyces cerevisiae. J Proteome Res 2006; 5: 2339- 2347.

[63] Rappsilber J, Ryder U, Lamond AI, Mann M. Large-scale proteomic analysis of the human spliceosome. Genome Res 2002; 12: 1231-1245.

[64] Ishihama Y, Oda Y, Tabata T, Sato T, Nagasu T, Rappsilber J, Mann M. Exponentially modified protein abundance index (emPAI) for estimation of absolute protein amount in proteomics by the number of sequenced peptides per protein. Mol Cell Prot 2005; 4: 1265- 1272.

[65] Gerber SA, Kettenbach AN, Rush J, Gygi SP. The absolute quantification strategy: application to phosphorylation profiling of human separase serine 1126. Methods Mol Biol 2007; 359: 71-86.

[66] Karp NA, Lilley KS. Design and analysis issues in quantitative proteomics studies. Proteomics 2007; 7: 42-50.

[67] Mortensen P, Gouw JW, Olsen JV, Ong SE, Rigbolt KTG, Bunkenborg J, Cox J, Foster L, Heck AJR, Blagoev B, Andersen JS, Mann M. MSQuant an open source platform for mass spectrometry-based quantitative proteomics. J Proteome Res 2010; 9: 393-403.

[68] Park SK, Venable JD, Xu T, Yates JR. A quantitative analysis software tool for mass spectrometry-based proteomics. Nat Methods 2008; 5: 319-322.

[69] MacCoss MJ, Wu CC, Liu H, Sadygov R, Yates JR. A correlation algorithm for the automated quantitative analysis of shotgun proteomics data. Anal Chem 2003; 75: 6912-6921.

[70] Carvalho PC, Hewel J, Barbosa VC, Yates JR. Identifying differences in protein expression levels by spectral counting and feature selection. Genet Mol Res 2008; 7: 342-356.

[71] Carvalho PC, Fischer JSG, Chen EI, Yates JR, Barbosa VC. PatternLab for proteomics: a tool for differential shotgun proteomics. BMC Bioinformatics 2008; 9: No 316.

[72] Pedrioli PGA, Eng JK, Hubley R, Vogelzang M, Deutsch EW, Raught B, Pratt B, Nilsson E, Angeletti RH, Apweiler R, Cheung K, Costello CE, Hermjakob H, Huang S, Julian RK, Kapp E, McComb ME, Oliver SG, Omenn G, Paton NW, Simpson R, Smith R, Taylor CF, Zhu W, Aebersold R. A common open representation of mass spectrometry data and its application to proteomics research. Nat Biotechnol 2004; 22: 1459-1466.

[73] Lin SM, Zhu L, Winter AQ, Sasinowski M, Kibbe WA. What is mzXML good for?. Exp Rev Prot 2005; 2: 839-845.

[74] Orchard S, Montechi-Palazzi L, Deutsch EW, Binz PA, Jones AR, Paton N, Pizarro A, Creasy DM, Wojcik J, Hermjakob H. Five years of progress in the standardization of proteomics data 4th annual spring workshop of the hupo-proteomics standards initiative April 23-25 2007 ecole nationale superieure (ens) Lyon France. Proteomics 2007; 7: 3436-3440.

[75] Katajamaa M, Miettinen J, Oresic M. MZmine: toolbox for processing and visualization of mass spectrometry based molecular profile data. Bioinformatics 2006; 22: 634-636.

[76] Sturm M, Kohlbacher O. TOPPview: an open-source viewer for mass spectrometry data. J Proteome Res 2009; 8: 3760-3763.

[77] Brusniak MY, Bodenmiller B, Campbell D, Cooke K, Eddes J, Garbutt A, Lau H, Letarte S, Mueller LN, Sharma V, Vitek O, Zhang N, Aebersold R, Watts JD. CORRA: Computational framework and tools for LC-MS discovery and targeted mass spectrometry-based proteomics. BMC Bioinformatics 2008; 9: 542.

[78] Braisted JC, Kuntumalla S, Vogel C, Marcotte EM, Rodrigues AR, Wang R, Huang S.T, Ferlanti ES, Saeed AI, Fleischmann RD, Peterson SN, Pieper R. The APEX quantitative proteomics tool: generating protein quantitation estimates from LC-MS/MS proteomics results. BMC Bioinformatics 2008; 9: 529.

[79] Shadforth IP, Dunkley TPJ, Lilley KS, Bessant C. i-Tracker: for quantitative proteomics using iTRAQ. BMC Genomics 2005; 6: 145.

[80] Mauri P, Deho G. A proteomic approach to the analysis of RNA degradosome composition in Escherichia coli. Methods Enzymol 2008; 447: 99-117.

[81] Kuster B, Schirle M, Mallick P, Aebersold R. Scoring proteomes with proteotypic peptide probes. Nat Rev Mol Cell Biol 2005; 6: 577-583.

[82] Malmström J, Lee H, Aebersold R. Advances in proteomic workflows for systems biology. Curr Opin Biotechnol 2007; 18: 378-384.

[83] Lange V, Picotti P, Domon B, Aebersold R. Selected reaction monitoring for quantitative proteomics: a tutorial. Mol Syst Biol 2008; 4: 222.

[84]   Drexler DM, Belcastro JV, Dickinson KE, Edinger KJ, Hnatyshyn SY, Josephs JL, Langish RA, McNaney CA, Santone KS, Shipkova PA, Tymiak AA, Zvyaga TA, Sanders M. An automated high throughput liquid chromatography-mass spectrometry process to assess the metabolic stability of drug candidates. Assay Drug Dev Technol 2007; 5: 247-264.

[85]   Shipkova P, Drexler DM, Langish R, Smalley J, Salyan ME, Sanders M. Application of ion trap technology to liquid chromatography/mass spectrometry quantitation of large peptides. Rapid Commun Mass Spectrom 2008; 22: 1359-1366.

[86]   Yang X, Lazar IM. MRM screening/biomarker discovery with linear ion trap ms: a library of human cancer-specific peptides. BMC Cancer 2009; 9: 96.

[87]   Gstaiger M, Aebersold R. Applying mass spectrometry-based proteomics to genetics genomics and network biology. Nat Rev Genet 2009; 10: 617-627.

[88]   Craig R, Cortens JP, Beavis RC. The use of proteotypic peptide libraries for protein identification. Rapid Commun Mass Spectrom 2005; 19: 1844-1850.

[89]   Desiere F, Deutsch EW, King NL, Nesvizhskii AI, Mallick P, Eng J, Chen S, Eddes J, Loevenich SN, Aebersold R. The PeptideAtlas project. Nucleic Acids Res 2006; 34: D655-D658.

[90]   Mathivanan S, Ahmed M, Ahn NG, Alexandre H, Amanchy R, Andrews PC, Bader JS, Balgley BM, Bantsche M, Bennett KL, Björling E, Blagoev B, Bose R, Brahmachari SK, Burlingame AS, Bustelo XR, Cagney G, Cantin GT, Cardasis HL, Celis JE, Chaerkady R, Chu F, Cole PA, Costello CE, Cotter RJ, Crockett D, DeLany JP, Marzo AMD, DeSouza LV, Deutsch EW, Dransfield E, Drewes G, Droit A, Dunn MJ, Elenitoba-Johnson K, Ewing RM, Eyk JV, Faca V, Falkner J, Fang X, Fenselau C, Figeys D, Gagne P, Gel C, Gevaert K, Gimble JM, Gnad F, Goel R, Gromov P, Hanash SM, Hancock WS, Harsha HC, Hart G, Hays F, He F, Hebbar P, Helsens K, Hermeking H, Hide W, Hjern K, Hochstrasser DF, Hofmann O, Horn DM, Hruban RH, Ibarrola N, James P, Jensen ON, Jensen PH, Jung P, Kandasamy K, Kheterpal I, Kikuno RF, Korf U, Körner R, Kuster B, Kwon MS, Lee HJ, Lee YJ, Lefevre M, Lehvaslaiho M, Lescuyer P, Levander F, Lim MS, Löbke C, Loo JA, Mann M, Martens L, Martinez-Heredia J, McComb M, McRedmond J, Mehrle A, Menon R, Miller CA, Mischak H, Mohan SS, Mohmood R, Molina H, Moran MF, Morgan JD, Moritz R, Morzel M, Muddiman DC, Nalli A, Navarro JD, Neubert TA, Ohara O, Oliva R, Omenn GS, Oyama M, Paik YK, Pennington K, Pepperkok R, Periaswamy B, Petricoin EF, Poirier GG, Prasad TSK, Purvine SO, Rahiman BA, Ramachandran P, Ramachandra YL, Rice RH, Rick J, Ronnholm RH, Salonen J, Sanchez JC, Sayd T, Seshi B, Shankari K, Sheng SJ, Shetty V, Shivakumar K, Simpson RJ, Sirdeshmukh R, Siu KWM, Smith JC, Smith RD, States DJ, Sugano S, Sullivan M, Superti-Furga G, Takatalo M, Thongboonkerd V, Trinidad JC, Uhlen M, Vandekerckhove J, Vasilescu J, Veenstra TD, Vidal-Taboada JM, Vihinen M, Wait R, Wang X, Wiemann S, Wu B, Xu T, Yates JR, Zhong J, Zhou M, Zhu Y, Zurbig P, Pandey A. Human proteinpedia enables sharing of human protein data. Nat Biotechnol 2008; 26: 164-167.

[91]   Martens L, Hermjakob H, Jones P, Adamski M, Taylor C, States D, Gevaert K, Vandekerckhove J, Apweiler R. PRIDE: the proteomics identifications database. Proteomics 2005; 5: 3537-3545.

[92]   Siepen JA, Belhajjame K, Selley JN, Embury SM, Paton NW, Goble CA, Oliver SG, Stevens R, Zamboulis L, Martin N, Poulovassillis A, Jones P, Côté R, Hermjakob H, Pentony MM, Jones DT, Orengo CA, Hubbard SJ. Ispider central: an integrated database web-server for proteomics. Nucleic Acids Res 2008; 36: W485-W490.

[93]   Tang H, Arnold RJ, Alves P, Xun Z, Clemmer DE, Novotny MV, Reilly JP, Radivojac P. A computational approach toward label-free protein quantification using predicted peptide detectability. Bioinformatics 2006; 22: e481-e488.

[94]   Webb-Robertson BJM, Cannon WR, Oehmen CS, Shah AR, Gurumoorthi V, Lipton MS, Waters KM. A support vector machine model for the prediction of proteotypic peptides for accurate mass and time proteomics. Bioinformatics 2008; 24: 1503-1509.

[95]   Sanders WS, Bridges SM, McCarthy FM, Nanduri B, Burgess SC. Prediction of peptides observable by mass spectrometry applied at the experimental set level. BMC Bioinformatics 2007; 8: Suppl 7 S23.

[96]   Everitt B. Cluster Analysis Halsted: New York 1980.

[97]   Mark S. Aldenderfer RKB. Cluster Analysis SAGE 1984.

[98]   Eisen MB, Spellman PT, Brown PO, Botstein D. Cluster analysis and display of genome-wide expression patterns. Proc Natl Acad Sci U S A 1998; 95: 14863-14868.

[99]   DeRisi JL, Iyer VR, Brown PO. Exploring the metabolic and genetic control of gene expression on a genomic scale. Science 1997; 278: 680-686.

[100]   Chu NF, Rimm EB, Wang DJ, Liou HS, Shieh SM. Clustering of cardiovascular disease risk factors among obese schoolchildren: the Taipei children heart study. Am J Clin Nutr 1998; 67: 1141-1146.

[101]   Altman RB, Raychaudhuri S. Whole-genome expression analysis: challenges beyond clustering. Curr Opin Struct Biol 2001; 11: 340-347.

[102]   Bhattacherjee V, Mukhopadhyay P, Singh S, Johnson C, Philipose JT, Warner CP, Greene RM, Pisano M.M. Neural crest and mesoderm lineage-dependent gene expression in orofacial development. Differentiation 2007; 75: 463-477.

[103]   Meunier B, Dumas E, Piec I, Bechet D, Hebraud M, Hocquette JF. Assessment of hierarchical clustering methodologies for proteomic data mining. J Proteome Res 2007; 6: 358-366.

[104] Harris RA, Yang A, Stein RC, Lucy K, Brusten L, Herath A, Parekh R, Waterfield MD, O'Hare MJ, Neville MA, Page MJ, Zvelebil MJ. Cluster analysis of an extensive human breast cancer cell line protein expression map database. Proteomics 2002; 2: 212-223.

[105] Dwek MV, Alaiya AA. Proteome analysis enables separate clustering of normal breast benign breast and breast cancer tissues. Br. J. Cancer 2003; 89: 305-307.

[106] Sodek KL, Evangelou AI, Ignatchenko A, Agochiya M, Brown TJ, Ringuette MJ, Jurisica I, Kislinger T. Identification of pathways associated with invasive behavior by ovarian cancer cells using Multidimensional Protein Identification Technology (MudPIT). Mol Biosyst 2008; 4: 762-773.

[107] Flikka K, Meukens J, Helsens K, Vandekerckhove J, Eidhammer I, Gevaert K, Martens L. Implementation and application of a versatile clustering tool for tandem mass spectrometry data. Proteomics 2007; 7: 3245-3258.

[108] Ru QC, Zhu LA, Silberman J, Shriver CD. Label-free semiquantitative peptide feature pro ling of human breast cancer and breast disease sera via two-dimensional liquid chromatography-mass spectrometry. Mol Cell Prot 2006; 5: 1095-1104.

[109] Cagney G, Park S, Chung C, Tong B, O'Dushlaine C, Shields DC, Emili A. Human tissue profiling with Multidimensional Protein Identification Technology. J Proteome Res 2005; 4: 1757-1767.

[110] Lawlor K, Nazarian A, Lacomis L, Tempst P, Villanueva J. Pathway-based biomarker search by high-throughput proteomics profiling of secretomes. J Proteome Res 2009; 8: 1489-1503.

[111] Kerr MK, Churchill GA. Bootstrapping cluster analysis: assessing the reliability of conclusions from microarray experiments. Proc Natl Acad Sci U S A 2001; 98: 8961-8965.

[112] Yeung KY, Fraley C, Murua A, Raftery AE, Ruzzo WL. Model-based clustering and data transformations for gene expression data. Bioinformatics 2001; 17: 977-987.

[113] Datta S, Datta S. Comparisons and validation of statistical clustering techniques for microarray gene expression data. Bioinformatics 2003; 19: 459-466.

[114] Barabasi AL, Oltvai ZN. Network biology: understanding the cell's functional organization. Nat Rev Genet 2004; 5: 101-113.

[115] Berger SI, Iyengar R. Network analyses in systems pharmacology. Bioinformatics 2009; 25: 2466-2472.

[116] Hood L. A personal view of molecular technology and how it has changed biology. J Proteome Res 2002; 1: 399-409.

[117] Albert R. Network inference analysis and modeling in systems biology. Plant Cell 2007; 19: 3327-3338.

[118] Khammash M. Reverse engineering: the architecture of biological networks. Biotechniques 2008; 44: 323-329.

[119] Schilling CH, Schuster S, Palsson BO, Heinrich R. Metabolic pathway analysis: basic concepts and scientific applications in the post-genomic era. Biotechnol Prog 1999; 15: 296-303.

[120] Patil KR, Nielsen J. Uncovering transcriptional regulation of metabolism by using metabolic network topology. Proc Natl Acad Sci U S A 2005; 102: 2685-2689.

[121] Joyce AR, Palsson BO. The model organism as a system: integrating 'omics' data sets. Nat Rev Mol Cell Biol 2006; 7: 198-210.

[122] Pavelka N, Fournier ML, Swanson SK, Pelizzola M, Ricciardi-Castagnoli P, Florens L, Washburn MP. Statistical similarities between transcriptomics and quantitative shotgun proteomics data. Mol Cell Prot 2008; 7: 631-644.

[123] Ewing RM, Chu P, Elisma F, Li H, Taylor P, Climie S, McBroom-Cerajewski L, Robinson MD, O'Connor L, Li M, Taylor R, Dharsee M, Ho Y, Heilbut A, Moore L, Zhang S, Ornatsky O, Bukhman YV, Ethier M, Sheng Y, Vasilescu J, Abu-Farha M, Lambert JP, Duewel HS, Stewart I.I, Kuehl B, Hogue K, Colwill K, Gladwish K, Muskat B, Kinach R, Adams SL, Moran MF, Morin GB, Topaloglou T, Figeys D. Large-scale mapping of human protein-protein interactions by mass spectrometry. Mol Syst Biol 2007; 3: 89.

[124] Mishra GR, Suresh M, Kumaran K, Kannabiran N, Suresh S, Bala P, Shivakumar K, Anuradha N, Reddy R, Raghavan T.M, Menon S, Hanumanthu G, Gupta M, Upendran S, Gupta S, Mahesh M, Jacob B, Mathew P, Chatterjee P, Arun KS, Sharma S, Chandrika KN, Deshpande N, Palvankar K, Raghavnath R, Krishnakanth R, Karathia H, Rekha B, Nayak R, Vishnupriya G, Kumar HGM, Nagini M, Kumar GSS, Jose R, Deepthi P, Mohan SS, Gandhi TKB, Harsha HC, Deshpande KS, Sarker M, Prasad TSK, Pandey A. Human protein reference database-2006 update. Nucleic Acids Res 2006; 34: D411-D414.

[125] Zanzoni A, Montecchi-Palazzi L, Quondam M, Ausiello G, Helmer-Citterich M, Cesareni G. MINT: a Molecular INTeraction database. FEBS Lett 2002; 513: 135-140.

[126] Kerrien S, Alam-Faruque Y, Aranda B, Bancarz I, Bridge A, Derow C, Dimmer E, Feuermann M, Friedrichsen A, Huntley R, Kohler C, Khadake J, Leroy C, Liban A, Lieftink C, Montecchi-Palazzi L, Orchard S, Risse J, Robbe K, Roechert B, Thorneycroft D, Zhang Y, Apweiler R, Hermjakob H. Intact-open source resource for molecular interaction data. Nucleic Acids Res 2007; 35: D561-D565.

[127] Joshi-Tope G, Gillespie M, Vastrik I, D'Eustachio P, Schmidt E, de Bono B, Jassal B, Gopinath GR, Wu GR, Matthews L, Lewis S, Birney E, Stein L. Reactome: a knowledgebase of biological pathways. Nucleic Acids Res 2005; 33: D428-D432.

[128] Xenarios I, Rice DW, Salwinski L, Baron MK, Marcotte EM, Eisenberg D. DIP: the database of interacting proteins. Nucleic Acids Res 2000; 28: 289-291.

[129] Stark C, Breitkreutz BJ, Reguly T, Boucher L, Breitkreutz A, Tyers M. Biogrid: a general repository for interaction datasets. Nucleic Acids Res 2006; 34: D535-D539.

[130] Brown KR, Jurisica I. Online predicted human interaction database. Bioinformatics 2005; 21: 2076-2082.

[131] Han K, Park B, Kim H, Hong J, Park J. HPID: the Human Protein Interaction Database. Bioinformatics 2004; 20: 2466-2470.

[132] Xia K, Dong D, Han JDJ. IntNetDB v1.0: an integrated protein-protein interaction network database generated by a probabilistic model. BMC Bioinformatics 2006; 7: 508.

[133] von Mering C, Jensen LJ, Kuhn M, Chaffron S, Doerks T, Krüger B, Snel B, Bork P. STRING 7-recent developments in the integration and prediction of protein interactions. Nucleic Acids Res. 2007; 35: D358-D362.

[134] Huang TW, Tien AC, Huang WS, Lee YCG, Peng CL, Tseng HH, Kao CY, Huang CYF. Point: a database for the prediction of protein-protein interactions based on the orthologous interactome. Bioinformatics 2004; 20: 3273-3276.

[135] Shah SP, Huang Y, Xu T, Yuen MMS, Ling J, Ouellette BFF. Atlas - a data warehouse for integrative bioinformatics. BMC Bioinformatics 2005; 6: 34.

[136] Birkland A, Yona G. Biozon: a hub of heterogeneous biological data. Nucleic Acids Res 2006; 34: D235-D242.

[137] Chang AN, McDermott J, Frazier Z, Guerquin M, Samudrala R. Integrator: interactive graphical search of large protein interactomes over the web. BMC Bioinformatics 2006; 7: 146.

[138] Shannon PT, Reiss DJ, Bonneau R, Baliga NS. The gaggle: an open-source software system for integrating bioinformatics software and data sources. BMC Bioinformatics 2006; 7: 176.

[139] Suderman M, Hallett M. Tools for visually exploring biological networks. Bioinformatics 2007; 23: 2651-2659.

[140] Shannon P, Markiel A, Ozier O, Baliga NS, Wang JT, Ramage D, Amin N, Schwikowski B, Ideker T. Cytoscape: a software environment for integrated models of biomolecular interaction networks. Genome Res 2003; 13: 2498-2504.

[141] Baitaluk M, Qian X, Godbole S, Raval A, Ray A, Gupta A. Pathsys: integrating molecular interaction graphs for systems biology. BMC Bioinformatics 2006; 7: 55.

[142] Demir E, Babur O, Dogrusoz U, Gursoy A, Nisanci G, Cetin-Atalay R, Ozturk M. Patika: an integrated visual environment for collaborative construction and analysis of cellular pathways. Bioinformatics 2002; 18: 996-1003.

[143] Hu Z, Hung JH, Wang Y, Chang YC, Huang CL, Huyck M, DeLisi C. VisANT 3.5: multi-scale network visualization analysis and inference based on the gene ontology. Nucleic Acids Res 2009; 37: W115-W121.

[144] Iragne F, Nikolski M, Mathieu B, Auber D, Sherman D. ProViz: protein interaction visualization and exploration. Bioinformatics 2005; 21: 272-274.

[145] Chalkley RJ, Baker PR, Huang L, Hansen KC, Allen NP, Rexach M, Burlingame AL. Comprehensive analysis of a multidimensional liquid chromatography mass spectrometry dataset acquired on a quadrupole selecting quadrupole collision cell time-of- flight mass spectrometer: II. new developments in protein prospector allow for reliable and comprehensive automatic analysis of large datasets. Mol Cell Proteomics 2005; 4: 1194-1204.

[146] Taylor JA, Walsh KA, Johnson RS. Sherpa: a macintosh-based expert system for the interpretation of electrospray ionization LC/MS and MS/MS data from protein digests. Rap Comm Mass Spec 1996; 10: 679-687.

[147] Tabb DL, Fernando CG, Chambers MC. Myrimatch: highly accurate tandem mass spectral peptide identification by multivariate hypergeometric analysis. J Proteome Res 2007; 6: 654-661.

[148] Bern M, Cai Y, Goldberg D. Lookup peaks: A hybrid of de novo sequencing and database search for protein identification by tandem mass spectrometry. Anal Chem 2007; 79: 1393-1400.

[149] Tanner S, Shu H, Frank A, Wang LC, Zandi E, Mumby M, Pevzner PA, Bafna V. Inspect: identification of posttranslationally modified peptides from tandem mass spectra. Anal Chem 2005; 77: 4626-4639.

[150] Liu J, Erassov A, Halina P, Canete M, Nguyen DV, Chung C, Cagney G, Ignatchenko A, Fong V, Emili A. Sequential interval motif search: unrestricted database surveys of global ms/ms data sets for detection of putative post-translational modifications. Anal Chem 2008; 80: 7846-7854.

[151] Slotta DJ, Barrett T, Edgar R. NCBI peptidome: a new public repository for mass spectrometry peptide identifications. Nat Biotechnol 2009; 27: 600-601.

[152] Suzuki R, Shimodaira H. Pvclust: an R package for assessing the uncertainty in hierarchical clustering. Bioinformatics 2006; 22: 1540-1542.

[153] Brock G, Pihur V, Datta S, Datta S. clValid an R package for cluster validation. J Stat Software 2007; 25: 1-22.

[154] Nagasaki M, Doi A, Matsuno H, Miyano S. Genomic object net: I. a platform for modelling and simulating biopathways. App Bioinformatics 2003; 2: 181-184.

[155] Doi A, Nagasaki M, Fujita S, Matsuno H, Miyano S. Genomic object net: II. modelling biopathways by hybrid functional petri net with extension. App Bioinformatics 2003; 2: 185-188.

[156] Funahashi A, Morohashi M, Kitano H, Tanimura N. Celldesigner: a process diagram editor for gene-regulatory and biochemical networks. BIOSILICO 2003; 1: 159-162.

[157] Kanehisa M, Goto S, Hattori M, Aoki-Kinoshita K.F, Itoh M, Kawashima S, Katayama T, Araki M, Hirakawa M. From genomics to chemical genomics: new developments in KEGG. Nucl Acids Res 2006; 34: D354-357.

[158] Mi H, Lazareva-Ulitsky B, Loo R, Kejariwal A, Vandergri J, Rabkin S, Guo N, Muruganujan A, Doremieux O, Campbell MJ, Kitano H, Thomas PD. THE PANTHER database of protein families sub-families functions and pathways. Nucl Acids Res 2005; 33: D284-288.

[159] Uhlen M, Bjorling E, Agaton C, Szigyarto CA, Amini B, Andersen E, Andersson A, Angelidou P, Asplund A, Asplund C, Berglund L, Bergstrom K, Brumer H, Cerjan D, Ekstrom M, Elobeid A, Eriksson C, Fagerberg L, Falk R, Fall J, Forsberg M, Bjorklund MG, Gumbel K, Halimi A, Hallin I, Hamsten C, Hansson M, Hedhammar M, Hercules G, Kampf C, Larsson K, Lindskog M, Lodewyckx W, Lund J, Lundeberg J, Magnusson K, Malm E, Nils-son P, Odling J, Oksvold P, Olsson I, Oster E, Ottosson J, Paavilainen L, Persson A, Rimini R, Rockberg J, Runeson M, Sivertsson A, Skollermo A, Steen J, Stenvall M, Sterky F, Stromberg S, Sundberg M, Tegel H, Tourle S, Wahlund E, Walden A, Wan J, Wernerus H, Westberg J, Wester K, Wrethagen U, Xu LL, Hober S, Ponten F. A Human Protein Atlas for normal and cancer tissues based on antibody proteomics. Mol Cell Proteomics 2005; 4: 1920-1932.

[160] Bader GD, Betel D, Hogue CWV. BIND: the Biomolecular Interaction Network Database. Nucl Acids Res 2003; 31; 248-250.

[161] Bowers P, Pellegrini M, Thompson M, Fierro J, Yeates T, Eisenberg D. Prolinks: a database of protein functional linkages derived from coevolution. Gen Bio 2004; 5: R35.

[162] Dennis G, Sherman B, Hosack D, Yang J, Gao W, Lane H, Lempicki R. David: Database for annotation visualization and integrated discovery. Genome Biology 2003; 4: R60.

[163] Wu CH, Apweiler R, Bairoch A, Natale DA, Barker WC, Boeckmann B, Ferro S, Gasteiger E, Huang H, Lopez R, Magrane M, Martin MJ, Mazumder R, O'Donovan C, Redaschi N, Suzek B. The Universal Protein Resource (UNIPROT): an expanding universe of protein information. Nucl Acids Res 2006; 34: D187-191.

[164] Bairoch A, Apweiler R. The Swiss-Prot protein sequence database and its supplement trEMBL in 2000. Nucl Acids Res 2000; 28: 45-48.

[165] Stoesser G, Tuli M, Lopez R, Sterk P. The EMBL nucleotide sequence database. Nucl Acids Res 1999; 27: 18-24.

[166] Wheeler DL, Barrett T, Benson DA, Bryant SH, Canese K, Chetvernin V, Church DM, DiCuccio M, Edgar R, Federhen S, Geer LY, Kapustin Y, Khovayko O, Landsman D, Lipman DJ, Madden TL, Maglott DR, Ostell J, Miller V, Pruitt KD, Schuler GD, Sequeira E, Sherry ST, Sirotkin K, Souvorov A, Starchenko G, Tatusov RL, Tatusova TA, Wagner L, Yaschenko E. Database resources of the National Center for Biotechnology Information. Nucl Acids Res 2006; 1031: D173-180.

CHAPTER 4

Current Methods for the Quantitative Analysis of Pharmaceutical Compounds from Whole Blood Matrix Using Liquid Chromatography Mass Spectrometry

Raymond Naxing Xu*, Matthew J. Rieser, and Tawakol A. El-Shourbagy

Abbott Laboratories, Department of Drug Analysis, 100 Abbott Park Road, Abbott Park, IL 60064-6126

Abstract: This article reviews current methods used in quantitative analysis of pharmaceutical compounds from whole blood matrix by liquid chromatography mass spectrometry. Whole blood matrix reviewed here includes traditional liquid whole blood and dried blood spot (DBS) on a collection paper. A number of bioanalytical methods have been reported over the years for the determination of pharmaceutical compounds from liquid whole blood matrix. These methods have been used to support drug discovery and development as well as therapeutic drug monitoring. Dried blood spot technique was initially developed for newborn screening, later adapted to therapeutic drug monitoring, and now expanding into pharmaceutical drug discovery and development. Sample pretreatment, extraction, chromatography, and mass detection procedures for sample analysis from both liquid whole blood and dried blood spot are summarized in this article. Factors influencing assay performance such as sampling and automation are discussed. Emerging techniques allowing direct analysis of blood samples using mass spectrometry technique are also included.

INTRODUCTION

In pharmaceutical discovery and development, biological fluids such as plasma, serum, whole blood, and urine are most commonly analyzed for pharmacokinetic parameter evaluation. Most bioanalytical applications are from plasma samples. But for many compounds such as cyclosporin A (CsA) that mainly distributes in the erythrocyte, whole blood rather than plasma or serum is the matrix of choice for the measurement of drug exposure in animal or human subjects. Whole blood samples present unique challenges in method development and validation because of the viscous nature of blood and complexity of its constituents. Of all the current techniques available, LC-MS/MS has been generally accepted as the preferred technique for quantitative and analysis of small-molecule drugs, metabolites, and other xenobiotic molecules in biological matrices including liquid whole blood samples due to its inherent specificity and sensitivity [1-2].

Recently, dried blood spot (DBS) technique has been attracting interest in bioanalytical field. Originated from newborn testing, DBS has gained acceptance in therapeutic drug monitoring, and now has been expanding into pharmaceutical discovery and development. Usually both liquid whole blood and dried blood samples are not directly compatible with LC–MS/MS analyses. In this review article, we will describe the current methods for quantitative analysis of pharmaceutical compounds from both liquid whole blood and dried whole blood samples. Developing techniques and future perspectives are also included in this review.

METHODS FOR LIQUID WHOLE BLOOD SAMPLE ANALYSIS

Whole blood samples are typically more difficult to work with than plasma samples because of their viscous nature. To start with, it takes more caution to accurately transfer whole blood samples. When using robotic liquid handler to pipette whole blood samples, the aspiration speed of the liquid handler has to be adjust properly to ensure accurate aliquoting of whole blood sample. Preparation of calibration standard and quality control samples in whole blood takes longer time than plasma samples. In whole blood samples collected from subjects dosed with active pharmaceutical ingredient analytes have already reached the distribution equilibrium (partition coefficient) between plasma and blood cells. Sufficient time and care are needed for spiked whole blood samples or serial diluted samples to reach the equilibrium so that they can approximate the unknowns. Prior to extraction of whole blood samples, a lysing step is usually needed since analytes may form molecular complexes in the red blood cells. This is often achieved through osmotic pressure to the whole blood sample. This step is often combined with addition of internal standard solution.

Analyte stability in the matrix is another factor to consider when developing a whole blood assay. For instance,

Address correspondence to this author Raymond Naxing Xu, Ph.D. at: Dept. R46W, Bldg. AP13A-2, Abbott Laboratories, 100 Abbott Park Road, Abbott Park, IL 60064-6126, USA; Tel: + 1-847-938-8158; Fax: + 1-847-938-7789. E-mail: raymond.xu@abbott.com

reminfentanil has been found to be unstable in both human whole blood and plasma [3]. To avoid hydrolysis of the *N*-substituted ester group of reminfentanil, whole blood analysis is preferred to plasma. Specific pre-treatment is required to prevent the decomposition of reminfentanil when blood samples are collected at clinics or calibration standard and quality controls are prepared. By adding a certain percentage of citric acid solution to these samples and after mixing and flash frozen at -20 °C, studies have shown that reminfentanil can be stable after three freeze-thaw cycles as well as 6 month storage in the -20 °C freezer and subsequent processing at room temperature [3].

Sample preparation techniques for whole blood assays are as diverse as those for plasma assays. Although simple sample preparation technique like protein precipitation (PPT) has been used in the whole blood analysis, whole blood analysis typically calls for more labor-intensive treatment such as liquid-liquid extraction (LLE) or solid-phase extraction (SPE). This might be attributed from the complexity of the nature of the whole blood matrix. Table **1** summaries representative applications of LC-MS/MS in liquid whole blood bioanalysis [3-28]. In some cases, PPT is used as a pretreatment method prior to LLE or SPE. Among them, a method for the estimation of indapamide in human whole blood has been developed with a sensitivity of 0.5 ng/ml as lower limit of quantification (LLOQ). The procedure for the extraction of indapamide and glimepiride as internal standard (IS) involves haemolysis and deproteination of whole blood using ZnSO4 followed by liquid–liquid extraction using ethyl acetate [7]. The sample extracts after drying were reconstituted and analyzed by LC–MS/MS. The mean recovery for indapamide was 82.40 and 93.23% for IS. The total run time was 2.5 min to monitor both indapamide and the IS. The method is fully validated over the range of 0.5–80.0 ng/ml and also applied to subject-sample analysis of bioequivalence study for 1.5 mg sustained-release formulations. Similarly, a highly specific LC-MS/MS method is reported for the determination in human whole blood of Aplidin, a novel depsipeptide under investigation in clinical studies [9]. Didemnin B was used as internal standard and, after protein precipitation with acetonitrile and liquid–liquid extraction with chloroform, APL was separated by liquid chromatography using a gradient program. A combination with PPT with SPE was used as the extraction approach in the simultaneous determination of 6 beta-blockers and 3 calcium-channel antagonists from human whole blood [8]. Sample clean-up was achieved by precipitation and solid phase extraction (SPE) with a mixed-mode column. Quantification was performed by reversed phase high performance liquid chromatography with positive electrospray ionization mass spectrometric detection (HPLC-MS). The method has been developed and robustness tested by systematically searching for satisfactory conditions using experimental designs including factorial and response surface designs.

While most of whole blood methods reported to date employed manual extraction, 96-well techniques have been developed for whole blood analysis. Ji et. al reported a quantitative method for the analysis of ABT-578 in human whole blood samples. Sample preparation was achieved by a semi-automated 96-well format liquid-liquid extraction (LLE) method [10]. Aluminum/polypropylene heat seal foil was used to enclose each well of the 96-well plate for the liquid-liquid extraction. A LC-MS/MS method with pre-column regeneration was developed for the analysis of sample extracts. The ammonium adduct ions generated from electrospray ionization were monitored as the precursor ions. The assay was validated for a linear dynamic range of 0.20-200.75 ng/ml. The correlation coefficient (r) was between 0.9959 and 0.9971. The intra-assay CV (%) was between 1.9 and 13.5% and the inter-assay CV (%) was between 4.7 and 11.3%. A 96-well based LC-MS/MS method for the determination of N-methyl-4-isoleucine-cyclosporin (NIM811) was developed and validated over the concentration range 1–2500 ng/mL in human whole blood using a 0.05 mL sample volume [11]. NIM811 and the internal standard, d12-cyclosporin A (d12-CsA), were extracted from blood using Methyl *tert*-butyl ether (MTBE) via liquid–liquid extraction in 96-well plate. After evaporation of the organic solvent and reconstitution, a 10 μL aliquot of the resulting extract was injected onto the LC-MS/MS system. Chromatographic separation of NIM811 and internal standard was performed using a Waters Symmetry column. The total run time was 3.5 min with a flow rate of 0.8 mL/min. The method has been used to measure the exposure of NIM811 in human subjects.

On-line SPE has also been reported for whole blood sample analysis. Klawitter et. al reported a semi-automated LC/LC-MS/MS assay for the quantification of imatinib in human whole blood and leukemia cells [12]. After protein precipitation, samples were injected into the HPLC system and trapped onto the enrichment column (flow 5mL/min); extracts were back-flushed onto the analytical column. A commercially available compound, trazodone, was used as internal standard. A calibration range of 0.03–75 ng/mL was validated. This semi-automated method was demonstrated to be simple with only one manual step and has proven to be robust in

Table 1: Representative applications of LC-MS/MS in liquid whole blood bioanalysis.

Compound	Matrix	Lysing/ pretreatment solution	Extraction	Chromatography	MS detection	LLOQ	Ref.
Remifentanil	Human whole blood	0.1 mol/L Phosphate buffer, pH 7.4	LLE with dichloromethane	Isocratic separation Mobile phase: a mixture of acetonitrile/chloroform (1:2 v/v), containing 2mmol/l ammonium acetate, 3 min runtime	ESI, MRM	0.1 ng/mL	[3]
Tacrolimus and Cyclosporin	Human whole blood	0.1 mol/L zinc sulfate	PPT with acetonitrile	Gradient separation with (A) 100% water and (B) 100% methanol, 1.5 min runtime	APCI, MRM	1.0 ng/mL for tacrolimus and 25 ng/mL for cyclosporine	[4]
Sirolimus (rapamycin) and cyclosporin	Human whole blood	a mixture of methanol and 0.4 mol/ 1 zinc sulfate (4:1, v/v) Served as PPT reagent as well	PPT followed by on line SPE with 30 x 4 mm C18 Nucleosil 100 10 μm particle extraction column	HPLC–UV methods have been developed for Isocratic separation Mobile Phase: methanol–water 90/10 (v/v), 15 min runtime	ESI, SIM of sodium adducts	0.2 ng/mL for sirolimus and 4 ng/mL for Cyclosporin	[5]
Cyclosporine A	Human whole blood	Not mentioned	PPT with methanol	Isocratic separation Mobile Phase: 90% methanol/10% water 5 min runtime	ESI, SIM of protonated molecule, sodium adducts, and potassium adducts	40 ng/mL	[6]
Indapamide	Human whole blood	5.0% ZnSO$_4$	LLE with ethyl acetate	Isocratic separation Mobile phase: 10:90 (v/v). ammonium acetate:acetonitrile, 2.5 min runtime	ESI, MRM	0.5 ng/mL	[7]
Six beta-blockers, 3 calcium-channel antagonists, 4 angiotensin-II antagonists and 1 antiarrhytmic drug	Human whole blood	Not mentioned	ice cold acetonitrile-methanol solution (85:15, v:v) followed by off-line SPE	Gradient separation with formic acid (A) and acetonitrile (B), 18 min runtime	ESI, SIM	1.0 μM	[8]
Aplidin	Human whole blood	Not mentioned	PPT with acetonitrile and LLE with chloroform	Gradient separation with acetonitrile and water (both containing 0.5% formic acid), 12 min runtime	ESI, MRM	1 ng/mL	[9]
ABT-578	Human whole blood	freshly prepared 4:1 (v/v) methanol:100mM ammonium acetate solution	LLE with ethyl acetate and hexane (50:50 v/v)	Isocratic separation 5mM ammonium acetate and 0.03% (v/v) formic acid in the solvent mixture of 80/20 (v/v) methanol/water, 10 min runtime	ESI, MRM, ammonium adducts as precursor ions	0.20 ng/mL	[10]
NIM811	Human whole blood	0.1% ammonium hydroxide solution	LLE with MTBE	Gradient separation with water and acetonitrile, 3.5 min runtime	ESI, MRM, ammonium adducts as precursor ions	1 ng/mL	[11]
Imatinib	Human whole blood	Methanol-0.2 M zinc sulfate (7:3 v/v)	PPT followed by on line SPE with Eclipse XDB-C$_8$ extraction column	Gradient separation with methanol, ammonium acetate, and TFA, 10.5 min runtime	ESI, MRM	0.03 ng/mL	[12]
Voclosporin	Human whole blood	Methanol-0.2 M zinc sulfate (8:2 v/v)	PPT followed by on line SPE with Zorbax SB-C8 2.1 x 12.5 mm extraction column	Isocratic separation with 80% MeOH containing 0.02% (v/v) acetic acid and 0.02mM sodium acetate, 2 min runtime	ESI, MRM, sodium adducts as precursor ions	1 ng/mL	[13]
Cocaine and metabolites	Human whole blood	Methanol-0.2 M zinc sulfate (8:2 v/v)	PPT followed on-line SPE with Hysphere MM anion SPE cartridge	Gradient separation with acetonitrile, water, and formic acid, 10 min runtime	ESI, MRM	8 to 47 ng/mL	[14]
Tetrahydrocannabinol and metabolites	Human whole blood	Not mentioned	PPT followed on-line SPE with Hysphere C8-EC exchange (10 x 2mm) SPE cartridge	Gradient separation with acetonitrile, water, and formic acid, 10 min runtime	ESI, MRM	2 to 8 ng/mL	[15]
Eight benzodiazepines	Human whole blood	25% ammonia solution	LLE with n-butylchloride	Iocratic separation with 5mM aqueous ammonium formate adjusted to pH 3 with formicv acid–acetonitrile (65:35, v/v),, runtime less than 5 min	APCI, SIM	2.5 ng/mL	[16]
4-Dimethylaminophenol	Dog whole blood	Not mentioned	PPT with acetonitrile Followed by chemical derivatization	Isocratic separation with mobile phase composed of acetonitrile–water–formic acid in the ratio of 85:15:0.1 (v/v/v), runtime approximately 4.5 min	ESI, MRM	2 ng/mL	[17]
β, β-Dimethylacrylshikonin (DASK)	Rat Whole Blood	Phosphate buffer (pH 7.4)	Chemical derivatization followed by LLE with cyclohexane	Gradient separation with methanol and water, 6.5 min runtime	ESI, MRM	3 ng/mL	[18]

larger studies. Similarly, a rapid LC-MS/MS method was developed and validated for the therapeutic drug monitoring of voclosporin in human whole blood [13]. Sample aliquots of 100μL were processed utilizing a protein precipitation procedure that contained a mixture of methanol, 0.2M ZnSO4, and deuterated voclosporin internal standard. Supernatant was injected onto a Zorbax SB-C8, 2.1×12.5mm column (at 60 °C), and washed with water–acetonitrile, supplemented with 0.02% glacial acetic acid and 0.02mM sodium acetate, to remove poorly retained components. After washing, voclosporin and internal standard were eluted to mass spectrometer for detection in multiple reaction monitoring (MRM) mode. Analytical performance was assessed in the range of 1–200 ng/ml in whole blood. Jagerdeo et. al reported a fully automated LC-MS/MS method for the analysis of cocaine and its metabolites (benzoylecgonine, ecgoninemethyl ester, ecgonine and cocaethylene) from whole blood [14]. The method utilizes an online solid-phase extraction (SPE) setup based on Spark Holland's Symbiosis system. Pretreatment of samples involve only protein precipitation and ultracentrifugation. An efficient online SPE procedure was developed using Hysphere MM anion sorbent. A gradient chromatography method was used for the complete separation of all components. For the analysis, two MRM transitions are monitored for each analyte and one transition is monitored for each internal standard. Linearity was analyte dependent but generally fell between 8 and 500 ng/mL. The limits of quantitation (LLOQs) ranged from 8 to 47 ng/mL. The same approach has been applied to Marijuana testing with an analytical method for the determination of tetrahydrocannabinol and metabolites in whole blood [15]. Overall, the major advantage of on-line SPE over off-line extraction techniques is that the sample preparation step is embedded into the chromatographic separation and thus eliminates most of the sample preparation time traditionally performed at the bench.

To speed up the analytical process and reduce run time, a simple and fast procedure based on monolithic chromatography was developed for the simultaneous determination of eight benzodiazepines in whole blood [16]. Sample pretreatment was carried out using a simple liquid–liquid extraction (LLE) with *n*-butylchloride, and chromatographic separation was performed using a monolithic silica column. APCI and electrospray ionization (ESI) were compared for the assay performance. Whereas both ionization techniques appeared suitable for BZDs, APCI was found to be slightly more sensitive, especially for the determination of frequently low-dosed compounds. The limit of quantification (LLOQ) was 2.5 ng/mL for all the compounds. Analysis of the eight compounds was accomplished in less than 5 min.

Chemical derivatization has also been reported for enhancement of sensitivity in detection and also for increasing the retention time of the analytes to avoid ion suppression effects. An LC–MS/MS method combined with precolumn dansyl-chloride derivatization was developed to determine dog blood 4-dimethylaminophenol (DMAP) concentrations [17]. The linearity of the method was observed within the concentration range of 2–2000 ng/mL. The precision, accuracy, stability, recovery and matrix effect of the method were also investigated and found to meet the requirements for pharmacokinetic studies of the drug. Similarly, an LC-MS/MS method was developed and validated for the determination of β, β-dimethylacrylshikonin (DASK) in rat whole blood. DASK was pretreated using pre-column derivatization with 2-mercaptoethanol followed by liquid–liquid extraction with cyclohexane [18]. Detection was performed by selected reaction monitoring mode and the linear range for the determination of DASK in rat whole blood ranged from 3 to 3000 ng/mL.

Reproducibility testing has profound impact on bioanalytical method development and validation process. Recently, such testing has been implemented as mandatory repeat experiment using incurred samples for regulated bioanalytical studies in pharmaceutical industry. The results from mandatory repeat experiment are often treated as a part of method validation. Currently, there is very limited publication on incurred sample reproducibility (ISR) for whole blood analytical methods. Xu et. al assessed method reproducibility of bioanalytical methods on concentration determination of a pharmaceutical compound and its metabolite in whole blood matrix by LC-MS/MS [29]. The analytical method was initially developed to satisfy typically requirements such as precision and accuracy, selectivity, stability, and matrix effect. The method was further evaluated with incurred samples in mandatory repeat mode. The evaluation was performed on both rat and dog whole blood methods. The result demonstrated that proper sample preparation procedures such as sample transfer and lysing of red blood cells are key to reproducible results. The volume of organic used for lysing needs to be controlled not causing significant protein precipitation.

There are a number of whole blood methods developed and validated for a class of compounds, immunosuppressants [4-6, 11, 13, 25-28, 30-31]. These compounds include cyclosporine A (CsA), tacrolimus (FK 506), sirolimus, and etc. A common characteristic of these methods is that they are can easily form adducts with ammonium, Na^+, K^+ ions during ionization process. While sodium-adduct ion was used in some assays [30-

31], it also has a deteriorating effect on mass spectrometer's sensitivity. Most of methods on immunosuppressant determination use ammonium adduct as precursor in multiple reaction monitoring (MRM). The formation of uncontrolled adduct ions would not only reduce the sensitivity of the target detection but also affect the accuracy and precision of the method. Chen et. al described such observations in an assay for the simultaneous determination of three isomeric metabolites of tacrolimus (FK506), 13-O-demethylated (M1), 31-O-demethylated (M2) and 15-O-demethylated (M3) tacrolimus in human whole blood [25]. They noted that the use of glassware for solutions and sample preparation would result in a highly variable intra- and inter-batch precision and accuracy. Exclusion of any potential sodium origination was found to be a necessary measure to have a reliable ammonium-adduct based LC–MS/MS method.

METHODS FOR DRIED BLOOD SPOT SAMPLE ANALYSIS

Dried blood spot (DBS) technique has been around for more than 40 years and has been adopted widely in newborn screening applications. A few drops of blood from several million newborn infants are screened annually throughout the world. In the United States, more than 95% of newborns are screened for inherited metabolic disorders by analyzing a range of compounds including amino acid, hormone, and RNA. There are a number of publications for newborn screening by LC-MS/MS on dried blood spot [32-39]. These reports will not be reported in this article. Instead, we will mainly focus on the use of DBS in therapeutic drug monitoring, pharmaceutical discovery and development.

In DBS method, whole blood samples are collected on filter paper and allowed to be dried. A portion of the dried spot then punched out or whole spot is or cut out. The spot is then extracted with solvent prior to LC-MS analysis. Table **1** summarizes some representative applications of dried-blood spot technique in LC-MS/MS bioanalysis [40-52].

One of the advantages of DBS is small blood volume needed for the analysis. This is well-suited for therapeutic drug monitoring in which blood samples can be obtained by finger prick. Edelbroek et. al recently reviewed dried blood spot sampling in therapeutic drug monitoring by using a wide range of assay techniques including high performance HPLC with UV and fluorescence detection, HPLC-MS/MS, polarization immunoassay, and radioimmunoassay [53]. In the reports reviewed by Edelbroek for therapeutic drug monitoring, sampling was done using 2 methods: The first method (A) involves the sampling of a drop of whole blood directly on the sampling paper within a premarked circle, and the second method (B) requires accurately pipetting capillary blood with a capillary pipette on sampling paper. In method A, a paper disk is punched out from the DBS with a smaller diameter than the blood spot itself. The punching technique must be reproducible, and a special punching apparatus is necessary. The disk represents a volumetric measurement comparable with a liquid measurement. If the spot has been pipetted on the paper, the whole blood spot has to be cut out; then, extraction takes place. In method B, part of the convenience and simplicity of method A is lost by introducing the pipetting step. Using the pipette, however, eliminates problems with sample volume variability caused by spreadability characteristics or an incorrect sampling technique. In therapeutic drug monitoring, the use of fresh normal blood for calibration standards and quality control standards is essential, especially for the DBS method using the paper disk as a volumetric measurement [52]. Hemolyzed blood or blood with a deviating hematocrit should not be used. Therefore, because the permeation of fresh blood in the paper is complete, there is less spreading and a smaller diameter. Edelbroek cautioned against the degree of potential errors introduced via the sampling method. He also concluded that LC-MS/MS to be generally preferable in specificity and sensitivity to other methods. Other details of Edelbroek's review are not included in this article. Some selected methods by using LC-MS/MS in therapeutic drug monitoring are included in Table **2** for comparison purpose.

The collection of whole blood samples as DBS for pharmacokinetic (PK) studies in drug discovery and development also offers a number of advantages over conventional plasma sampling. The small blood volumes required for DBS samples (less than 100 μL, compared to >0.5 mL blood which are usually obtained for conventional plasma analyses) make this a particularly suitable approach for the collection of blood samples for pediatric studies. In addition, it offers the advantage of less invasive sampling (finger or heel prick, rather than conventional venous cannula) which enables recruitment of subjects for clinical studies. Further, the simpler matrix preparation and transfer (no refrigerated centrifugation to produce plasma) and easy storage and shipment to analytical laboratories (no requirement for freezers and dry ice) offer further benefits. In addition, these requirements lead to notable environmental benefits. The transport and storage of samples is further simplified by the antimicrobial properties of the DBS sample, removing the requirements for special biohazard arrangements [54]. The significant reduction in blood volume required for DBS, allows for the simplification of current approaches to the determination of drug exposure (toxicokinetics, i.e., TK) in pre-clinical animal studies and leads to significant benefits in the reduction of animal use in drug development. More consistent data can be obtained through more serial sampling as DBS from individual animals and less reliance on composite data.

Table 2: Representative applications of dried blood spot technique in LC-MS/MS bioanalysis.

Compound	Blood Spot Card /Paper	Blood Volume for calibration standards (µL)	Dried Blood Spot size (ID)	Extraction solvent	Chromatography	MS detection	LLOQ	Ref.
Dextromethorphan and its metabolite dextrorphan	Whatman FTA Elute cards	50	3mm	Methyl tert-butyl ether (MTBE)	Gradient separation with a mobile phase of Acetic acid, formic acid, water, and acetonitrile, 3.5 min runtime	ESI, MRM	0.2 ng/mL for both analytes	[40]
Acetaminophen	FTA Card by Whatman	15	3mm	1:1 MeOH:H2O (v/v)	Gradient separation with a mobile phase of ammonium acetate and methanol, 2 min runtime	ESI, MRM	25 ng/mL	[41]
Acetaminophen	FTA Elute Card by Whatman	15	3mm	MeOH	Gradient separation with a mobile phase of ammonium acetate and methanol, 1.5 min runtime	ESI, MRM	100 ng/mL	[42]
Amprenavir, nelfinavir, indinavir, lopinavir, saquinavir, ritonavir, atazanavir, nevirapine, and efavirenz	Testkarten 76×108 mm; Schleicher & Schuell	5	5mm	50:50 MeOH/0.2M ZnSO4 (v/v)	Gradient separation with a mobile phase of ammonium acetate, acetic acid, water, and methanol, 8 min runtime	ESI, MRM	41–102 ng/mL	[43]
Atazanavir, darunavir, lopinavir, ritonavir, efavirenz, and nevirapine	Whatman 903 protein saver Cards	40	0.25 inch	MeOH:ACN :0.2M ZnSO4 (1:1:2, v/v/v)	Stepwise gradient separation with a mobile phase of ammonium acetate and methanol, 10 min runtime	APCI, MRM	50-100 ng/mL	[44]
Etravirine (TMC125)	Whatman 903 protein saver cards	25	0.25 inch	MeOH:ACN :0.2M ZnSO4 (1:1:2, v/v/v)	Stepwise gradient separation with a mobile phase of acetate buffer and methanol, 10 min runtime	ESI, MRM	50 ng/mL	[45]
Four discovery compounds synthesized at Merck Frosst and Co.	Schleicher and Schuell 903 paper	40	3.2mm	1:1 ACN:H2O (v/v)	Gradient separation with a mobile phase of formic acid and acetonitrile, 3 min runtime	ESI, MRM	50 ng/mL as demonstrated	[46]
Raltegravir	Whatman 903 protein saver Cards	15	0.25 inch	MeOH:ACN :0.2M ZnSO4 (1:1:2, v/v/v)	Stepwise gradient separation with a mobile phase of acetate buffer and methanol, 10 min runtime	ESI, MRM	50 ng/mL	[47]
Cyclosporin A	Whatman 903 protein saver Cards	50	8mm	1:1 MeOH:H2O (v/v)	Gradient separation with a mobile phase of formic acid, ammonium acetate and methanol, 3.5 min runtime	ESI, MRM	25 ng/mL	[48]
Everolimus	Protein saver 903, Whatman	30	NA	MeOH	Online SPE, isocratic separation with a mobile phase of formic acid, ammonium acetate and methanol, 6.5 min runtime	ESI, MRM	2 ng/mL	[49]
Topiramate	Whatman 903	20	3.2mm	30:70 H2O:CAN with 0.05% of formic acid	Isocratic separation with a mobile phase of formic acid, water and acetonitrile, 3 min runtime	ESI, MRM	500 ng/mL	[50]
Corticosterone, deoxycorticosterone, progesterone, 17alpha-hydroxyprogesterone, 11-deoxycortisol, 21-deoxycortisol, androstenedione, testosterone, dihydrotestosterone and cortisol	Whatman Schleicher & Schuell paper	25	6mm	50:50 (v/v) ACN/MeOH	Gradient separation with a mobile phase of formic acid, water, and methanol, 6 min runtime	ESI, MRM	5 ng/mL for cortisol and 12.5 ng/mL for other steroids	[51]
Tacrolimus	Whatman Schleicher & Schuell paper	30	7.5mm	40:10 (v/v) MeOH:ACN	Online SPE, isocratic separation with a mobile phase of formic acid, ammonium acetate and methanol, 6.5 min runtime	ESI, MRM	1 ng/mL	[52]

Beaudette et. al applied DBS technique for discovery stage pharmacokinetic determination [46]. Four compounds, each from a different structural class, were investigated with rats being the animal chosen for dosing. The calculated log P for these compounds ranged from 1.61 to 7. Blood samples were collected from rats by tail-bleeding onto the blood collection cards at the specified timepoints. When the study was completed, the cards were air dried overnight. After punching out 3.2mm discs of the dried blood spots and placing in a 96-well plate, extraction of the compounds was performed by using 1:1 acetonitrile/water containing internal standard as the extraction solvent. A portion of the particulate-free extract was transferred for LC-MS/MS analysis which employed a 3-min HPLC gradient program combined with simultaneous MRM detection for the 4 testing compounds. The unknown samples were measured against calibration standards that were pipetted on the blood collection card. The method was demonstrated to be both precise and accurate for the compounds tested with acceptable inter and intra-assay variability. The compounds tested were also found to be stable for up to one month at room temperature.

Barfield et. al reported first application of DBS analysis to a TK study in support of a safety assessment study [42]. A reversed phase HPLC-MS/MS method has been developed and validated for the quantitative bioanalysis of acetaminophen in dried blood spots prepared from small volumes (15µL) of dog blood. Samples were extracted with methanol prior to analysis. Detection was performed in positive ion mode with selected reaction monitoring. The analytical concentration range was 0.1–50 µg/mL. The intra-day precision and bias values were both less than 15%. Acetaminophen was stable in DBS stored at room temperature for at least 10 days. The methodology was applied in a toxicokinetic (TK) study where the data obtained from DBS samples was physiologically comparable with results from duplicate blood samples (diluted 1:1 (v/v) with water) analyzed using identical HPLC-MS/MS conditions. The authors demonstrated that quantitative analysis of a drug extracted from DBS could provide high quality TK data while minimizing the volume of blood withdrawn from experimental animals, to an order of magnitude lower than is current practice in the pharmaceutical industry. The similar approach was later used in Spooner's report for the quantitative determination of circulating drug concentrations in clinical studies using acetaminophen as a tool compound [41]. An assay with a range from 25 to 5000 ng/mL in human blood was validated by aliquoting 15 µL of sample onto DBS card with a repeater pipette. The assay employed simple solvent extraction of a punch taken from the DBS sample, followed by reversed phase HPLC separation, combined with selected reaction monitoring mass spectrometric detection. In addition to performing routine experiments to establish the validity of the assay to internationally accepted criteria (precision, accuracy, linearity, sensitivity, selectivity), a number of experiments were performed to specifically demonstrate the quality of the quantitative data generated using DBS sample format. The authors have demonstrated that a volume change from 10 µL to 20 µL for blood spotted, a device change from pipette to glass capillary for spotting the blood, or a temperature change from 0 to 37 Celsius for blood spotted had minimal impact on the assay. DBS and whole blood samples spiked individually with acetaminophen glucuronide and sulfate metabolite standards showed no detectable formation of acetaminophen after storage for 24 and 6 h at room temperature, respectively. Further, a qualitative assessment of the peak areas for the metabolites showed no notable decrease with time. This indicates that these metabolites are stable in DBS on the collection paper and as whole blood samples and will therefore not interfere with the quantification of acetaminophen. The validated DBS approach was successfully applied to a clinical study

Both Barfield and Spooner's reports called for simple extraction of dried blood spot samples using either 1:1 MeOH:H$_2$O (v/v) or MeOH as extraction solvent. Recently, DBS technology was evaluated in an assay for the quantitation of dextromethorphan (DM) and its metabolite, dextrorphan (DT), in human whole blood using methyl tert-butyl ether (MTBE) as extraction solvent. Both the parent drug and metabolite were spiked in the blood matrix and subsequently allowed to dry on a specimen collection card. The dried blood spots were removed using a manual punch and then subjected to extraction by MTBE. The organic supernatant was transferred and evaporated and the residue was reconstituted in 20% acetonitrile. The overall method recovery of DM and DT was 87.8% and 95.4%, respectively. The assay was linear over the concentration range of 0.2–200 ng/mL for both analytes. Several factors that potentially affect DBS assay quantitation were investigated, such as punch size, DBS sample punch-out location, and the volume of the blood sample pipetted on the specimen collection cards. The study determined that punch size did not affect assay quantitation accuracy. Sampling from different location on the specimen collection cards showed no significant variation for both drugs. The results showed that acceptable results could be achieved with some variation of the sample volume, which allows a simple blood sampling procedure at the test sites. By stacking several blood spots at the same concentration level together and performing the same extraction, the authors demonstrated a similar lower limit of quantitation (LLOQ) at 0.01 ng/mL for DM can be achieved as the plasma assay.

DEVELOPING TECHNIQUES FOR WHOLE BLOOD SAMPLE ANALYSIS

Déglon et. al demonstrated the feasibility of an on-line DBS procedure for bioanalysis of saquinavir, imipramine, and verapamil [55]. The authors designed an inox cell for receiving a blood sample (10µL) that was spotted on a filter paper. The cell was then integrated into LC/MS system where the analytes are desorbed out of the paper towards a column switching system ensuring the purification and separation of the compounds before their detection on a single quadrupole MS coupled to atmospheric pressure chemical ionization (APCI) source. The authors showed that no pretreatment is necessary in spite the analysis is based on whole blood sample. This conceptual on-line DBS technique allowed the analyses of these three compounds over their therapeutic concentrations from 50 to 500 ng/mL for imipramine and verapamil and from 100 to 1000 ng/mL for saquinavir. Good selectivity was obtained and no endogenous or chemical components interfered with the quantitation of the analytes. The method also showed good repeatability with relative standard deviation (RSD) lower than 15% based on two levels of concentration (low and high). Function responses were found to be linear over the therapeutic concentration for each compound and were used to determine the concentrations of real patient samples for saquinavir. Comparison of the founded values with those of a validated method used routinely in a reference laboratory showed a good correlation between the two methods.

Lately, desorption electrospray ionization (DESI) has been gaining interest from analytical field as a potential tool for direct and quantitative analysis of small molecules from various substrates [56-58]. DESI is a member of the family of ambient ionization methods which are characterized by the ability to record mass spectra on ordinary samples in their native environment, without sample preparation or pre-separation, by creating ions outside the instrument. DESI can be applied for quantitative measurements of spotted samples deposited on artificial surfaces. A solvent is electrosprayed to generate charged droplets, which are directed at the analyte surface. The secondary droplets are then directed through the ion interface of a mass spectrometer and mass analyzed. The

DESI mechanism involves a droplet pick-up process followed by ESI-like desolvation of the secondary droplets and formation of gas-phase ions. Droplet pick-up involves an interaction in which the impacting solvent droplet makes contact with the wet sample surface, causing ejection of solvent droplet(s) containing the dissolved analyte. DESI has been applied to pharmaceutical cleaning validation, *in vivo* recognition of *Bacillus Subtilis*, trace analysis of agrochemicals in food, analysis of diterpene glycosides from *Stevia* leaves, imaging drugs and metabolites in tissues, and etc. A newly developed DESI source was just characterized in terms of its performance for quantitative analysis [58]. A 96-sample array, containing pharmaceuticals in various matrices, was analyzed in a single run with a total analysis time of 3 min. Chemical background-free samples of propranolol (PRN) and carbamazepine (CBZ) were examined. So were two other sample sets consisting of the analytes at varying concentration in a biological milieu of 10% urine or porcine brain total lipid extract. The lower limit of detection (LOD) for PRN and CBZ when analyzed without chemical background was 10 and 30 fmol, respectively. The LOD of PRN increased to 400 fmol analyzed in 10% urine, and 200 fmol when analyzed in the brain lipid extract. Although still at its early application stage, the further advancement of DESI technique may make rapid and direct analysis of whole blood samples, especially DBS samples, a reality for routine pharmacokinetic studies and therapeutic monitoring use in the future.

CONCLUSIONS

Various approaches in sample pretreatment, extraction, chromatography, and mass spectrometric detection have been developed for liquid whole blood sample analysis to satisfy sensitivity, selectivity and other assay requirements. Some of the methods have been automated in 96-well format for pharmaceutical development. Meanwhile, dried blood spot (DBS) technique has shown great potential to be used in not only therapeutic drug monitoring, but also preclinical and clinical studies because of the many advantages it offers in reduced sample volume, shipping, storage, and etc. The success of DBS techniques in preclinical and clinical studies has led to growing intent to apply DBS technology as the recommended analytical approach for the assessment of pharmacokinetics for new oral small molecule drug candidates. Further improvements in areas like sampling, automation, and sensitivity will be keys to a wide adaption to DBS technique in bioanalysis for pharmaceutical discovery and development.

ABBREVIATIONS

MTBE	=	methyl *tert*-butyl ether
MS	=	mass spectrometry
HPLC	=	high-performance liquid chromatography
MeOH	=	methanol
ACN	=	acetonitrile
ESI	=	electrospray ionization
MRM	=	multiple reaction monitoring

PPT	=	protein precipitation
LLE	=	liquid-liquid extraction
SPE	=	solid-phase extraction
$ZnSO_4$	=	zinc sulfate
DESI	=	desorption electrospray ionization
DBS	=	dried-blood spot
CV	=	coefficient of variation
LLOQ (LOQ)	=	lower limit of quantitation
LOD	=	low limit of detection
MS/MS	=	tandem mass spectrometry

REFERENCES

[1]   Jemal M, Xia YQ. LC-MS development strategies for quantitative bioanalysis. Curr Drug Metabol 2006; 7: 491-502.

[2]   Xu RN, Fan L, Rieser MJ, El-Shourbagy TA. Recent advances in high-throughput quantitative bioanalysis by LC-MS/MS. J Pharm Biomed Anal 2007; 44: 342-355.

[3]   Selinger K, Lanzo C, Sekut A. Determination of remifentanil in human and dog blood by HPLC with UV detection. J Pharm Biomed Anal 1994; 12: 243-248.

[4]   Salm P, Taylor PJ, Rooney F. A high-performance liquid Chromatography-mass spectrometry method using a novel atmospheric pressure chemical ionization approach for the rapid simultaneous measurement of tacrolimus and cyclosporin in whole blood. Ther Drug Monit 2008; 30: 292-300.

[5]   Kirchner GI, Vidal C, Jacobsen W, Franzke A, Hallensleben K, Christians U, Sewing KF. Simultaneous on-line extraction and analysis of sirolimus (rapamycin) and ciclosporin in blood by liquid chromatography-electrospray mass spectrometry. J chromatogr B Biomed Sci. & Appl 1999; 721: 285-94.

[6]   Zhou L, Tan D, Theng J, Lim L, Liu YP, Lam KW. Optimized analytical method for cyclosporin A by high-performance liquid chromatography-electrospray ionization mass spectrometry. J Chromatogr B, Biomed Sci & Appl 2001; 754: 201-207.

[7]   Jain DS, Subbaiah G, Sanyal M, Pande UC, Shrivastav P. Liquid chromatography-tandem mass spectrometry validated method for the estimation of indapamide in human whole blood. J Chromatogr B 2006; 834: 149-154.

[8]   Kristoffersen L, Øiestad EL, Opdal MS,, Krogh M, Lundanes E, Christophersen AS. Simultaneous determination of 6 beta-blockers, 3 calcium-channel antagonists, 4 angiotensin-II antagonists and 1 antiarrhythmic drug in post-mortem whole blood by automated solid phase extraction and liquid chromatography mass spectrometry. Method development and robustness testing by experimental design. J Chromatogr B 2007; 850: 147-160.

[9]   Celli N, Mariani B, Di Carlo F, Zucchetti M, Lopez-Lazaro L, D'Incalci M, Rotilio D., Determination of aplidin, a marine-derived anticancer drug, in human plasma, whole blood and urine by liquid chromatography with electrospray ionisation tandem mass spectrometric detection. J Pharm Biomed Anal 2004; 34: 619-630.

[10] Ji QC, Todd RM, El-Shourbagy TA. 96-Well liquid-liquid extraction liquid chromatography-tandem mass spectrometry method for the quantitative determination of ABT-578 in human blood samples. J Chromatogr B 2004; 805: 67-75.

[11] Li W, Luo S, Hayes M, He H, Tse F.L. Determination of N-methyl-4-isoleucine-cyclosporin (NIM811) in human whole blood by high performance liquid chromatography-tandem mass spectrometry. Biomed Chromatogr 2007; 21: 249-256.

[12] Klawitter J, Zhang YL, Klawitter J, Anderson N, Serkova NJ, Christians U. Development and validation of a sensitive assay for the quantification of imatinib using LC/LC-MS/MS in human whole blood and cell culture. Biomed Chromatogr. 2009; 23: 1251-1258

[13] Handy R, Trepanier D, Scott G, Foster R, Freitag D. Development and validation of a LC/MS/MS method for quantifying the next generation calcineurin inhibitor, voclosporin, in human whole blood. J Chromatogr B 2008; 874: 57-63.

[14] Jagerdeo E, Montgomery MA, Lebeau MA, Sibum M. An automated SPE/LC/MS/MS method for the analysis of cocaine and metabolites in whole blood. J Chromatogr B 2008; 874: 15-20.

[15] Jagerdeo E, Schaff J.E, Montgomery M.A, LeBeau M.A. A semi-automated solid-phase extraction liquid chromatography/tandem mass spectrometry method for the analysis of tetrahydrocannabinol and metabolites in whole blood. Rapid Commun Mass Sp 2009; 23: 2697-705.

[16] Bugey A, Rudaz S, Staub C. A fast LC-APCI/MS method for analyzing benzodiazepines in whole blood using monolithic support. J Chromatogr B 2006; 832: 249-55.

[17] Zhuang XM, Yuan M, Zhang ZW, Wang XY, Zhang ZQ, Ruan JX. Determination of 4-dimethylaminophenol concentrations in dog blood using LC-ESI/MS/MS combined with precolumn derivatization. J Chromatogr B 2008; 876: 76-82.

[18] Tian H, Sun D, Dou G, Yuan D, Meng Z. Quantitative determination of beta,beta-dimethylacrylshikonin (DASK) in rat whole blood by liquid chromatography-tandem mass spectrometry with pre-column derivation and its pharmacokinetic application. Biomed Chromatogr 2009; 23: 365-70.

[19] Naik H, Imming P, Schmidt MS, Murry DJ, Fleckenstein L. Development and validation of a liquid chromatography-mass spectrometry assay for the determination of pyronaridine in human blood for application to clinical pharmacokinetic studies. J Pharm Biomed Anal 2007; 45: 112-119.

[20] Ariffin MM, Anderson RA. LC/MS/MS analysis of quaternary ammonium drugs and herbicides in whole blood. J Chromatogr B 2006; 842: 91-97.

[21] Johansen SS, Bhatia HM. Quantitative analysis of cocaine and its metabolites in whole blood and urine by high-performance liquid chromatography coupled with tandem mass spectrometry J Chromatogr B 2007; 852: 338-344.

[22] Johansen SS, Jensen JL. Liquid chromatography-tandem mass spectrometry determination of LSD, ISO-LSD, and the main metabolite 2-oxo-3-hydroxy-LSD in forensic samples and application in a forensic case. J Chromatogr B 2005; 825: 21-28.

[23] Yeh LT, Nguyen M, Dadgostari S, Bu W, Lin CC. LC-MS/MS method for simultaneous determination of viramidine and ribavirin levels in monkey red blood cells J Pharm Biomed Anal 2007; 43: 1057-64.

[24] Kontrimaviciūte V, Breton H, Mathieu O, Mathieu-Daudé JC, Bressolle FM. Liquid chromatography-electrospray mass spectrometry determination of ibogaine and noribogaine in human plasma and whole blood. Application to a poisoning involving Tabernanthe iboga root. J Chromatogr B 2006; 843: 131-141.

[25] Chen YL, Hirabayashi H, Akhtar S, Pelzer M, Kobayashi M. Simultaneous determination of three isomeric metabolites of tacrolimus (FK506) in human whole blood and plasma using high performance liquid chromatography-tandem mass spectrometry. J Chromatogr B 2006; 830: 330-341.

[26] Li JL, Wang XD, Wang CX, Fu Q, Liu LS, Huang M, Zhou SF. Rapid and simultaneous determination of tacrolimus (FK506) and diltiazem in human whole blood by liquid chromatography-tandem mass spectrometry: application to a clinical drug-drug interaction study. J Chromatogr B 2008; 867: 111-118.

[27] Bogusz MJ, Enazi EA, Hassan H, Abdel-Jawaad J, Ruwaily JA, Tufail MA. Simultaneous LC-MS-MS determination of cyclosporine A, tacrolimus, and sirolimus in whole blood as well as mycophenolic acid in plasma using common pretreatment procedure. J Chromatogr B 2007; 850: 471-80.

[28] Alak AM, Moy S, Cook M, Lizak P, Niggebiugge A, Menard S, Chilton A. An HPLC/MS/MS assay for tacrolimus in patient blood samples. Correlation with results of an ELISA assay. J Pharm Biomed Anal 1997; 16: 7-13.

[29] Xu RN, Rieser M, and El-Shourbagy TA. Reproducibility Testing of Bioanalytical Method for Regulated Samples in Whole Blood Matrix by LC/MS/MS. 2009; March 8-13, Chicago, IL.

[30] Lensmeyer GL, Poquette MA. Therapeutic monitoring of tacrolimus concentration in blood: semi-automated extraction and liquid chromatography–electrospray ionization mass spectrometry. Ther Drug Monit 2001; 23: 239-249

[31] Poquette MA, Lensmeyer GL, Doran TC. Effective use of liquid chromatography–mass spectrometry (LC/MS) in the routine clinical laboratoryfor monitoring sirolimus, tacrolimus, and cyclosporine. Ther Drug Monit 2005; 27: 144-150.

[32] Cavicchi C, Malvagia S, la Marca G, Gasperini S, Donati MA, Zammarchi E, Guerrini R, Morrone A, Pasquini E. Hypocitrullinemia in expanded newborn screening by LC-MS/MS is not a reliable marker for ornithine transcarbamylase deficiency. J Pharm Biomed Anal 2009; 49: 1292-1295.

[33] Higashi T, Nishio T, Uchida S, Shimada K, Fukushi M, Maeda M. Simultaneous determination of 17alpha-hydroxypregnenolone and 17alpha-hydroxyprogesterone in dried blood spots from low birth weight infants using LC-MS/MS. J Pharm Biomed Anal 2008; 48: 177-82.

[34] Lai CC, Tsai CH, Tsai FJ, Lee CC, Lin WD. Rapid monitoring assay of congenital adrenal hyperplasia with microbore high-performance liquid chromatography/electrospray ionization tandem mass spectrometry from dried blood spots. Rapid Commun Mass Sp 2001; 15: 2145-51

[35] Malvagia S, Pasquini E, Innocenti M, Donati MA, Zammarchi E., Rapid 2nd-tier test for measurement of 3-OH-propionic and methylmalonic acids on dried blood spots: reducing the false-positive rate for propionylcarnitine during expanded newborn screening by liquid chromatography-tandem mass spectrometry. la Marca G Clin Chem 2007; 53: 1364-1369.

[36] deWilde A, Sadilkova K, Sadilek M, Vasta V, Hahn SH, Tryptic peptide analysis of ceruloplasmin in dried blood spots using liquid chromatography-tandem mass spectrometry: application to newborn screening. Clin Chem 2008; 54: 1941-1942.

[37] Oglesbee D, Sanders KA, Lacey JM, Magera MJ, Casetta B, Strauss KA, Tortorelli S, Rinaldo P, Matern D. Second-tier test for quantification of alloisoleucine and branched-chain amino acids in dried blood spots to improve newborn screening for maple syrup urine disease (MSUD). Clin Chem 2008; 54: 542-549.

[38]   Zoppa M, Gallo L, Zacchello F, Giordano G. Method for the quantification of underivatized amino acids on dry blood spots from newborn screening by HPLC-ESI-MS/MS. J Chromatogr B 2006; 831: 267-73.

[39]   Chace DH. Mass spectrometry in newborn and metabolic screening: historical perspective and future directions. J Mass Spectrom 2009; 44: 163–170.

[40]   Liang X, Li Y, Barfield M, Ji QC. Study of dried blood spots technique for the determination of dextromethorphan and its metabolite dextrorphan in human whole blood by LC-MS/MS. J Chromatogr B 2009; 877: 799-806.

[41]   Spooner N, Lad R, Barfield M. Dried blood spots as a sample collection technique for the determination of pharmacokinetics in clinical studies: considerations for the validation of a quantitative bioanalytical method. Anal. Chem. 2009; 81: 1557-63.

[42]   Barfield M, Spooner N, Lad R, Parry S, Fowles S. Application of dried blood spots combined with HPLC-MS/MS for the quantification of acetaminophen in toxicokinetic studies. J Chromatogr B 2008; 870: 32-7

[43]   Koal T, Burhenne H, Römling R, Svoboda M, Resch K, Kaever V. Quantification of antiretroviral drugs in dried blood spot samples by means of liquid chromatography/tandem mass spectrometry. Rapid Comm Mass Sp 2005; 9: 2995-3001.

[44]   ter Heine R, Davids M, Rosing H, van Gorp EC, Mulder JW, van der Heide YT, Beijnen JH, Huitema AD. Quantification of HIV protease inhibitors and non-nucleoside reverse transcriptase inhibitors in peripheral blood mononuclear cell lysate using liquid chromatography coupled with tandem mass spectrometry. J Chromatogr B 2009; 877: 575-80.

[45]   ter Heine R, Rosing H, van Gorp EC, Mulder JW, Beijnen JH, Huitema AD. Quantification of etravirine (TMC125) in plasma, dried blood spots and peripheral blood mononuclear cell lysate by liquid chromatography tandem mass spectrometry. J Pharm Biomed Anal 2009; 49: 393-400.

[46]   Beaudette P, Bateman KP, Discovery stage pharmacokinetics using dried blood spots. J Chromatogr B 2004; 809: 153-158.

[47]   ter Heine R, Hillebrand MJ, Rosing H, van Gorp EC, Mulder JW, Beijnen JH, Huitema AD. Quantification of the HIV-integrase inhibitor raltegravir and detection of its main metabolite in human plasma, dried blood spots and peripheral blood mononuclear cell lysate by means of high-performance liquid chromatography tandem mass spectrometry. J Pharm Biomed Anal 2009; 49: 451-8.

[48]   Wilhelm AJ, den Burger JC, Vos RM, Chahbouni A, Sinjewel A. Analysis of cyclosporin A in dried blood spots using liquid chromatography tandem mass spectrometry. J Chromatogr B 2009; 877: 1595-1598.

[49]   van der Heijden J, de Beer Y, Hoogtanders K, Christiaans M, de Jong GJ, Neef C, Stolk L. Therapeutic drug monitoring of everolimus using the dried blood spot method in combination with liquid chromatography-mass spectrometry. J Pharm Biomed Anal 2009; 50: 664-70.

[50]   la Marca G, Malvagia S, Filippi L, Fiorini P, Innocenti M, Luceri F, Pieraccini G, Moneti G, Francese S, Dani FR, Guerrini R. Rapid assay of topiramate in dried blood spots by a new liquid chromatography-tandem mass spectrometric method. J Pharm Biomed Anal 2008; 48: 1392-1396.

[51]   Janzen N, Sander S, Terhardt M, Peter M, Sander J. Fast and direct quantification of adrenal steroids by tandem mass spectrometry in serum and dried blood spots. J Chromatogr B 2008; 861: 117-122.

[52]   Hoogtanders K, van der Heijden J, Christiaans M, Therapeutic drug monitoring of tacrolimus with the dried blood spot method. J Pharm Biomed Anal 2007; 44: 658–664.

[53]   Edelbroek PM, van der Heijden J, Stolk LML. Dried Blood Spot Methods in Therapeutic Drug Monitoring: Methods, Assays, and Pitfalls. Ther Drug Monit 2009; 31: 327-336.

[54]   Knudsen RC, Slazyk WE, Richmond JY, Hannon WH. CDC Guidelines for the Shipment of Dried Blood Spot Specimens; http:// www.cdc.gov/od/ohs/biosfty/driblood.html 1995.

[55]   Déglon J, Thomas A, Cataldo A, Mangin P, Staub C. On-line desorption of dried blood spot: A novel approach for the direct LC/MS analysis of micro-whole blood samples. J Pharm Biomed Anal 2009; 49: 1034-1039.

[56]   Wiseman JM, Ifa DR, Zhu Y, Kissinger CB, Manicke NE, Kissinger PT, Cooks RG. Desorption electrospray ionization mass spectrometry: Imaging drugs and metabolites in tissues. P NATL ACAD SCI USA 2008; 105: 18120-18125.

[57]   Dill AL, Ifa DR, Manicke NE, Ouyang Z, Cooks RG. Mass spectrometric imaging of lipids using desorption electrospray ionization. J Chromatogr B 2009; 877: 2883-2889.

[58]   Manicke NE, Kistler T, Ifa DR, Cooks RG, Ouyang Z. High-throughput quantitative analysis by desorption electrospray ionization mass spectrometry. J Amer Soc Mass Spectr 2009; 20: 321-5.

Reviews in Pharmaceutical and Biomedical Analysis, 2010, 63-75 63

Microbial Cells and Biosensing: A Dual Approach - Exploiting Antibodies and Microbial Cells as Analytical/Power Systems.

Sushrut Arora[1,2,4], **Gabriele Pastorella**[1,4], **Barry Byrne**[1,3,4], **Enrico Marsili**[1,4] and **Richard O'Kennedy**[1,2,3,4]

[1]*School of Biotechnology, Dublin City University, Dublin 9, Ireland;* [2]*Biomedical Diagnostics Institute (BDI), Dublin City University, Dublin 9, Ireland;* [3]*Centre for Bioanalytical Sciences (CBAS), Dublin City University, Dublin 9, Ireland and* [4] *National Centre for Sensor Research (NCSR), Dublin City University, Dublin 9, Ireland*

Abstract: The primary focus of this review is the discussion of how biosensor-based platforms can be used in conjunction with microbial cells for monitoring, environmental and industrial applications. Two approaches will be comprehensively discussed. The first of these will examine how immunosensors can be used for the sensitive and selective detection of bacterial pathogens in a range of diverse and complex sample matrices. Secondly, we discuss the implementation of free and immobilised microbial cells for facilitating the analysis of chemicals and metabolites in cost-effective devices that, in turn, are directly applicable to environmental monitoring. Further examples, relating to the uses and advantages of microbial fuel cells are also discussed, with particular emphasis on recent and innovative developments.

INTRODUCTION

Biosensors play key roles in the sensitive and selective detection of a plethora of biologically important and structurally diverse analytes. These include proteinaceous biomarkers, toxic metabolites and whole microbial cells. Biosensing analytical devices use a target-specific biorecognition element, such as an antibody, lectin or deoxyribonucleic acid (DNA) probe, which is suitably immobilised on the sensor surface so as to promote efficient and functional interaction with its cognate target. The resultant biorecognition event introduces a physicochemical change that is converted (via a transducer) to a signal that can be interpreted and further quantified by the end user. This is typically facilitated by a dedicated computer-based readout system (Fig. **1**). The versatility of selecting biosensor-based detection is demonstrated by the availability of a selection of different formats that are based on the monitoring of changes in electrochemical, optical, mass, magnetic and thermometric properties [1,2], so that the operator can tailor the experimental design to suit their objectives by selecting the most appropriate platform.

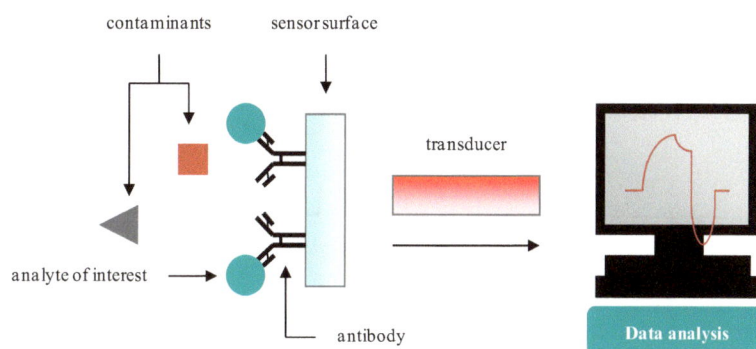

Figure 1: A schematic representation of a biosensor, utilising an immobilised full-length immunoglobulin for the detection of a target molecule in a complex sample matrix.

BIOLIGAND-BASED PATHOGEN DETECTION

Microorganisms, specifically bacterial and fungal cells, are ubiquitous in nature and are found in a selection of diverse ecosystems (e.g. soil, water and plant litter). It is well established that many microbes have a positive impact on human life through their participation in a variety of biologically important events. Some bacterial strains, for example, contribute towards the assimilation of polysaccharides and suppression of growth of invasive microorganisms on the epithelium of the colon by blocking adhesion sites [3]. Furthermore, a diverse array of medicinally and economically important metabolites are synthesised by bacterial and fungal cells. These include enzymes (xylanases, proteases, streptokinases), non-ribosomal peptides (vancomycin), antibiotics (penicillin) and polyketides (erythromycin A, amphotericin B). Finally, as will be discussed later, microorganisms may convert chemical energy into electrical energy in microbial fuel cells. These positive attributes contrast sharply with the inherent ability of selected bacterial and fungal species to have a deleterious effect on human well-being by acting as pathogens. The origin of these harmful microorganisms ranges from contaminated food-sources (cheese, coleslaw, meat) to hospital surfaces (as is the case with nosocomial-related infections, such as those caused by *Clostridium difficile*). In addition, many toxins are produced by bacterial strains, such as *Staphylococcus aureus*, and thus, accurate detection is of paramount importance for maintaining public health and ensuring compliance with legislative standards [1].

Biosensor-based analytical platforms greatly facilitate the rapid and sensitive detection of pathogens. Bioligands used in these formats include nucleic acid probes, as recently implemented by Liao and Ho [4] for the sensitive detection of the foodborne pathogen *Escherichia coli* O157:H7, and by Prabhakar and colleagues for monitoring the presence of *Mycobacterium tuberculosis* [5]. Lectins, specific for mono- and oligosaccharide elements of bacterial polysaccharide structures (glycocalyx), are also applicable. Ertl and Mikkelsen [6] devised an electrochemical lectin-based biosensor array for the detection of five bacterial strains (*Bacillus cereus*, *Enterobacter aerogenes*, *Proteus vulgaris*, *S. aureus* and *E. coli*) and *Saccharomyces cerevisiae*, an opportunistic fungal pathogen. More recently, this methodology was applied by Gamella and colleagues for analysis of *E. coli*, *S. aureus* and *Mycobacterium phlei* by immobilising bacterial cell-bound biotinylated lectins on a screen-printed gold electrode and subsequently monitoring changes in impedance. The authors were also able to differentiate between viable *E. coli* and *S. aureus* cells by monitoring β-galactosidase activity that is routinely associated with the metabolism of the former strain and other *Enterobacteriaceae* [7].

The detection of bacterial and fungal pathogens by biosensing is also facilitated by the diverse array of epitopes that are presented on the exterior of these cells which, in turn, can be selected as targets for antibody-based biorecognition. These antigenic targets are typically flagellar, capsular or surface-bound proteinaceous antigens [1] which are preferentially selected over carbohydrate-based epitopes, such as the polysaccharide elements [8,9] commonly used for lectin-based biosensor recognition, since carbohydrates typically have lower immunogenic potential [10]. Monoclonal, polyclonal and recombinant antibodies can be raised against almost any target through the development of a carefully designed and rigorous screening protocol, and a schematic representation of a full-length antibody and antibody fragments is shown in Fig. **2**. The generation and screening of these biorecognition elements against a target of interest on a pathogenic bacterial or fungal microorganism is comprehensively outlined in references [1] and [2].

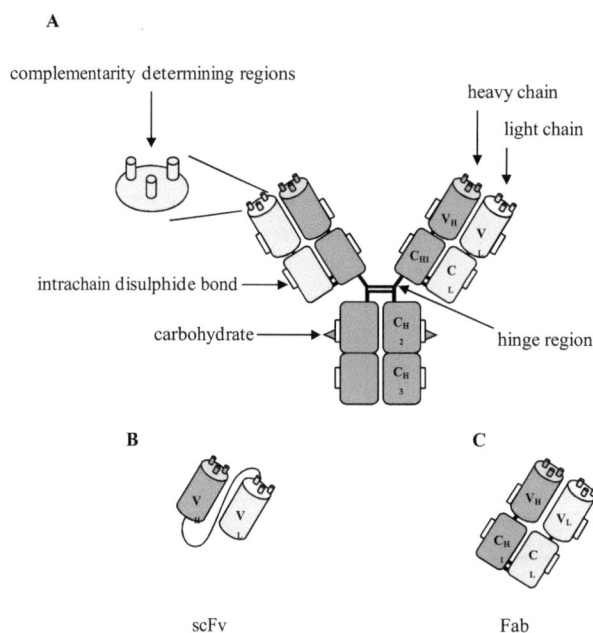

Figure 2: (A) A typical full-length immunoglobulin, ideal for immunosensor-based pathogen analysis. The three complementarity determining regions (CDR) shown above, located in the variable domains of the antibody, are the points of contact with the analyte of interest (e.g. pathogenic cell). Two recombinant antibody fragments, namely the single chain variable fragment (scFv; B) and fragment antigen binding (Fab; C) unit are shown, and their use in immunosensing is discussed in the text.

The efficacy of using an antibody-based biosensor (immunosensor) platform for pathogen monitoring is dependent on the quality of the two main components of the system, namely the transducer and the biorecognition element (in this case, the antibody). A key consideration here relates to identifying an antibody, be it monoclonal, polyclonal or recombinant, that is suitable for immunosensing. This is readily facilitated by rigorously screening candidates on advanced analytical platforms, such as Biacore™ (produced by GE Healthcare), to positively identify antibodies that have high-affinity for the target epitope and have sufficient sensitivity for identifying low cell numbers, which may often be present in a food sample contaminated with bacterial strains such as *Listeria monocytogenes*, *Salmonella typhimurium* or *E. coli* O157:H7. Furthermore, as several bacterial strains may often reside in a single analytical matrix, selectivity is another important consideration. Therefore, where a pathogen of interest is to be detected, it is beneficial to identify an antigen that is highly-specific to a bacterial (or fungal) strain and is constitutively expressed, and raise an antibody against this target [1]. Subsequent cross-reactivity analysis with unrelated microorganisms can validate this approach by ensuring that the selected antibody does not recognise antigens on other cells, which is understandably problematic where pathogen identification is an absolute necessity. The efficacy of using immunosensor-based methods for monitoring bacterial pathogens is summarised in Table 1. Examples relating to the detection of *Trichophyton rubrum* and *Puccinia striiformis* are also included to demonstrate that similar methodologies can be applied for monitoring of fungal pathogens.

Table 1: Immunosensor-based biorecognition of microbial cells. CFU=colony forming units.

Sensor format	Pathogen detected	Antibody type	Sensitivity	Reference
Electrochemical				
Amperometric	*S. typhimurium*	Polyclonal	8×10^3 CFU/ml	[95]
	E. coli O157:H7	Monoclonal	5×10^3 CFU/ml	[96]
	Trichophyton rubrum	Polyclonal	1×10^{-15} mg/ml surface antigen	[97]
Impedimetric	*E. coli* O157:H7	Polyclonal	1×10^4 CFU/ml	[98]
	L. monocytogenes	Monoclonal	1×10^2 CFU/ml	[99]
Potentiometric	*E. coli* O157:H7	Polyclonal	2.5×10^4 cells/ml (live)	[100]
			7.1×10^2 cells/ml (heat-treated)	
Conductimetric	*E. coli* O157:H7	Polyclonal	79 CFU/ml	[101]
	S. typhimurium		83 CFU/ml	
	E. coli O157:H7	Polyclonal	0.5 CFU/ml	[102]
Optical	*L. monocytogenes*	Monoclonal	1×10^7 CFU/ml	[103]
	L. monocytogenes	Polyclonal	1×10^5 CFU/ml	[104]
	Campylobacter jejuni	Polyclonal	1×10^3 CFU/ml	[105]
	Puccinia striiformis	Monoclonal	3.1×10^5 CFU/ml	[106]
Piezoelectric	*Salmonella paratyphi*	Monoclonal	1.7×10^2 CFU/ml	[107]
	E. coli O157:H7	Polyclonal	1×10^6 CFU/ml	[108]
	Candida albicans	Monoclonal	1×10^6 CFU/ml	[109]
Magnetic	*E. coli* O157:H7	Polyclonal	1×10^5 CFU/ml	[110]

MICROBIAL CELL-BASED SENSOR FORMATS

Whole cell-based biosensors (WCBs) (Fig. **3**) monitor physiological changes in reporter cells exposed to biological or industrial samples containing pathogens, pollutants, biomolecules or drugs [11]. This methodology has been applied for over twenty years [12], with specific examples of analytes that can be monitored including organic compounds [13,14], primary and secondary metabolites [15], xenobiotic compounds [16] and heavy metals [17,18]. The selection of these platforms allows basic metabolic responses, such as viability and growth rate, to be determined and also allows the monitoring of oxygen consumption rates [19]. Arguably the most important feature of WCBs is their ability to provide functional information about biologically active agents [20-24], with particular emphasis on determining the bioavailable fraction and the effect that these compounds have on the growth rate of microbial cells. This, in turn, allows the quantification of complex parameters such as ecotoxicity [25] and biochemical oxygen demand (BOD) [26-28], which is discussed in more detail later. Finally, WCBs can also be used to evaluate biochemical stress resulting from the presence of non-characterised chemicals/mixtures [29,30].

Both of the aforementioned analytical platforms (e.g. immunosensors for pathogens and WCBs) combine sensors and microbial cells. When comparing both formats, it is worth noting that the former are excellent candidates for the sensitive detection of pathogens and chemicals (e.g. toxins), where specific, sensitive and high-affinity/engineered antibodies are available as biorecognition elements. In contrast, WCBs are generally

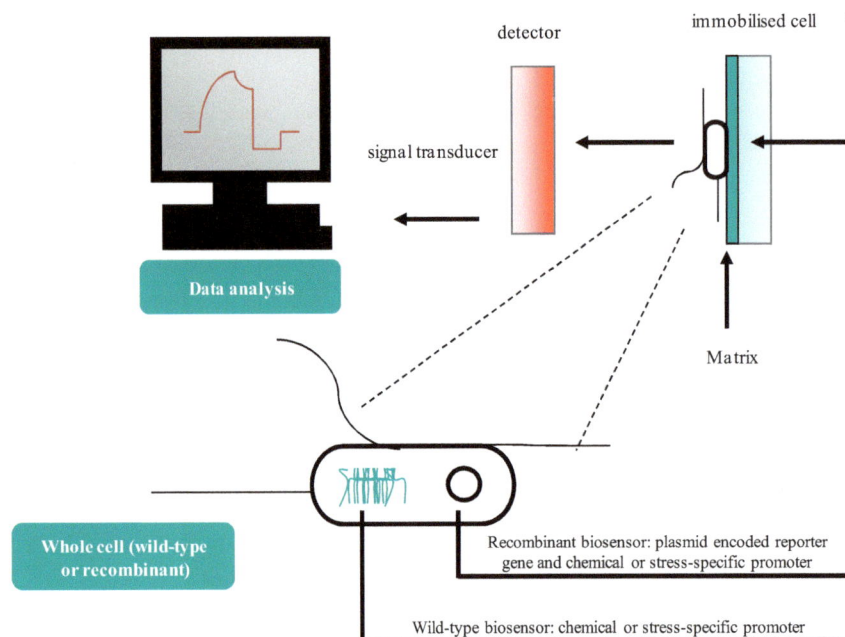

Figure 3: A schematic representation of a whole cell biosensor. For illustration purposes, a whole microbial cell immobilised on a solid support matrix (e.g. agar) is shown which, in turn, is located in close proximity with a suitable detector. Suppression or gene expression, resulting from the presence of an analyte, can be used for monitoring purposes. Examples of applications using wild-type or recombinant microbial cells are discussed in the text.

less selective and sensitive than their antibody-based counterparts, although recent advances in recombinant DNA technology have allowed bacterial cells to be engineered to enhance their biosensing attributes. The implementation of these mutant strains in WCB formats [17] have generated assays with comparable sensitivities and selectivities to immunosensor platforms, with relevant examples discussed later. Viable microorganisms (wild-type or mutant) selected for use can be either suspended, immobilised in a polymeric matrix or directly attached to the detector, either through adsorption or covalent bonds [25,26]. When viable cells are provided with necessary nutrients and organic substrates, they are capable of self-sustaining and self-repairing, resulting in a longer shelf-life. In addition, the ability to rapidly produce microbial cells for implementation represents a more cost-effective alternative to the production of antibodies which is often expensive and requires costly screening and purification methods.

WCBs BASED ON MICROBIAL CELLS

Whole cell biosensors employing suspended microbial cells have been developed and commercialised in recent years [20,29,31,32]. Typically, these formats are used for "proof-of-concept" purposes before immobilisation on a suitable matrix (see below), but nonetheless have many uses. To illustrate these, we focus on recent advances involving the implementation of recombinant bacterial strains. In a typical configuration, a reporter gene for a luminescent molecule (e.g. luciferase or green-fluorescent protein; GFP) that is activated in the presence of a specific analyte is transformed into a suitable microbial strain. In response to the presence of this analyte, an optical signal is generated, detected and quantified (the precise mechanisms involved here are discussed in detail later in this review). The high quantum efficiency of optical detectors enables the detection of inducer molecules in quantities of between 10^{-18} and 10^{-21}M, levels which are significantly lower than those of most spectroscopic techniques [33]. As an added advantage over conventional colourimetric, fluorescent and electrochemical techniques, WCBs based on luminescence do not require external sources of radiation [32]. The ability to non-destructively measure optical responses also enables *in situ* and "real-time" monitoring of selected targets [34]. At the genome level (Fig. **3**), the aforementioned reporter genes (e.g. *lux*) are fused with an appropriate transcriptional response element that is sensitive to the presence of an analyte or to a change in an environmental parameter (such as pH). This approach can be extended to target additional genes within recombinant hosts, thereby enabling multiple analytes to be detected in parallel [17]. Engineered cells selected for use in these applications can be further classified as being either constitutively-expressed ("light-off") or inducibly-expressed ("light-on"), depending on the nature of the promoter fused to the reporter gene [17].

Ecotoxicity assays based on "light-off" cells have been already commercialised [35]. "Light-off" WCBs typically utilise a strong promoter that is expressed under normal conditions, resulting in high basal-level expression of the reporter genes. In the presence of unknown toxic chemical pollutants or other stress factors, the level of expression decreases, allowing the toxicity of a sample to be approximately quantified by correlating with the degree of under-expression of the reporter gene. Thus, the rate of growth of the bacterial cell or the intensity of the reporter signal (referenced against a control signal) can indirectly be used to monitor the total concentration of the toxic chemical(s) or the intensity of the stress factor [36]. "Light-on" systems are based on the use of microbial cells that express an open reading frame (ORF), through an inducible promoter, that has low basal levels of expression in the absence of an inducer. For example, isopropyl-β-D-1-thiogalactopyranoside (IPTG) enables transcriptional activation (induction) of the galactose operon promoter. WCBs based on "light-on" luminescence can detect a wide range of analytes which are of great relevance to bioprocess and environmental monitoring [16,17,37]. However, since many bacterial transcription regulators need to be triggered by the relative effector, the number of compounds that can be analysed is limited by the availability of an appropriate chemically-activated transcription factor. In future, transcription factor design through molecular engineering and directed evolution of individual genes will enhance the applicability of "light-on" whole cell biosensing by enabling particular open reading frames to be targeted and engineered [38].

Numerous applications of the "light-on" / "light-off" methodologies have been described. Ivask and colleagues [17] developed a platform to determine the bioavailable fraction of mercury (Hg), cadmium (Cd), copper (Cu) and lead (Pb) in drinking water. Here, Gram positive and Gram negative recombinant strains (*Staphylococcus aureus*, *Bacillus subtilis*, *E. coli* and *Pseudomonas fluorescens*) were engineered to express luminescence-encoding genes (*luxCDABE*) from *Photorhabdus luminescens* in response to these bioavailable metals. They constructed both 'lights-on' (containing metal-response elements) and 'lights-off' (constitutively expressed) biosensors, the former having a sensitivity of sub-μg/l levels. WCBs are in general not suitable for the detection of aromatic compounds at low concentrations due to their very low solubility in water. Keane and co-workers [16] developed a bioluminescent *Pseudomonas putida* WCB employing non-ionic surfactants, such as Triton X100, Brij 30 and Brij 35, to enhance the aromatic hydrocarbon bioavailability. The resultant platform allowed the detection of toluene, naphthalene, and phenanthrene at sub-mg/l concentrations. Kulakova and colleagues [23] described a system for the detection of the metabolic intermediate phosphonoacetate, based on the *LysR*-like transcriptional regulator (*PhnR*) from *Pseudomonas fluorescens*. *Escherichia coli* DH5α cells, containing a fusion of *PhnR* and the structural gene-encoding promoter, exhibited phosphonoacetate-dependent GFP fluorescence in response to threshold concentrations as low as 0.5μM which is 100 times lower than the detection limit of currently available non-biological analytical methods [23]. In other examples, viable *Saccharomyces cerevisiae* cells were used in an indirect amperometric sensor for the detection of low μg/l levels of copper (Cu) [40]. Neufeld and co-workers [41] employed recombinant cells harbouring plasmids that carry *fabA* and *fabR* genes and with high-resolution amperometric responses to membrane-damaging chemicals, such as phenol and toluene. These WCBs (e.g. amperometric) may be used also to indentify single chemicals, rather than to determine toxicity effects, provided that a selective enzyme is expressed under normal growth conditions. For example, *Gluconobacter* contains several redox enzymes, not all of which are available commercially at reasonable purities. This bacterial strain was used in an amperometric whole cell biosensing system to detect, for example, mono- and poly-alcohols and disaccharides, and a comprehensive overview of other recent developments in *Gluconobacter*-based biosensing is provided by Švitel and colleagues [42].

KEY ISSUES IN USE OF MICROBIAL WCBS

Microbial cells can be immobilised in a variety of biocompatible materials which differ in their relative stabilities, electrical properties, costs, and preparation procedures. Examples include chitosan [43], polyacrylamide [44], and agarose [45]. Alternatively, it is possible to immobilise cells in biofilms, which are commonly studied in environmental and clinical microbiology. While the former strategy for biosensing is commonly used and reasonably successful, only a handful of studies have been published on natural biofilm-based WCBs [46]. In general, cells employed as environmental sensors require an encapsulation matrix that is strong enough to endure the rigours of the outside environment, yet resilient enough to maintain the fragile cells viability without inhibiting the response to analytes - with efficient signal transduction [47]. Also, the design of a microbial biosensor often requires immobilisation on transducers in order to enhance signal detection. The efficacy of this immobilisation is one of the most critical factors in developing an applicable WCB, as this has a direct influence on response time, detection limits and the shelf-life of the biosensor [48]. Similar considerations need to be addressed for immunosensors, where incorrectly immobilised or orientated antibodies have a deleterious effect on the assay performance [1]. The immobilisation of microbial cells can be performed by chemical or physical methods. The most common chemical methods are covalent binding and cross-linking, but these are poorly utilised due to the resultant harsh treatment of bacterial cells which is often to their detriment for biosensing purposes due to significant reductions in cell viability and structural damage to cell walls and membranes [7,49]. Milder physical methods, such as adsorption and entrapment, are therefore commonly used. Adsorption is the simplest method for capture of microbial cells [49]. Here, immobilisation results from low-energy interactions, such as hydrogen and hydrophobic bonds. Adsorption generally does not suit long-term stability because of desorption of microbes [19]. However, the ability to form new links between immobilised cells and the matrices *in lieu* of those damaged suggests that this may be used for long-term applications. Moreover, immobilisation of microbial cells in matrices can also introduce additional substrate and analyte diffusion resistances that reduce sensitivity and increase response times [49].

Numerous WCBs based on immobilised cells have been reported [43,50-56], and, as with those using suspended cells, the most recent applications employ recombinant strains [51,52,57-60]. The microbial cells used in these biosensors are bacterial [26,37,54,62], fungal [56,63] or algal [18,64] and contain promoter and reporter genes [65-67]. Similarly to their suspension-based counterparts, detection can be based on optical [45,55,61] or electrical-based signals [62,68]. They can also be further classified as being either enzyme activity-based [18,56,64] or physiological response-based WCBs [43,63], although this is dependent on the analyte or response involved.

WCBs based on immobilised cells have been proposed for environmental monitoring of heavy metals, pesticide residues, organic pollutants, and for bioprocess monitoring (Table **2**). For example, recombinant *E. coli* that express luciferase under the control of the "SOS" promoter control were immobilised on polyethyleneimine fibers and maintained in non-growth medium. When exposed to ethanol, the cells generated a stable signal for a few days. In another application, Kumlanghan and colleagues [27] developed a microbial biochemical oxygen demand (BOD) biosensor for monitoring treatment of wastewater from an industry processing concentrated rubber latex. The BOD biosensor used immobilised mixed cultures of *E. coli* as biological sensing elements and an oxygen electrode as the transducer. The sensor had a short response time and good stability, and was applied to "on-line" BOD determination in wastewater from anaerobic processes.

MICROBIAL FUEL CELL-POWERED BIOSENSORS

In this review, we have already described how viable microorganisms can be used in amperometric biosensors. For field work, however, there is a requirement for an independent power source for operation and this does not allow analysis in remote environments, such as groundwater and sediments, where the substitution of batteries is not a feasible option. A solution to this problem is to use microorganisms to produce the electrical energy needed for biosensor operation. Microbial fuel cells (MFCs) transform chemical energy into electrical energy via electrochemical reactions involving viable cells [69,70]. These formats consist of an anode, cathode, electrolyte and an external electrical circuit, with microbial cells in contact with one or both of the electrodes. Electrons are produced at the anode through the oxidation of organic carbon substrates by anodic microorganisms and transferred through the external circuit to the cathode, while charge-balancing ions are transferred between anode and cathode through a semi-permeable membrane [71]. Microorganisms used in MFCs have a metabolic pathway enabling them to accept or donate electrons via an electrode *in lieu* of a soluble or mineral electron acceptor or donor, thereby supporting the oxidation of electron donors such as lactate, glucose, acetate and a panel of xenobiotic compounds [72-74]. Microorganisms employed in MFCs include mainly γ- and δ-Proteobacteria, such as *Geobacter* spp. [75,76] and *Shewanella* spp. [77]. Fewer reports exist about the use of other bacterial strains, such as *Firmicutes* [78] and *Bacillus* spp. [79], in these formats. Currently, MFCs can produce only small amounts of electrical power (< 40 W/m^2 and less than 500 W/m^3) [80, 81]. While such energy outputs are too small for large-scale electrical energy production, they are sufficient to power environmental sensors and related data logging and transmission devices (Fig. **4**) [82,83]. Since power output of MFCs may be insufficient for continuous sensor operation, the use of a microcapacitor to store electrical energy was proposed to allow the operation of MFC-based sensors in discontinuous mode [84].

Table 2: Different applications of whole-cell biosensors (WCB) based on immobilised microbial cells.

Microbial cell used	Immobilisation method / matrix	Target agent	Analyte and / or response	Transducer	Limit of detection	Response time	Active life-time	Ref/
WCBs used in environmental monitoring								
Chlorella vulgaris	Self-assembled monolayers (SAMs) of alkanethiolate	Cadmium	Alkaline phosphatase (AP) activity	Conductimetric transducer	1ppb	30 mins	17 days	[64]
Chlorella vulgaris	Porous silica matrix	Diuron [3-(3,4-dichlorophenyl)-1,1-dimethylurea] (herbicide)	Chlorophyll fluorescence detection	Spex Fluorolog 2 fluorometer	1µg/l	5 mins	5 weeks	[55]
Chlorella vulgaris	Cells entrapped in a bovine serum albumin (BSA) membrane	Cadmium and Zinc ions (Zn^{2+} and Cd^{2+})	AP activity	Diamond electrode	0.1ppb	5 mins	-	[18]
Circinella spp.	Carbon paste electrode	Cupric ions (Cu^{2+})	Voltammetric determination of Cu^{2+}	Carbon paste electrode	5.4×10^{-8}M (0.0034mg/l)	-	-	[68]
E. coli	Carbon electrode	Heavy metallic ions (Hg^{2+}, Cu^{2+}, Zn^{2+} and Ni^{2+})	Change in bacterial respiratory activity	Carbon electrode	-	30 mins	-	[37]
E. coli	Carbon electrode	Organic pollutants (o-chlorophenol, 2,4-dichlorophenol and p-nitrophenol)	Change in bacterial respiratory activity	Carbon electrode	-	30 mins		
E. coli MC1061 harbouring parsluxCDABE plasmid	Multimode optical fibers	Bioavailable Arsenic (III) ions (As^{3+})	Luminescence in presence of As^{3+}	Fiber optic	0.012mg/l	-	2 months at -80°C in $CaCl_2$ solution	[61]
E. coli MC1061 harbouring parsluxCDABE plasmid	Multimode optical fibers	Bioavailable Arsenic (V) ions (As^{5+})	Luminescence in presence of As^{5+}	Fiber optic	0.014mg/l	-	2 months at -80°C in $CaCl_2$ solution	
E. coli MC1061 harbouring pmerRluxCDABE plasmid	Multimode optical fibers	Bioavailable mercuric ions (Hg^{2+})	Luminescence in the presence of Hg^{2+}	Fiber optic	0.0026mg/l	-	2 months at -80°C in $CaCl_2$ solution	
Flavobacterium spp.	Cells trapped in glass fiber filter	Methyl parathion (pesticide)	*p*-nitrophenol produced by hydrolysis of methyl parathion by organophosphorus hydrolase enzyme	Optical fiber spectrophotometer	0.3µM	< 3 mins	> 1 month	[54]
Lyophilised biomass of *Brevibacterium ammoniagenes*	Polystyrene sulphonate-polyaniline (PSS-PANI) conducting polymer	Urea	Urease activity	Platinum twin wire electrode	0.125mM	-	7 days	[19]
Pseudomonas aeruginosa	Polyethersulphone	Acrylamide	Amidase activity	Ammonium ion-selective electrode	4.48×10^{-5}M	55 secs	48 days	[44]
Pseudomonas aeruginosa	Porablot NY plus (nylon)	Acrylamide	Amidase activity	Ammonium ion-selective electrode	4.48×10^{-5}M	55 secs	54 days	[44]
Pseudomonas fluorescens HK44	Agar hydrogels	Naphthalene in air	Bioluminescence	Photomultiplier tube-based photon counting device	20nmol/l	-	-	[45]
Pseudomonas putida	Screen-printed graphite electrodes (SPGE)	2,4-dichloro phenoxy acetic acid (2,4-D)	Metabolic oxygen consumption	SPGE electrodes	10µM	10 mins	-	[62]
Pseudomonas putida	Clark oxygen probe	2,4-dichloro phenoxy acetic acid (2,4-D)	Metabolic oxygen consumption	Dissolved oxygen-meter	20µM	200 secs	-	
Pseudomonas spp.	Polycarbonate membrane	*p*-nitrophenol	Change in oxygen concentration	Clark oxygen electrode	10µM	7 mins	-	[111]
Pseudomonas spp. MB58 strain	Entrapped bacterial cells in agarose, carrageenan and alginate	2,2-dichloro propionic and D-2-chloro propionic acids	Chloride content	Chloride selective electrode	0.1mg/dm³ (2,2-dichloropropionate) and 0.05mg/dm³ (D-2-chloropropionate)	-	-	[112]
Pseudomonas syringae	Highly porous micro-cellular polymer disk	Biochemical oxygen demand (BOD)	Respiration rate of microbial cell	Dissolved oxygen electrode	3.30mg/dm³	3-5 mins	95 days	[26]
Recombinant *S. cerevisiae* SEY6210/YEp352-FUSI strain	Capillary membrane	Copper ions (Cu^{2+})	Induction of *CUP1* promoter gene by Cu^{2+} results in β-galactosidase expression by	Oxygen electrode	0.0067mg/l	-	2 months at 4°C (dry) 2 weeks at 4°C in 0.1M phosphate	[40]

MFCs are a promising alternative to conventional batteries for the aforementioned devices, because they overcome the need to change the ephemeral chemical battery and alleviate related environmental concerns. A further application of this methodology relates to their used in powering sensors immersed in freshwater and saltwater sediments [82,85]. A typical sediment microbial fuel cell (SMFC) consists of a sediment-submerged anode and a water-submerged cathode (Fig. **4**) [82,83]. Microorganisms in the sediment colonise the anode surface and oxidise organic compounds in seawater [83]. The resultant electrons travel through the electrical circuit and enable oxygen reduction at the cathode.

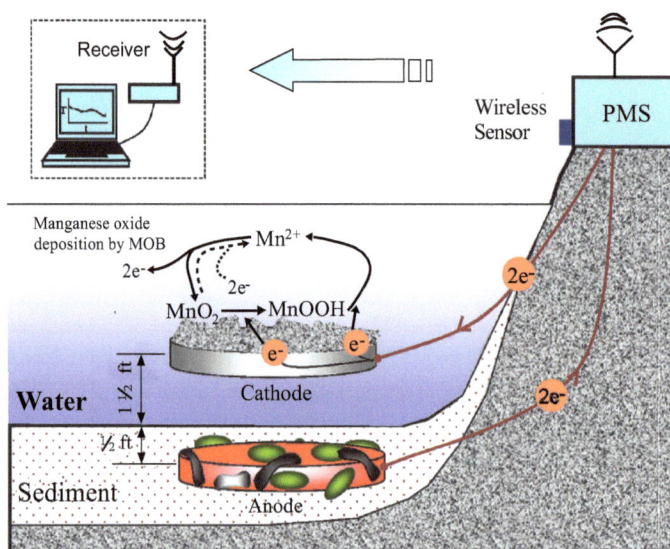

Figure 4: A sediment microbial fuel cell (SMFC) with microbial anode and cathode providing energy for a wireless sensor. MOB = Manganese-oxidising bacteria; PMS = power management system. (Reproduced with permission from [82]).

In other applications, MFCs may be used in parallel as energy sources and biosensors [86]. When exposed to toxic contaminants or where there is a lack of organic carbon substrates, microorganisms in MFCs decrease their metabolic activities, thereby reducing the oxidation current at the anode or the reduction current at the cathode. Consequently, the electrical power output may be correlated to the concentration of pollutants and nutrients in the environment surrounding the MFCs. This strategy was proposed for remote monitoring of anaerobic bioremediation processes, where the bioremediation rate is controlled through the concentration of organic substrates delivered to the sub-surface anaerobic soil. To our knowledge, only a few applications of this idea have been proposed. A MFC biosensor may be also used to measure the *in situ* respiration rate in anaerobic sub-surface environments during a bioremediation process for heavy metals and organic contaminants [87]. In such conditions, the MFC power output can provide "real-time" data for electron donor availability and biological activity of the microorganisms involved in the process. The application of this concept to field studies requires further investigation, since the power output is affected by cell growth at the electrodes and by the presence of other contaminants.

Finally, MFCs have been used as biological oxygen demand (BOD) sensors. BOD provides an indication of the concentration of labile organic carbon, i.e., the organic carbon which is rapidly oxidised by microorganisms available in wastewater. BOD is one of the most common parameters in wastewater treatment, and the availability of fast and reliable methods for its measurement is crucial to the correct operation of wastewater treatment plants. The conventional BOD test (BOD5) measures the molecular oxygen utilised during biological degradation of organic material over five days at 20°C in the dark [88]. This test is therefore not suitable for process control and "real-time" monitoring where rapid analysis is a prerequisite [46]. Hence, the use of a MFC for BOD measurement is a convenient alternative. Preliminary studies adopting a two-chambered MFC resulted in the development of a very large apparatus with rather limited reliability and sensitivity [27,28,89]. More recently, however, a single chambered, continuous-flow MFC was developed and implemented, and excellent agreement between BOD measured through MFC and conventional methods has been reported [90]. Although MFC-based BOD sensors need improvement to increase their sensitivity, the current linearity range encompasses the labile organic carbon concentration commonly encountered in municipal wastewater treatment plants.

CONCLUSIONS

Biosensor-based analysis has emerged in recent years as a popular area of interdisciplinary research, with active contributions made by chemists, engineers and microbiologists in developing analytical platforms such as those described in this review. More than 700 original papers and 50 patents have been published or filed in the last five years (source: ISI Web of Science) in the areas of immunosensing and the application of WCBs, demonstrating the applicability of these methodologies. With reference to pathogen detection, recent developments in high-throughput antibody screening have greatly facilitated the identification of suitable antibody candidates. While many of the platforms described use monoclonal or polyclonal antibodies, recombinant antibodies are extremely useful alternatives that can readily be produced in *E. coli* and engineered to improve their sensitivity and specificity [1,2]. With reference to whole cell biosensing, a key factor in the development of such platforms relates to the requirement for sensitive environmental sensing networks in vast freshwater or seawater areas, in addition to having suitable platforms for the monitoring of remediation and bioremediation processes. The maturation of recombinant technology and device miniaturisation/integration has allowed some of these goals to be reached, as demonstrated by the observation of comparable sensitivities between WCB-based analytical platforms and conventional chemical and physical sensors [91].

The use of viable cells as transducers in biosensors has numerous advantages. Due to their ubiquity and their rapid evolution, microorganisms have the metabolic machinery to recognise and detect virtually every chemical or biochemical species in nature. Furthermore, they are able to adapt readily in unfavorable circumstances and metabolise new analytes [92]. Consequently, WCBs are easy to operate and do not require bulky, expensive or fragile instrumentation [93] and, hence, they may be applied in "on-site" or *in situ* analysis. Moreover, WCBs can provide unique functional information, such as the concentration of the bioavailable fraction of the analyte, *in lieu* of the total concentration of the analyte that is provided by conventional methods. Thus, WCBs permit user-friendly and "real-time" measurements of ecotoxicity [29,34].

Despite the cited advantages and high potential, WCBs have not yet passed the "proof-of-concept" phase [94] and their performance in real biological and environmental samples remains largely uninvestigated. Furthermore, the development of sensitive and reliable WCBs faces many challenges, with response performances dependent on the temperature, pH and chemical composition of the samples selected for analysis, while the lack of repeatability requires frequent calibration and complicates their use in the field. It is worth noting that the most sensitive and selective WCBs are based on recombinant technology. Since viable microorganisms are exposed to indigenous strains, they may be rapidly outcompeted or lose their genetic make-up (e.g. by horizontal gene transfer). This may result in lower sensitivity and a lack of repeatability. Finally, changes in cell densities and cell-cell interactions change the sensor response with time.

In addition to these problems, the immobilisation strategies for capturing microbial cells poses additional problems for substrate diffusion, sensor response and stability with time following cell growth. In theory, it is possible to adopt natural immobilisation strategies for cells on biofilms implementing, for example, nanoparticle-modified supports. These technologies are still in the developmental phases, so it is yet to be determined if this will result in improved whole cell biosensing.

In summary, immunosensing and WCB-based analysis are still attractive options for pathogen detection and for environmental monitoring. The use of microbial cells in conjunction with sensor-based analytical platforms should permit rapid and reliable formats to be developed for the desired applications in future.

ACKNOWLEDGEMENTS

We gratefully acknowledge the support of Science Foundation Ireland (CSET Grant no. 05/CE3/B754), the Irish Research Council for Science, Engineering and Technology (IRCSET) Embark Scholarship (to Sushrut Arora), the Environmental Protection Agency Ireland (EPA) Doctoral Fellowship (to Gabriele Pastorella), the Biomedical Diagnostics Institute (BDI) and the Centre for Bioanalytical Sciences (CBAS).

REFERENCES

[1]   Byrne B, Stack E, Gilmartin N, O'Kennedy R. Antibody-based sensors: principles, problems and potential for detection of pathogens and associated toxins. Sensors 2009; 9(6): 4407-4445.

[2]   Conroy PJ, Hearty S, Leonard P, O'Kennedy RJ. Antibody production, design and use for biosensor-based applications. Semin Cell Dev Biol 2009; 20(1): 10-26.

[3]   Sears CL. A dynamic partnership: celebrating our gut flora. Anaerobe 2005; 11(5): 247-251.

[4]   Liao W, Ho JA. Attomole DNA electrochemical sensor for the detection of *Escherichia coli* O157. Anal Chem 2009; 81(7): 2470-2476.

[5]   Prabhakar N, Arora K, Singh H, Malhotra BD. Polyaniline based nucleic acid sensor. J Phys Chem B 2008; 112(15): 4808-4816.

[6]   Ertl P, Mikkelsen SR. Electrochemical biosensor array for the identification of microorganisms based on lectin-lipopolysaccharide recognition. Anal Chem 2001; 73(17): 4241-4248.

[7]   Gamella M, Campuzano S, Parrado C, Reviejo AJ, Pingarrón JM. Microorganisms recognition and quantification by lectin adsorptive affinity impedance. Talanta 2009; 78(4-5): 1303-1309.

[8]   Meyer MHF, Krause H, Hartmann M, Miethe P, Oster J, Keusgen M. *Francisella tularensis* detection using magnetic labels and a magnetic biosensor based on frequency mixing. J Magn Magn Mater 2007; 311(1): 259-263.

[9]   Garcia-Ojeda PA, Hardy S, Kozlowski S, Stein KE, Feavers IM. Surface plasmon resonance analysis of antipolysaccharide antibody specificity: responses to meningococcal group C conjugate vaccines and bacteria. Infect Immun 2004; 72(6): 3451-3460.

[10]   Byrne B, Donohoe GG, O'Kennedy R. Sialic acids: carbohydrate moieties that influence the biological and physical properties of biopharmaceutical proteins and living cells. Drug Discov Today 2007; 12(7-8): 319-326.

[11]   El-Ali J, Sorger PK, Jensen KF. Cells on chips. Nature 2006; 442(7101): 403-411.

[12]   Racek J, Musil J. Biosensor for lactate determination in biological fluids. I. Construction and properties of the biosensor. Clin Chim Acta 1987; 162(2): 129-139.

[13]   Naessens M, Tran-Minh C. Whole-cell biosensor for determination of volatile organic compounds in the form of aerosols. Anal Chim Acta 1998; 364(1-3): 153-158.

[14]   DeAngelis KM, Firestone MK, Lindow SE. Sensitive whole-cell biosensor suitable for detecting a variety of N-acyl homoserine lactones in intact rhizosphere microbial communities. Appl Environ Microbiol 2007; 73(11): 3724-3727.

[15]   Gerald A, Urban G, Jobst I. Chemo- and biosensor microsystems for clinical applications. Proc SPIE 1998; 3539: 46-50.

[16]   Keane A, Lau PC, Ghoshal S. Use of a whole-cell biosensor to assess the bioavailability enhancement of aromatic hydrocarbon compounds by non-ionic surfactants. Biotechnol Bioeng 2008; 99(1): 86-98.

[17]   Ivask A, Rolova T, Kahru A. A suite of recombinant luminescent bacterial strains for the quantification of bioavailable heavy metals and toxicity testing. BMC Biotechnol 2009; 9: 41.

[18]   Chong KF, Loh KP, Ang K, Ting YP. Whole cell environmental biosensor on diamond. Analyst 2008; 133(6): 739-743.

[19]   Jha SK, Kanungo M, Nath A, D'Souza SF. Entrapment of live microbial cells in electropolymerized polyaniline and their use as urea biosensor. Biosens Bioelectron 2009; 24(8): 2637-2642.

[20]   Bousse L. Whole cell biosensors. Sens Actuators B 1996; 34(1-3): 270-275.

[21]   Deuschle K, Okumoto S, Fehr M, Looger LL, Kozhukh L, Frommer WB. Construction and optimization of a family of genetically encoded metabolite sensors by semirational protein engineering. Protein Sci 2005; 14(9): 2304-2314.

[22]   Wang P, Xu G, Qin L, Xu Y, Li Y, Li R. Cell-based biosensors and its application in biomedicine. Sens Actuators B 2005; 108(1-2): 576-584.

[23]   Kulakova AN, Kulakov LA, McGrath JW, Quinn JP. The construction of a whole-cell biosensor for phosphonoacetate, based on the *LysR*-like transcriptional regulator *PhnR* from *Pseudomonas fluorescens* 23F. Microb Biotechnol 2009; 2(2): 234-240.

[24]   Park TJ, Zheng S, Kang YJ, Lee SY. Development of a whole-cell biosensor by cell surface display of a gold-binding polypeptide on the gold surface. FEMS Microbiol Lett 2009; 293(1): 141-147.

[25]   Shitanda I, Takamatsu S, Watanabe K, Itagaki M. Amperometric screen-printed algal biosensor with flow injection analysis system for detection of environmental toxic compounds. Electrochim Acta 2009; 54(21): 4933-4936.

[26]   Kara S, Keskinler B, Erhan E. A novel microbial BOD biosensor developed by the immobilization of *P. syringae* in micro-cellular polymers. J Chem Technol Biotechnol 2009; 84: 511-518.

[27]   Kumlanghan A, Kanatharana P, Asawatreratanakul P, Mattiasson B, Thavarungkul P. Microbial BOD sensor for monitoring treatment of wastewater from a rubber latex industry. Enzyme Microb Technol 2008; 42(6): 483-491.

[28]   Kim BH, Chang IS, Cheol Gil G, Park HS, Kim HJ. Novel BOD (Biological Oxygen Demand) sensor using mediator-less microbial fuel cell. Biotechnol Lett 2003; 25(7): 541-545.

[29]   Hansen LH, Sørensen SJ. The use of whole-cell biosensors to detect and quantify compounds or conditions affecting biological systems. Microb Ecol 2001; 42(4): 483-494.

[30]   Lee JH, Youn CH, Kim BC, Gu MB. An oxidative stress-specific bacterial cell array chip for toxicity analysis. Biosens Bioelectron 2007; 22(9-10): 2223-2229.

[31]   Gu M, Mitchell R, Kim B. Whole-cell-based biosensors for environmental biomonitoring and application. Biomanufacturing 2004; 87: 269-305.

[32]   Nivens DE, McKnight TE, Moser SA, Osbourn SJ, Simpson ML, Sayler GS. Bioluminescent bioreporter integrated circuits: potentially small, rugged and inexpensive whole-cell biosensors for remote environmental monitoring. J Appl Microbiol 2004; 96(1): 33-46.

[33] Roda A, Pasini P, Mirasoli M, Michelini E, Guardigli M. Biotechnological applications of bioluminescence and chemiluminescence. Trends Biotechnol 2004; 22(6): 295-303.

[34] Werlen C, Jaspers MC, van der Meer JR. Measurement of biologically available naphthalene in gas and aqueous phases by use of a *Pseudomonas putida* biosensor. Appl Environ Microbiol 2004; 70(1): 43-51.

[35] Parvez S, Venkataraman C, Mukherji S. A review on advantages of implementing luminescence inhibition test (*Vibrio fischeri*) for acute toxicity prediction of chemicals. Environ Int 2006; 32(2): 265-268.

[36] Ben-Yoav H, Biran A, Pedahzur R, Belkin S, Buchinger S, Reifferscheid G, *et al.* A whole cell electrochemical biosensor for water genotoxicity bio-detection. Electrochim Acta 2009; In Press, Corrected Proof.

[37] Wang H, Wang XJ, Zhao JF, Chen L. Toxicity assessment of heavy metals and organic compounds using CellSense biosensor with *E. coli*. Chinese Chem Lett 2008; 19(2): 211-214.

[38] Galvão TC, de Lorenzo V. Transcriptional regulators à la carte: engineering new effector specificities in bacterial regulatory proteins. Curr Opin Biotechnol 2006; 17(1): 34-42.

[39] Rawson DM, Willmer AJ, Turner AP. Whole-cell biosensors for environmental monitoring. Biosen 1989; 4(5): 299-311.

[40] Tag K, Riedel K, Bauer H, Hanke G, Baronian KHR, Kunze G. Amperometric detection of Cu^{2+} by yeast biosensors using flow injection analysis (FIA). Sens Actuators B 2007; 122(2): 403-409.

[41] Neufeld T, Biran D, Popovtzer R, Erez T, Ron EZ, Rishpon J. Genetically engineered *pfabA pfabR* bacteria: an electrochemical whole cell biosensor for detection of water toxicity. Anal Chem 2006; 78(14): 4952-4956.

[42] Švitel J, Tkáč J, Voštiar I, Navrátil M, Štefuca V, Bučko M, *et al.* Gluconobacter in biosensors: applications of whole cells and enzymes isolated from *gluconobacter* and *acetobacter* to biosensor construction. Biotechnol Lett 2006; 28(24): 2003-2010.

[43] Odaci D, Timur S, Telefoncu A. A microbial biosensor based on bacterial cells immobilized on chitosan matrix. Bioelectrochem 2009; 75(1): 77-82.

[44] Silva N, Gil D, Karmali A, Matos M. Biosensor for acrylamide based on an ion-selective electrode using whole cells of *Pseudomonas aeruginosa* containing amidase activity. Biocatal Biotransform 2009; 27(2): 143-151.

[45] Valdman E, Gutz IGR. Bioluminescent sensor for naphthalene in air: cell immobilization and evaluation with a dynamic standard atmosphere generator. Sens Actuators B 2008; 133(2): 656-663.

[46] Liu J, Mattiasson B. Microbial BOD sensors for wastewater analysis. Water Res 2002; 36(15): 3786-3802.

[47] Trogl J, Ripp S, Kuncova G, Sayler GS, Churava A, Parik P, *et al.* Selectivity of whole cell optical biosensor with immobilized bioreporter *Pseudomonas fluorescens* HK44. Sens Actuators B 2005; 107(1): 98-103.

[48] Rodriguez-Mozaz S, Lopez deAlda MJ, Barceló D. Biosensors as useful tools for environmental analysis and monitoring. Anal Bioanal Chem 2006; 386(4): 1025-1041.

[49] Lei Y, Chen W, Mulchandani A. Microbial biosensors. Anal Chim Acta 2006; 568(1-2): 200-210.

[50] Kitagawa Y, Ameyama M, Nakashima K, Tamiya E, Karube I. Amperometric alcohol sensor based on an immobilised bacteria cell membrane. Analyst 1987; 112(12): 1747-1749.

[51] Galindo E, Bautista D, García JL, Quintero R. Microbial sensor for penicillins using a recombinant strain of *Escherichia coli*. Enzyme Microb Technol 1990; 12(9): 642-646.

[52] Heitzer A, Malachowsky K, Thonnard JE, Bienkowski PR, White DC, Sayler GS. Optical biosensor for environmental on-line monitoring of naphthalene and salicylate bioavailability with an immobilized bioluminescent catabolic reporter bacterium. Appl Environ Microbiol 1994; 60(5): 1487-1494.

[53] Beyersdorf-Radeck B, Riedel K, Karlson U, Bachmann TT, Schmid RD. Screening of xenobiotic compounds degrading microorganisms using biosensor techniques. Microbiol Res 1998; 153(3): 239-245.

[54] Kumar J, Jha SK, D'Souza SF. Optical microbial biosensor for detection of methyl parathion pesticide using *Flavobacterium* sp. whole cells adsorbed on glass fiber filters as disposable biocomponent. Biosens Bioelectron 2006; 21(11): 2100-2105.

[55] Nguyen-Ngoc H, Tran-Minh C. Fluorescent biosensor using whole cells in an inorganic translucent matrix. Anal Chim Acta 2007; 583(1): 161-165.

[56] Voronova EA, Iliasov PV, Reshetilov AN. Development, investigation of parameters and estimation of possibility of adaptation of *Pichia angusta*-based microbial sensor for ethanol detection. Anal Lett 2008; 41(3): 377-391.

[57] Rainina EI, Efremenco EN, Varfolomeyev SD, Simonian AL, Wild JR. The development of a new biosensor based on recombinant *E. coli* for the direct detection of organophosphorus neurotoxins. Biosens Bioelectron 1996; 11(10): 991-1000.

[58] Hansen LH, Sørensen SJ. Versatile biosensor vectors for detection and quantification of mercury. FEMS Microbiol Lett 2000; 193(1): 123-127.

[59] Gu MB, Chang ST. Soil biosensor for the detection of PAH toxicity using an immobilized recombinant bacterium and a biosurfactant. Biosens Bioelectron 2001; 16(9-12): 667-674.

[60] Gil GC, Kim YJ, Gu MB. Enhancement in the sensitivity of a gas biosensor by using an advanced immobilization of a recombinant bioluminescent bacterium. Biosens Bioelectron 2002; 17(5): 427-432.

[61] Ivask A, Green T, Polyak B, Mor A, Kahru A, Virta M, *et al.* Fibre-optic bacterial biosensors and their application for the analysis of bioavailable Hg and As in soils and sediments from Aznalcollar mining area in Spain. Biosens Bioelectron 2007; 22(7): 1396-1402.

[62] Odaci D, Sezgintürk MK, Timur S, Pazarlioğlu N, Pilloton R, Dinçkaya E, *et al. Pseudomonas putida* based amperometric biosensors for 2,4-D detection. Prep Biochem Biotechnol 2009; 39(1): 11-19.

[63] Seo K, Choo K, Chang H, Park J. A flow injection analysis system with encapsulated high-density *Saccharomyces cerevisiae* cells for rapid determination of biochemical oxygen demand. Appl Microbiol Biotechnol 2009; 83(2): 217-223.

[64] Guedri H, Durrieu C. A self-assembled monolayers based conductometric algal whole cell biosensor for water monitoring. Microchim Acta 2008; 163(3): 179-184.

[65] Simpson ML, Sayler GS, Applegate BM, Ripp S, Nivens DE, Paulus MJ, *et al.* Bioluminescent-bioreporter integrated circuits form novel whole-cell biosensors. Trends Biotechnol 1998; 16(8): 332-338.

[66] Sayler GS, Simpson ML, Cox CD. Emerging foundations: nano-engineering and bio-microelectronics for environmental biotechnology. Curr Opin Microbiol 2004; 7(3): 267-273.

[67] Islam SK, Vijayaraghavan R, Zhang M, Ripp S, Caylor SD, Weathers B, *et al.* Integrated circuit biosensors using living whole-cell bioreporters. IEEE Trans Circuits Syst I-Regul Pap 2007; 54(1): 89-98.

[68] Alpat Ş, Alpat SK, Çadırcı BH, Yaşa İ, Telefoncu A. A novel microbial biosensor based on *Circinella* sp. modified carbon paste electrode and its voltammetric application. Sens Actuators B 2008; 134(1): 175-181.

[69] Allen RM, Bennetto HP. Microbial fuel-cells. Electricity production from carbohydrates. Appl Biochem Biotechnol 1993; 37(39/40): 27-40.

[70] Bullen RA, Arnot TC, Lakeman JB, Walsh FC. Biofuel cells and their development. Biosens Bioelectron 2006; 21(11): 2015-2045.

[71] Zhao F, Slade RCT, Varcoe JR. Techniques for the study and development of microbial fuel cells: an electrochemical perspective. Chem Soc Rev 2009; 38(7): 1926-1939.

[72] Bond DR, Lovley DR. Electricity production by *Geobacter sulfurreducens* attached to electrodes. Appl Environ Microbiol 2003; 69(3): 1548-1555.

[73] Chaudhuri SK, Lovley DR. Electricity generation by direct oxidation of glucose in mediatorless microbial fuel cells. Nat Biotechnol 2003; 21(10): 1229-1232.

[74] Clauwaert P, van der Ha D, Boon N, Verbeken K, Verhaege M, Rabaey K, *et al.* Open air biocathode enables effective electricity generation with microbial fuel cells. Environ Sci Technol. 2007; 41(21):7564-7569.

[75] Biffinger JC, Ray R, Little BJ, Fitzgerald LA, Ribbens M, Finkel SE, *et al.* Simultaneous analysis of physiological and electrical output changes in an operating microbial fuel cell with *Shewanella oneidensis*. Biotechnol Bioeng 2009; 103(3):524-531.

[76] Borole AP, Hamilton CY, Vishnivetskaya TA, Leak D, Andras C, Morrell-Falvey J, *et al.* Integrating engineering design improvements with exoelectrogen enrichment process to increase power output from microbial fuel cells. J Power Sources 2009; 191(2):520-527.

[77] Nevin KP, Kim B, Glaven RH, Johnson JP, Woodard TL, Methé BA, *et al.* Anode biofilm transcriptomics reveals outer surface components essential for high density current production in *Geobacter sulfurreducens* fuel cells. PLoS ONE 2009; 4(5):e5628.

[78] Kumlanghan A, Liu J, Thavarungkul P, Kanatharana P, Mattiasson B. Microbial fuel cell-based biosensor for fast analysis of biodegradable organic matter. Biosens Bioelectron 2007; 22(12):2939-2944.

[79] Nimje VR, Chen C, Chen C, Jean J, Reddy AS, Fan C, *et al.* Stable and high energy generation by a strain of *Bacillus subtilis* in a microbial fuel cell. J Power Sources 2009; 190(2):258-263.

[80] Kim B, Chang I, Gadd G. Challenges in microbial fuel cell development and operation. Appl Microbiol Biotechnol 2007; 76(3): 485-494.

[81] Kargi F, Eker S. High power generation with simultaneous COD removal using a circulating column microbial fuel cell. J Chem Technol Biotechnol 2009; 84(7): 961-965.

[82] Donovan C, Dewan A, Heo D, Beyenal H. Batteryless, wireless sensor powered by a sediment microbial fuel cell. Environ Sci Technol 2008; 42(22): 8591-8596.

[83] Tender LM, Gray SA, Groveman E, Lowy DA, Kauffman P, Melhado J, *et al.* The first demonstration of a microbial fuel cell as a viable power supply: powering a meteorological buoy. J Power Sources 2008; 179(2): 571-575.

[84] Dewan A, Beyenal H, Lewandowski Z. Intermittent energy harvesting improves the performance of microbial fuel cells. Environ Sci Technol 2009; 43(12): 4600-4605.

[85] Lowy DA, Tender LM, Zeikus JG, Park DH, Lovley DR. Harvesting energy from the marine sediment-water interface II: kinetic activity of anode materials. Biosens Bioelectron 2006; 21(11): 2058-2063.

[86] Yi H, Nevin KP, Kim B, Franks AE, Klimes A, Tender LM, *et al.* Selection of a variant of *Geobacter sulfurreducens* with enhanced capacity for current production in microbial fuel cells. Biosens Bioelectron 2009; 24(12):3498-3503.

[87] Tront JM, Fortner JD, Plötze M, Hughes JB, Puzrin AM. Microbial fuel cell biosensor for *in situ* assessment of microbial activity. Biosens Bioelectron 2008; 24(4): 586-590.

[88] Clesceri LS, Eaton AD, Greenberg AE, Franson MAH, American Public Health Association, American Water Works Association, *et al.* Standard methods for the examination of water and wastewater: 19th edition supplement. Washington, DC: American Public Health Association; 1996.

[89]   Chang IS, Jang JK, Gil GC, Kim M, Kim HJ, Cho BW, *et al.* Continuous determination of biochemical oxygen demand using microbial fuel cell type biosensor. Biosens Bioelectron 2004; 19(6): 607-613.

[90]   Di Lorenzo M, Curtis TP, Head IM, Scott K. A single-chamber microbial fuel cell as a biosensor for wastewaters. Water Res 2009; In Press, Corrected Proof.

[91]   Petänen T, Romantschuk M. Use of bioluminescent bacterial sensors as an alternative method for measuring heavy metals in soil extracts. Anal Chim Acta 2002; 456(1): 55-61.

[92]   Durrieu C, Tran-Minh C, J.M. Chovelon, Barthet L, Chouteau C, Védrine C. Algal biosensors for aquatic ecosystems monitoring. Eur Phys J Appl Phys 2006; 36(2): 205-209.

[93]   Akyilmaz E, Yaşa İ, Dinçkaya E. Whole cell immobilized amperometric biosensor based on *Saccharomyces cerevisiae* for selective determination of vitamin B1 (thiamine). Anal Biochem 2006; 354(1): 78-84.

[94]   Harms H, Wells M, van der Meer JR. Whole-cell living biosensors—are they ready for environmental application? Appl Microbiol Biotechnol 2006; 70(3): 273-280.

[95]   Gehring AG, Gerald Crawford C, Mazenko RS, Van Houten LJ, Brewster JD. Enzyme-linked immunomagnetic electrochemical detection of *Salmonella typhimurium*. J Immunol Methods 1996; 195(1-2): 15-25.

[96]   Lin Y, Chen S, Chuang Y, Lu Y, Shen TY, Chang CA, *et al.* Disposable amperometric immunosensing strips fabricated by Au nanoparticles-modified screen-printed carbon electrodes for the detection of foodborne pathogen *Escherichia coli* O157: H7. Biosens Bioelectron 2008; 23(12): 1832-1837.

[97]   Medyantseva EP, Khaldeeva EV, Glushko NI, Budnikov HC. Amperometric enzyme immunosensor for the determination of the antigen of the pathogenic fungi *Trichophyton rubrum*. Anal Chim Acta 2000; 411(1-2): 13-18.

[98]   Radke SM, Alocilja EC. A high density microelectrode array biosensor for detection of *E. coli* O157: H7. Biosens Bioelectron 2005; 20(8): 1662-1667.

[99]   Wang R, Ruan C, Kanayeva D, Lassiter K, Li Y. TiO$_2$ Nanowire bundle microelectrode based impedance immunosensor for rapid and sensitive detection of *Listeria monocytogenes*. Nano Lett 2008; 8(9): 2625-2631.

[100] Gehring AG, Patterson DL, Tu S. Use of a light-addressable potentiometric sensor for the detection of *Escherichia coli* O157: H7. Anal Biochem 1998; 258(2): 293-298.

[101] Muhammad-Tahir Z, Alocilja EC. A conductometric biosensor for biosecurity. Biosens Bioelectron 2003; 18(5-6): 813-819.

[102] Hnaiein M, Hassen WM, Abdelghani A, Fournier-Wirth C, Coste J, Bessueille F, *et al.* A conductometric immunosensor based on functionalized magnetite nanoparticles for *E. coli* detection. Electrochem Commun 2008; 10(8): 1152-1154.

[103] Leonard P, Hearty S, Quinn J, O'Kennedy R. A generic approach for the detection of whole *Listeria monocytogenes* cells in contaminated samples using surface plasmon resonance. Biosens Bioelectron 2004; 19(10): 1331-1335.

[104] Hearty S, Leonard P, Quinn J, O'Kennedy R. Production, characterisation and potential application of a novel monoclonal antibody for rapid identification of virulent *Listeria monocytogenes*. J Microbiol Methods 2006; 66(2): 294-312.

[105] Wei D, Oyarzabal OA, Huang T, Balasubramanian S, Sista S, Simonian AL. Development of a surface plasmon resonance biosensor for the identification of *Campylobacter jejuni*. J Microbiol Methods 2007; 69(1): 78-85.

[106] Skottrup P, Hearty S, Frøkiaer H, Leonard P, Hejgaard J, O'Kennedy R, *et al.* Detection of fungal spores using a generic surface plasmon resonance immunoassay. Biosens Bioelectron 2007; 22(11): 2724-2729.

[107] Fung YS, Wong YY. Self-assembled monolayers as the coating in a quartz piezoelectric crystal immunosensor to detect *Salmonella* in aqueous solution. Anal Chem 2001; 73(21): 5302-5309.

[108] Pohanka M, Skládal P, Pavliš O. Label-free piezoelectric immunosensor for rapid assay of *Escherichia coli*. J Immunoassay Immunochem 2008; 29(1): 70-79.

[109] Muramatsu H, Kajiwara K, Tamiya E, Karube I. Piezoelectric immune-sensor for the detection of *Candida albicans* microbes. Anal Chim Acta 1986; 188: 257-261.

[110] Mujika M, Arana S, Castaño E, Tijero M, Vilares R, Ruano-López JM, *et al.* Magnetoresistive immunosensor for the detection of *Escherichia coli* O157: H7 including a microfluidic network. Biosens Bioelectron 2009; 24(5): 1253-1258.

[111] Banik RM, Mayank, Prakash R, Upadhyay SN. Microbial biosensor based on whole cell of *Pseudomonas* sp. for online measurement of p-nitrophenol. Sens Actuators B 2008; 131(1): 295-300.

[112] Maliszewska I, Wilk KA. Detection of some chloroorganic compounds by a microbial sensor system. Mater Sci 2008; 26(2): 451-458.

Electroanalytical Methods as Tools for Predictive Drug Metabolism Studies

B. Blankert[1,2,*] and J.-M. Kauffmann[1]

[1]*Université Libre de Bruxelles – Institute of Pharmacy, Laboratory of Instrumental Analysis and Bioelectrochemistry, Belgium and* [2]*University of Mons, Faculty of Medicine & Pharmacy, Lab. of Pharmaceutical Analysis, Belgium*

Abstract: The search for new *in vitro* screening tools for early metabolite profiling and identification is becoming a major focus of interest in the pharmaceutical industry. This is motivated by the hope to avoid late failure in drug development and ultimately to launch safer drugs with fewer side effects. Electroanalytical methods alone or coupled on-line with mass spectrometry, can find a niche in this context as they may be readily implemented for the electrically driven synthesis and characterization of xenobiotics oxidized or reduced form(s). Intimately integrated in a dual electrode-enzyme configuration, electroanalysis offers also a mean to study electron transfer at the redox active center of the enzyme in the presence of a substrate and/or an inhibitor.

INTRODUCTION

With the ever increasing need to anticipate the *in vivo* pattern of newly synthesized and potentially pharmacologically active compounds there is currently a regain of activity towards the development of investigating tools capable of mimicking biotransformation events[1]. This is particularly of great concern for the pharmaceutical industry given the risk of drug failure in late clinical trials or after a drug has been released. Drug induced adverse reactions are the consequence of complex and multiple biochemical pathways. It is well accepted that this bioactivation passes by reactive intermediates and metabolites[2]. In the drug discovery schedule, and in contrast to former step by step research strategies, it is nowadays advised to initiate parallel studies for new lead compound screening and for metabolite identification. This strategy, however, imposes new and strong experimental constrains given that a larger number of assays need to be performed in short periods of time. High throughput and miniturisation are thus key criteria for modern analytical tools in pharmaceutical research.

Drug metabolism studies are routinely performed using laboratory animals but due to metabolic interspecies differences when compared to man, they suffer from relatively poor accuracy to anticipate the metabolic profile of a drug in humans. In addition, animal models use is costly, and is lacking of the required throughtput required in early discovery. There are several *in vitro* strategies that may be implemented as complementary tools for drug biotranformation studies. *In silico* predictions can be performed by making use of databases on molecular structure and metabolic transformations or by exploiting the 3D structure of enzymes and calculated drug conformations into the active sites.

The implementation of biological systems *in vitro* is a prerequiste in any drug development scheme. Isolated human microsomes (HM) and human hepatocytes (HM) are preferably applied as they provide informations on Phase I reactions (IM) or Phase I reactions and Phase II conjugations (HH) very similar to that found *in vivo*. The former reactions are often redox biotransformations such as biocatalytic oxidations governed by the cytochrome P450 (P450) isoenzyme family. Human tissues slices, and cultivated cells as well as c-DNA expressed P450s are also commercially available for biotransformation and inhibition studies. Besides being expensive and difficult to store and handle, these bioassays need also to be designed for allowing high throughput and automation capabilities for profiling large compound libraries[3-9].

As most of the bioactivation events occur *in vivo* by redox processes, electrochemical methods occupy a place of choice in the pharmaceutical analytical tools arsenal. A great number of approved pharmaceutical compounds exhibit electroactivity at solid electrodes [10, 11]. Some drug compounds, however, require a too high oxidation potential for being synthetized or detected in aqueous solutions [12]. Electrochemistry (EC) can be applied early in the drug development stage for the synthesis [13], quantification and identification of oxidized or reduced

*Address correspondence to this author at: University of Mons (UMONS), Fac. of Medicine & Pharmacy, Lab. of Pharmaceutical Analysis Av. Maistriau 19, Bât. Mendeleiev, 7000 Mons, Belgium - +32 65 37 3592 – E-mail: bertrand.blankert@umons.ac.be

C.K. Zacharis and P.D. Tzanavaras (Eds)

forms of drug compounds. The structure of the electrolyzed species and the mechanistic informations provided may be of great utility in later metabolite and toxicity screening [14]. Nowadays electrochemical methods are in an advanced and mature stage allowing for their implementation in routine analysis. Ultra low currents can be recorded at microelectrodes, different potential waveforms can be applied, and a variety of solid electrodes and cell configurations are available. Investigations can be performed in aqueous and non aqueous media.

CYCLIC VOLTAMMETRY (CV)

Electrochemistry offers a variety of technique for studying redox reactions. Among them, cyclic voltammetry is the most popular method as a mean of studying redox states [15-30]. It is often the first experiment performed in an electrochemical study of a compound. It consists of cycling the potential of an electrode, which is immersed in an unstirred solution, and measuring the resulting current. It enables a wide potential range to be scanned rapidly. The electrochemical cell comprises generally three electrodes: a working, a reference and an auxiliary electrode. The experiments may be performed in a few ml or μl volumes of solution. Instrumentation is readily available and at relatively low cost. By CV, the informations provided allow quantification of the studied analyte but the most useful aspect is its application to the qualitative diagnosis of electrode reactions which are coupled to homogeneous chemical reactions. This method can provide unique information both in terms of kinetics and thermodynamic parameters in redox processes. An interesting review article was recently provided by M.Goulart, C. Amatore *et al.* illustrating the usefulness of CV in investigations devoted to reactive oxygen species (when combined with the use of microelectrodes), biooxidative/bioreductive activation of pro-drugs and DNA alkylation with particular emphasis on quinones and related compounds [28] An illustration of the power of the informations provided by cyclic voltammetry is obtained by studying phenothiazine drug compounds (Figs. **1** and **2**) [31]. Here, the two phenothiazine molecules are of identical molecular weight and possess a similar configuration but they give a distinct CV pattern. Scanning towards the positive direction and starting at a potential of 0.0 V, promazine (PMZ) gives two oxidation peaks, the first one is reversible (peak A/A') and the second is irreversible (peak B). The second and subsequent scanning shows no additional peak in the voltammogram. Under identical conditions, promethazine (PMTZ) gives only one oxidation peak. This peak comprises several electrochemical steps (A + B etc...) and occurs at slightly more positive potentials than for PMZ. This peak has no reduction peak as partner (irreversible oxidation) but new redox couples appear at low potentials on subsequent scanning (C/C', D/D'). Such a dramatic difference in the voltammogram between two structurally related compounds is due to a distinct oxidation behavior. It was confirmed, by help of electrochemistry coupled on-line with mass spectrometry, that this distinct pattern is related to a breaking of the lateral chain of PMTZ by electrooxidation [31]. This peculiar phenomenon was not observed for PMZ. It was also studied by chemical and enzymatical oxidation and a general oxidation pattern has been suggested for the phenothiazine compounds namely: phenothiazines with two carbons between the two nitrogens in the lateral side chain (2C phenothiazines) give a break of the lateral chain upon oxidation and those with three carbons between the two nitrogens in the lateral chain (3C phenothiazines) are oxidized to the corresponding sulfoxide without cleavage of the lateral chain. The literature on phenothiazine based compounds metabolisation generally points out the formation of the corresponding sulfoxide as a main metabolite along with many minor components but the *in vivo* removal of the aminoalkyl side chain of phenothiazines was seldom reported. Of additional interest is the fact that the CV experiments permit to detect a reversible behavior of PMZ (peaks A/A') attributed to the formation of a relatively unstable cation radical. The high reactivity of the latter is often pointed out in pharmacological and toxicological studies on phenothiazines.

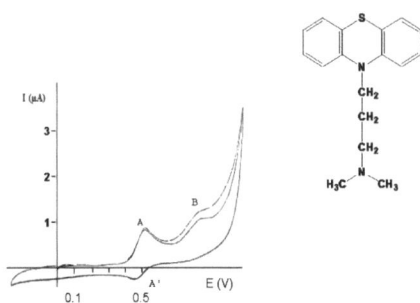

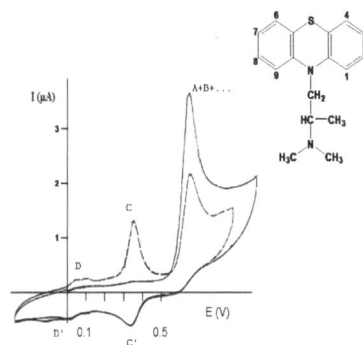

Figure 1: Left   **Figure 2:** Right

Cyclic voltammogram (two scans) of promazine (left) and promethazine (right): 0.1 mM, acetate buffer pH 4.6; methanol 10%, glassy carbon electrode, Volt vs Ag/AgCl 3 M KCl, scan rate 50 mV.s^{-1}.

This example highlights that such CV screenings performed early in the drug R&D program can offer key information on the redox patterns of potential drug candidates. Such data can advantageously be exploited to help for the search of metabolites difficult to isolate and/or identify. It will also allow to better understand the drug stability (oxidability) and to predict drug biotransformation mechanisms.

Interestingly, thiols such as glutathione (GSH) exhibit no electroactivity at carbon based electrodes in a broad potential range, a property which can be exploited to perform CV of a drug compound in the presence of GSH. Endogenous GSH adducts are often encountered during drug bioactivation, this has been reported for example for the metabolization of the neuroleptic clozapine (CLZ). By CV, it was confirmed that CLZ is oxidized to a relatively stable intermediate as inferred from the reversible profile of the oxidation peak[32]. This intermediate was identified in the literature as a reactive iminium cation of CLZ [33, 34]. In the presence of GSH, the reversibility was suppressed and new oxidation peaks were detected attributed to GS-CLZ adducts as confirmed by electrochemistry coupled on-line with mass spectrometry [35]. It is also worth mentioning that the potential at which electrooxidation occurs can be of interest in stability studies and in the early screening for new safer clozapine-like analogues[21, 36]. EC of drug compounds is not limited to oxidation processes, for example reduction potentials of anthelmintic drugs have possibly been related to their pharmacological activity [37].

ELECTROCHEMISTRY-MASS SPECTROMETRY (EC-MS)

Cyclic voltammetry, however, demands the support of additional tools for accurate identification of reaction mechanisms. The innovative combination of flow-through electrochemical cells on-line with mass spectrometry (Fig. **3**) fulfills this need. This hyphenated and purely instrumental configuration has allowed pushing electrochemistry several steps forward in the arsenal of analytical tools for drug biotransformation studies. With modern MS, structural identifications with a high degree of certainty can be achieved and better mechanistic informations are obtained about the oxidation or reduction products at solid electrodes [38]. Integrated in a flow injection analysis (FI) set up the EC-MS configuration is gaining much interest as it allows high throughput and automated capabilities [38-42]. As stated above, it permitted clear identification of promethazine oxidative cleavage and promazine cation radical formation [31] as well as CLZ iminium cation and GS-CLZ adducts formation [35]. Depending on the working electrode configuration, 100% electrolysis efficiency can be achieved, low amounts of sample can be studied and nucleophilic species such as thiols may be injected along with the drug compound to study possible conjugation reactions. By the proper choice of the applied potential, different oxidation patterns can be observed such as hydroxylation, sulfoxidation, N-and O-dealkylation etc. Careful interpretation of the mass spectra must be realized especially taking into account the risk of redox reactions occuring in the electrospray itself. Likewise care must be taken for the selection of the electrochemical cell design in order to avoid cross contamination by redox processes taking place at the working and auxiliary electrodes [43]. A thorough review article of the different oxidations performed by cytochrome P450 and how they correlate to EC oxidation was recently reported by Jurva *et al.* [12]. Thanks to specific oxidations (comprising specific cleavage and distinction between phosphorylated and unphosphorylated tyrosine residues) EC/MS belongs to the possible techniques for on-line protein digestion and peptides mapping system [44, 45].

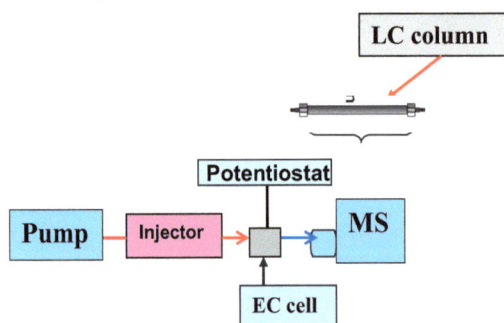

Figure 3: Flow injection (FI) setup with on line electrochemical flow-through cell (EC cell) and mass spectrometer (MS). Also shown is the possibility to insert an LC column (post or pre) the EC cell.

ELECTROCHEMISTRY-LIQUID CHROMATOGRAPHY-MASS SPECTROMETRY (EC-LC-MS)

Coupling a separation step after electrochemical activation is a judicious mean for refining the identification of the oxidation products. The analysis takes more time comparing to FI-EC-MS but mass spectra are less complex. Additional information regarding the polarity of the species is provided and isomers may be detected separately. It permits also to infer if a redox reaction has occurred in the electrospray. The coupling of FI-EC-MS or EC-LC-MS was recently described for the assay of several drug compounds such as estradiol [41], tamoxifen and amitriptyline [39], toremifene [46], acetaminophen [39, 41, 47, 48], phenothiazine derivatives[31, 49], clozapine [35, 48], zotepine and chlorpromazine[50], boscalid [51], amodiaquine [48, 52], amsacrine and mitoxantrone[52], dopamine [53], olsalazine [54], lidocaine and 7-ethoxy coumarin [55], trimethoprim [48], metoprolol [56], catechol derivatives [48, 57] and tetrazepam [58]. Analogies between electrochemically generated products and metabolites are often encountered but some differences exist though [55]. Better understanding of the origin of such subtle differences should be achieved by compiling the data from the various strategies for metabolism predictive studies.

ELECTROCHEMICAL BIOSENSORS

Biosensors are analytical tools which combine, in an intimate contact, a biological compound of recognition with physical transducers and which can be of high interest in drug discovery and drug analysis [59]. Their ability to overcome the disadvantage of conventional methods represents a trump in terms of potential development [60, 61]. An interesting concept is to integrate metabolically competent enzymes as biorecognition element. Among them, enzymes from the CYP450 family due to their major *in vivo* metabolic involvement were immobilized on an amperometric biosensor. Two major difficulties appear: (i) the complexity to measure the enzymatic activity and (ii) the enzymatic stability. Several authors have solved these problems through different electrode surface modifications, for fast electrons transfer to be obtained, and through ingenious immobilization techniques [60-65]. This relatively recent trend concerning the immobilization of the enzyme cytochrome P450 onto an electrode surface allows drug metabolism predictions. The enzyme immobilization is preferably realized within a thin film of polycation layers coated onto gold [64] or glassy carbon and Pt electrodes [66] or into a colloidal gold /chitosan layer onto the electrode [67]. This allows direct reversible electron transfer of P450 to be observed by cyclic voltammetry and the current changes in the presence of a suitable substrate. The CYP450 biosensors allow the identification of drugs or drug candidates as substrates or inhibitors to the attached enzyme. Thorough reviews on cytochrome P450 biosensors have recently been published [63, 68]. In addition, the higher degree of miniaturization, performances and advantages proposed by the screen printed electrode technology increase the appeal and the advent for metabolically competent biosensors [69, 70]. Complementary, two new concepts of biosensors emerge: cell-based biosensors and DNA-based biosensors [71-74]. The first one represents an advantageous compromise between a purely *in vitro* model and the whole organism. The use of cells takes into account a more faithful perspective of the complexity of the *in vivo* environment. In comparison with isolated enzymes, they include (i) a wide supply of biocatalysts, (ii) the entire and natural metabolic pathways, (iii) optimal environmental conditions for the implicated enzyme, (iv) access to enzymes not available in an isolated form [73, 75]. DNA-based immobilized electrodes have recently emerged as useful tools as useful tools for drug - DNA interactions studies [76-80] and for detecting damages caused by metabolically generated species which can affect the integrity of the genetic material [72, 81, 82].

The enzyme Horseradish Peroxidase (HRP) is an other heme containing molecule which can be immobilised onto electrodes for the peroxidation or CYP450 metabolisation mimicking of drugs in the presence of hydrogen peroxide [32, 83-86]. The biosensor usually detects the enzymatic oxidation product(s) at low applied potentials. Sub-micromolar drug concentrations are determined and short lived intermediates may be detected due to the electrode surface confined reaction. HRP is a readily available enzyme and its behavior simulates to some extent *in vivo* biotransformation (Fig. **4**). Identification of the oxidation product(s) and study of HRP activator and inhibitors may thus be of predictive value in drug metabolism [85, 86].

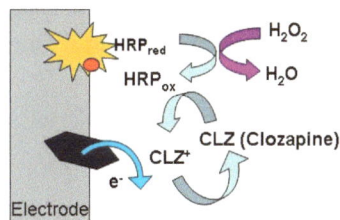

Figure 4: Cartoon of an HRP immobilized carbon paste electrode illustrating the biocatalytic oxidation, in the presence of hydrogen peroxide of clozapine (CLZ) to CLZ nitrenium (CLZ^+) and its subsequent electroreduction at the electrode surface [83, 85].

Conclusion

It is widely accepted that no unique methodology will comply with all the requirements imposed in drug metabolic profiling. Electrosynthesis and product identification as outlined above is only a small part of the analytical arsenal. Its implementation in early drug R&D can, however, considerably help to orientate and speed up the search for many oxidized or reduced metabolites and possible thio-conjugates for a closer understanding of the drug *in vivo* fate. The restricted applicability to electroactive compounds is a slight limitation which can be solved by the use of new electrode material with wider available potential range in aqueous media and less surface fouling (e.g. boron doped diamond). The development of electrochemical biosensing arrays with immobilized CYP450 isoforms is still in its infancy, it should accelerate studies on drug-enzyme and drug-drug interactions as well as on enzyme inhibition.

REFERENCES

[1]   Chen Y, Monshouwer M, Fitch W. Analytical Tools and Approaches for Metabolite Identification in Early Drug Discovery. Pharm Res 2007; 24(2): 248-257.

[2]   Walgren J L, Mitchell M D, Thompson D C. Role of Metabolism in Drug-Induced Idiosyncratic Hepatotoxicity. Crit Rev Toxicol 2005; 35(4): 325 - 361.

[3]   Wang J, Urban L. The impact of early ADME profiling on drug discovery and development strategy. Drug Discov World 2004; fall: 73-86.

[4]   Armer R E, Morris I D. Trends in early drug safety. Drug News Perspect 2004; 17(2): 143-148.

[5]   Vanparys P. ECVAM and pharmaceuticals. ATLA 2002; 30: 221-223.

[6]   Ekins S, Ring B J, Grace J, McRobie-Belle D J, Wrighton S A. Present and future *in vitro* approaches for drug metabolism. J Pharmacol Toxicol Methods 2000; 44(1): 313-324.

[7]   Lee M-Y, Park C B, Dordick J S, Clark D S. From the Cover: Metabolizing enzyme toxicology assay chip (MetaChip) for high-throughput microscale toxicity analyses. PNAS 2005; 102(4): 983-987.

[8]   Ekins S, Nikolsky Y, Nikolskaya T. Techniques: Application of systems biology to absorption, distribution, metabolism, excretion and toxicity. Trends Pharmacol Sci 2005; 26(4): 202-209.

[9]   Brandon E F A, Raap C D, Meijerman I, Beijnen J H, Schellens J H M. An update on *in vitro* test methods in human hepatic drug biotransformation research: pros and cons. Toxicol Appl Pharmacol 2003; 189(3): 233-246.

[10]   de Carvalho R M, Freire R S, Rath S, Kubota L T. Effects of EDTA on signal stability during electrochemical detection of acetaminophen. J Pharm Biomed Anal 2004; 34(5): 871-878.

[11]   Jane I, Mckinnon A, Flanagan R J. High-performance liquid chromatographic analysis of basic drugs on silica columns using non-aqueous ionic eluents : II. Application of UV, fluorescence and electrochemical oxidation detection. J Chromatogr 1985; 323(2): 191-225.

[12]   Jurva U, Wikström H V, Weidolf L, Bruins A P. Comparison between electrochemistry/mass spectrometry and cytochrome P450 catalyzed oxidation reactions. Rapid Commun Mass Spectrom 2003; 17(8): 800-810.

[13]   Felim A, Neudörffer A, Monnet F P, Largeron M. Environmentally Friendly Expeditious One-Pot Electrochemical Synthesis of Bis-Catechol-Thioether Metabolites of Ecstasy: *in vitro* Neurotoxic Effects in the Rat Hippocampus. Int J Electrochem Sci 2008; 3: 266-281.

[14]   Felim A, Urios A, Neudorffer A, Herrera G, Blanco M, Largeron M. Bacterial Plate Assays and Electrochemical Methods: An Efficient Tandem for Evaluating the Ability of Catechol - Thioether Metabolites of MDMA ("Ecstasy") to Induce Toxic Effects through Redox-Cycling. Chem Res Toxicol 2007; 20(4): 685-693.

[15]   Kissinger P T, Heineman W R. Cyclic voltammetry. J Chem Educ 1983; 60: 772-776.

[16]   Campanella L, Bonanni A, Bellantoni D, Tomassetti M. Biosensors for determination of total antioxidant capacity of phytotherapeutic integrators: comparison with other spectrophotometric, fluorimetric and voltammetric methods. J Pharm Biomed Anal 2004; 35(2): 303-320.

[17]   Campanella L, Bonanni A, Bellantoni D, Favero G, Tomassetti M. Comparison of fluorimetric, voltammetric and biosensor methods for the determination of total antioxidant capacity of drug products containing acetylsalicylic acid. J Pharm Biomed Anal 2004; 36(1): 91-99.

[18]   Chevion S, Roberts M A, Chevion M. The use of cyclic voltammetry for the evaluation of antioxidant capacity. Free Radic Biol Med 2000; 28(6): 860-870.

[19]   Sadeghi S J, Gilardi G, Cass A E G. Mediated electrochemistry of peroxidases-effects of variations in protein and mediator structures. Biosens Bioelectron 1997; 12: 1191-1198.

[20]   Ruffien-Ciszak A, Gros P, Comtat M, Schmitt A-M, Questel E, Casas C, Redoules D. Exploration of the global antioxidant capacity of the stratum corneum by cyclic voltammetry. J Pharm Biomed Anal 2006; 40(1): 162-167.

[21]   Kauffmann J-M, Vire J-C, Patriarche G J. Tentative correlation between the electrochemical oxidation of neuroleptics and their pharmacological properties. Bioelectrochem Bioenerg 1984; 12: 413-420.

[22]   Dakova B, Kauffmann J-M, Evers M, Lamberts L, Patriarche G J. Electrochemical behaviour of pharmacologically interesting seleno-organic compounds I. N-alkyl- and N-aryl-1,2-benzisoselenazol-3(2H)-one. Electrochim Acta 1990; 35: 1133-1138.

[23]   Lawrence N S, Beckett E L, Davis J, Compton R G. Advances in the Voltammetric Analysis of Small Biologically Relevant Compounds. Anal Biochem 2002; 303(1): 1-16.

[24]   Hotta H, Sakamoto H, Nagano S, Osakai T, Tsujino Y. Unusually large numbers of electrons for the oxidation of polyphenolic antioxidants. Biochim Biophys Acta 2001; 1526(2): 159-167.

[25]   Abo El-Maali N. Voltammetric analysis of drugs. Bioelectrochem 2004; 64(1): 99-107.

[26]   Rapta P, Vargová A, Polovková J, Gatial A, Omelka L, Majzlík P, Breza M. A variety of oxidation products of antioxidants based on N,N'-substituted p-phenylenediamines. Polym Degrad Stab 2009; 94(9): 1457-1466.

[27]   Gomes A, Fernandes E, Garcia M B Q, Silva A M S, Pinto D C G A, Santos C M M, Cavaleiro J A S, Lima J L F C. Cyclic voltammetric analysis of 2-styrylchromones: Relationship with the antioxidant activity. Bioorg Med Chem 2008; 16(17): 7939-7943.

[28]   He J-B, Yuan S-J, Du J-Q, Hu X-R, Wang Y. Voltammetric and spectral characterization of two flavonols for assay-dependent antioxidant capacity. Bioelectrochem 2009; 75(2): 110-116.

[29]   Seradilla Razola S, Blankert B, Quarin G, Kauffmann J-M. Phenothiazines drugs as redox mediators in horseradish peroxidase bioelectrocatalysis. Anal. Lett. 2003; 36(9): 1819-1833.

[30]   Hillard E A, Caxico de Abreu F, Ferreira D C M, Jaouen G, Goulart M O F, Amatore C. Electrochemical parameters and techniques in drug development, with an emphasis on quinones and related compounds. Chem Commun 2008; 23: 2612-2628.

[31]   Blankert B, Hayen H, van Leeuw S M, Karst U, Bodoki E, Lotrean S, Sandulescu R, Mora Diez N, Dominguez O, Arcos J, Kauffmann J-M. Electrochemical, chemical and enzymatic oxidations of phenothiazines. Electroanal 2005; 17: 1501-1510.

[32]   Yu D, Blankert B, Kauffmann J-M. Development of amperometric horseradish peroxidase based biosensors for clozapine and for the screening of thiol compounds. Biosens Bioelectron 2007; 22(11): 2707-2711.

[33]   Tafazoli S, O'Brien P J. Peroxidases: a role in the metabolism and side effects of drugs. Drug Discovery Today 2005; 10(9): 617-625.

[34]   Williams D P, Park B K. Idiosyncratic toxicity: the role of toxicophores and bioactivation. Drug Discovery Today 2003; 8(22): 1044-1050.

[35]   van Leeuw S M, Blankert B, Kauffmann J-M, Karst U. Prediction of clozapine metabolism by on-line electrochemistry/liquid chromatography/mass spectrometry. Anal. Bioanal.Chem 2005; 382: 742-750.

[36]   Liegeois J-F, Bruhwyler J, Petit C, Damas J, Delarge J, Geczy J, Kauffmann J-M, Lamy M, Meltzer H, Mouithys-Mickalad A. Oxidation Sensitivity May Be a Useful Tool for the Detection of the Hematotoxic Potential of Newly Developed Molecules: Application to Antipsychotic Drugs. Arch Biochem Biophys 1999; 370(1): 126-137.

[37]   Kovacic P, Ames J R, Rector D L, Jawdosiuk M, Ryan M D. Reduction potentials of anthelmintic drugs: Possible relationship to activity. Free Radic Biol Med 1989; 6(2): 131-139.

[38]   Diehl G, Karst U. On-line electrochemistry - MS and related techniques. Anal Bioanal Chem 2002; 373(6): 390-398.

[39]   Gamache P, Smith R, McCarthy R, Waraska J, Acworth I. ADME/TOx Profiling using coulometric electrochemistry and electrospray ionization mass spectrometry. Spectroscopy 2003; 18(6): 14-21.

[40]   Gamache P, McCarthy R, Waraska J, Acworth I. Pharmaceutical oxidative stability profiling with high- throughput voltammetry. American Laboratory 2003; 35: 21-25.

[41]   Gamache P H, Meyer D F, Granger M C, Acworth I N. Metabolomic applications of electrochemistry/Mass spectrometry. J Am Soc Mass Spectrom 2004; 15(12): 1717-1726.

[42]   Asa D. Drug metabolism studies aided by redox chemistry. Genetic Engineering News 2004; 24(12).

[43]   Zettersten C, Lomoth R, Hammarström L, Sjöberg P J R, Nyholm L. The influence of the thin-layer flow cell design on the mass spectra when coupling electrochemistry to electrospray ionisation mass spectrometry. J Electroanal Chem 2006; 590(1): 90-99.

[44]   Permentier H P, Jurva U, Barroso B, Bruins A P. Electrochemical oxidation and cleavage of peptides analyzed with on-line mass spectrometric detection. Rapid Commun Mass Spectrom 2003; 17(14): 1585-1592.

[45]   Permentier H P, Bruins A P, Bischoff R. Electrochemistry-Mass Spectrometry in Drug Metabolism and Protein Research. Mini Rev Med Chem 2008; 8(1): 46-56.

[46]   Lohmann W, Karst U. Electrochemistry meets enzymes: instrumental on-line simulation of oxidative and conjugative metabolism reactions of toremifene. Anal Bioanal Chem 2009; 394(5): 1341-1348.

[47]   Lohmann W, Karst U. Simulation of the detoxification of paracetamol using on-line electrochemistry/liquid chromatography/mass spectrometry. Anal Bioanal Chem 2006; 386(6): 1701-1708.

[48]   Madsen K G, Olsen J, Skonberg C, Hansen S H, Jurva U. Development and Evaluation of an Electrochemical Method for Studying Reactive Phase-I Metabolites: Correlation to *in Vitro* Drug Metabolism. Chem Res Toxicol 2007; 20(5): 821-831.

[49]   Hayen H, Karst U. Analysis of Phenothiazine and Its Derivatives Using LC/Electrochemistry/MS and LC/Electrochemistry/Fluorescence. Anal Chem 2003; 75(18): 4833-4840.

[50]   Nozaki K, Kitagawa H, Kimura S, Kagayama A, Arakawa R. Investigation of the electrochemical oxidation products of zotepine and their fragmentation using on-line electrochemistry/electrospray ionization mass spectrometry. J Mass Spectrom 2006; 41(5): 606-612.

[51]   Lohmann W, Dötzer R, Gütter G, Van Leeuwen S M, Karst U. On-Line Electrochemistry/Liquid Chromatography/Mass Spectrometry for the Simulation of Pesticide Metabolism. J Am Soc Mass Spectrom 2009; 20(1): 138-145.

[52]   Lohmann W, Karst U. Generation and Identification of Reactive Metabolites by Electrochemistry and Immobilized Enzymes Coupled On-Line to Liquid Chromatography/Mass Spectrometry. Anal Chem 2007; 79(17): 6831-6839.

[53]   Deng H, Berkel G J V. A Thin-Layer Electrochemical Flow Cell Coupled On-Line with Electrospray-Mass Spectrometry for the Study of Biological Redox Reactions. Electroanal 1999; 11(12): 857-865.

[54]   Bokman C F, Zettersten C, Sjober P J R, Nyholm L. A Setup for the Coupling of a Thin-Layer Electrochemical Flow Cell to Electrospray Mass Spectrometry. Anal Chem 2004; 76(7): 2017-2024.

[55]   Jurva U, Wikström H V, Bruins A P. *In vitro* mimicry of metabolic oxidation reactions by electrochemistry/mass spectrometry. Rapid Commun Mass Spectrom 2000; 14(6): 529-533.

[56]   Johansson T, Weidolf L, Jurva U. Mimicry of phase I drug metabolism - novel methods for metabolite characterization and synthesis. Rapid Commun Mass Spectrom 2007; 21(14): 2323-2331.

[57]   Arakawa R, Yamaguchi M, Hotta H, Osakai T, Kimoto T. Product analysis of caffeic acid oxidation by on-line electrochemistry/electrospray ionization mass spectrometry. J Am Soc Mass Spectrom 2004; 15(8): 1228-1236.

[58]   Baumann A, Lohmann W, Schubert B, Oberacher H, Karst U. Metabolic studies of tetrazepam based on electrochemical simulation in comparison to *in vivo* and *in vitro* methods. J Chromatogr A 2009; 1216(15): 3192-3198.

[59]   Yu D, Blankert B, Viré J-C, Kauffmann J-M. Biosensors in drug discovery and drug analysis. Anal. Lett. 2005; 38: 1687-1701.

[60]   Turner A P F. Biosensors: Sense and sensitivity. Science 2000; 290: 1315-1317.

[61]   Malhotra B D, Singhal R, Chaubey A, Sharma S K, Kumar A. Recent trends in biosensors. Curr Appl Phys 2005; 5(2): 92-97.

[62]   Cooper M A. Biosensor profiling of molecular interactions in pharmacology. Curr Opin Pharmacol 2003; 3(5): 557-562.

[63]   Hara M. Application of P450s for biosensing: combination of biotechnology and electrochemistry. Mater Sci Eng, C 2000; 12(1-2): 103-109.

[64]   Joseph S, Rusling J F, Lvov Y M, Friedberg T, Fuhr U. An amperometric biosensor with human CYP3A4 as a novel drug screening tool. Biochem Pharmacol 2003; 65(11): 1817-1826.

[65]   Pearson J, Gill A, Vadgama P. Analytical aspects of biosensors. Ann. Clin. Biochem. 2000; 37: 119-145.

[66]   Iwuoha E I, Joseph S, Zhang Z, Smyth M R, Fuhr U, Ortiz de Montellano P R. Drug metabolism biosensors: electrochemical reactivities of cytochrome P450cam immobilised in synthetic vesicular systems. J Pharm Biomed Anal 1998; 17: 1101-1110.

[67]   Liu S, Peng L, Yang X, Wu Y, He L. Electrochemistry of cytochrome P450 enzyme on nanoparticle-containing membrane-coated electrode and its applications for drug sensing. Anal Biochem 2008; 375(2): 209-216.

[68]   Bistolas N, Wollenberger U, Jung C, Scheller F W. Cytochrome P450 biosensors-a review. Biosens Bioelectron 2005; 20(12): 2408-2423.

[69]   Shumyantseva V V, Bulko T V, Archakov A I. Electrochemical reduction of cytochrome P450 as an approach to the construction of biosensors and bioreactors. J Inorg Biochem 2005; 99(5): 1051-1063.

[70]   Hart J P, Crew A, Crouch E, Honeychurch K C, Pemberton R M. Some recents designs and developments of screen-printed carbon electrochemical sensors/biosensors for biomedical, environmental, and industrial analyses. Ann. Lett. 2004; 37(5): 1-42.

[71]   Wang P, Xu G, Qin L, Xu Y, Li Y, Li R. Cell-based biosensors and its application in biomedicine. Sens Actuators, B 2005; 108(1-2): 576-584.

[72]   Zhou L, Yang J, Estavillo C, Stuart J D, Schenkman J B, Rusling J F. Toxicity screening by electrochemical detection of DNA damage by metabolites generated in situ in ultrathin DNA-enzyme films. J Am Chem Soc 2003; 125(1431-1436).

[73]   Durick K, Negulescu P. Cellular biosensors for drug discovery. Biosens Bioelectron 2001; 16(7-8): 587-592.

[74]   Bousse L. Whole cell biosensors. Sens Actuators, B 1996; 34(1-3): 270-275.

[75]   Wijesuriya D C, Rechnitz G A. Biosensors based on plant and animal tissues. Biosens Bioelectron 1993; 8: 155-160.

[76]   Krizkova S, Adam V, Petrlova J, Zitka O, Stejskal K, Zehnalek J, Sures B, Trnkova L, Beklova M, Kizek R. A Suggestion of Electrochemical Biosensor for Study of Platinum(II)-DNA Interactions. Electroanal 2007; 19(2-3): 331-338.

[77]   Szpakowska I, Krassowska-Swiebocka B, Maciejewska D, Kazmierczak P, Jemielita W, Konrad M, Trykowska J, Maj-Zurawska M. Electrochemical DNA Biosensor for Testing Pentamidine and Its Analogues as Potential Chemotherapeutics. Electroanal 2006; 18(13-14): 1422-1430.

[78]   Erdem A, Ozsoz M. Electrochemical DNA Biosensors Based on DNA-Drug Interactions. Electroanal 2002; 14(14): 965-974.

[79]   Bagni G, Osella D, Sturchio E, Mascini M. Deoxyribonucleic acid (DNA) biosensors for environmental risk assessment and drug studies. Anal Chim Acta 2006; 573-574: 81-89.

[80]   Rauf S, Gooding J J, Akhtar K, Ghauri M A, Rahman M, Anwar M A, Khalid A M. Electrochemical approach of anticancer drugs-DNA interaction. J Pharm Biomed Anal 2005; 37(2): 205-217.

[81]   Mbindyo J, Zhou L, Zhang Z, Stuart J D, Rusling J F. Detection od chemically induced DNA damage by drivative square wave voltammetry. Anal Chem 2000; 72: 2059-2065.

[82]   Rusling J F. Applications of polyions filsm containing biomolecules to sensing toxicity. Faraday Discuss 2000; 116: 77-87.

[83]   Blankert B, Dominguez O, El Ayyas W, Arcos J, Kauffmann J-M. Horseradish peroxidase electrode for the analysis of clozapine. Anal. Lett. 2004; 37(5): 917-928.

[84]   C.Petit, K.Murakami, A.Erdem, E.Kilinc, G.O.Borondo, Liégeois J-F, Kauffmann J-M. Horseradish peroxidase immobilized electrode for phenothiazine analysis. Electroanal 1998; 10(18): 1241-1248.

[85]   Yu D, Blankert B, Bodoki E, Bollo S, Viré J-C, Sandulescu R, Nomura A, Kauffmann J-M. Amperometric biosensor based on horseradish peroxidase-immobilised magnetic microparticles. Sens Actuators B 2005; 113: 749-754.

[86]   Yu D, Renedo O D, Blankert B, Sima V, Sandulescu R, Arcos J, Kauffmann J-M. A Peroxidase-Based Biosensor Supported by Nanoporous Magnetic Silica Microparticles for Acetaminophen Biotransformation and Inhibition Studies. Electroanal 2006; 18(17): 1637-1642.

84   *Reviews in Pharmaceutical and Biomedical Analysis*, 2010, 84-107

Sample Preparation Overview for the Chromatographic Determination of 1,4-Benzodiazepines in Biological matrices

Mohammad Nasir Uddin, Victoria F. Samanidou* and Ioannis N. Papadoyannis

Laboratory of Analytical Chemistry, Department of Chemistry, Aristotle University of Thessaloniki, Thessaloniki, GR-541 24, Greece

Abstract: Benzodiazepines due to their sedative, anti-depressive, muscle relaxant, tranquilizer, hypnotic and anticonvulsant properties, have become common worldwide prescribed medicines in the therapy of anxiety, sleep disorders and convulsive attacks, as relatively safe, with mild side effects. The availability of rapid, sensitive and selective analytical methods is essential for the determination of these drugs in clinical and forensic cases. Benzodiazepines are usually present at trace levels (μgmL^{-1} or $ngmL^{-1}$) in a complex biological matrix, and the potentially interfering compounds need to be removed prior to analysis. Therefore, a sample preparation technique is often mandatory both to extract the drugs of interest from the matrix and concentrate them. An extended and comprehensive review on sample preparation giving emphasis on extraction techniques for the chromatographic determination of major benzodiazepines and their metabolites in biological samples is presented providing important physicochemical and bio-pharmacological data to be useful for the development of procedure.

Keywords: Benzodiazepines, chromatography, sample preparation, biological matrices.

INTRODUCTION

Benzodiazepines comprise an important class of psychotherapeutic agents acting on the central nervous system. Since the introduction of chlordiazepoxide in 1960, over 50 of these have been investigated worldwide. They have become the most frequently prescribed drugs for the treatment of anxiety, sleep disturbance and status epileptics due to their tranquilizer, anti-depressive and sedative properties.[1-3] Benzodiazepines continue to be developed, evaluated and introduced for clinical use as the world-wide demand for benzodiazepine anxiolytics and hypnotics is extremely large. They are also used in the treatment of alcohol withdrawal, as well as to relieve tension in the pre-operative period and to induce amnesia in surgical procedures. They are often abused by the young illicit drug users and as "date-rape" drugs to render a victim incapable of resisting an attack.[4] Because of their drug abuse potential, BDZs are frequently present in the blood of drivers involved in traffic accidents.[5] In large doses benzodiazepines often cause profound behavioral effects and if misused they may also cause or contribute to sudden death. They are often subject to overdose in suicide attempts. The older populations are not immune to using benzodiazepines, and their continuous abuse leads to dependence [6-8].

The benzodiazepines represent a large range of potencies at low doses ranging from submilligram (1-30 mg) to over 100 mg, resulting in blood concentrations in the range from subnanogram per mL (10-500 $ngmL^{-1}$) to near-microgram per mL.[4] They undergo extensive metabolism often giving pharmacologically active metabolites. Hence assay methods are essential to be selective, sensitive and specific, *i.e.* capable of separating and determining the parent drug as well as major metabolites, to evaluate their pharmacokinetics, bioavailability and clinical pharmacology, and to detect and identify them in toxicological and forensic samples. In recent years a large number of analytical methods for the determination of 1,4-benzodiazepines and their metabolites in biological samples have been described. The dominant assay methods include mainly chromatography: High Performance Liquid Chromatography (HPLC), Gas Chromatography (GC), Micellar Liquid Chromatography (MLC), Micellar Electrokinetic Capillary Chromatography (MEKC) with Ultraviolet (UV), Electron Capture (ECD) or Mass Spectrometry (MS) detectors [9-11].

A review of the published methods for the determination of benzodiazepines in biological samples and pharmaceutical formulations in the time span 1996-2008 covering chromatographic conditions: column, mobile phase and detection used in HPLC, is presented in a recent work by Samanidou *et al.* [12].

The aim of current review is to cover sample preparation methodologies prior to the chromatographic determi-

---

*Address correspondence to this author Victoria Samanidou at the:** Aristotle University of Thessaloniki. GR-54 124 Thessaloniki, Greece; Tel:+30231997698, FAX: +302310997719; E-mail: samanidu@chem.auth.gr

nation of benzodiazepines, since sample preparation is usually the most critical, time-consuming and laborious step and it is often the bottleneck of any analytical process. Moreover, analytical methods are frequently hampered by impurities causing severe inconvenience in the quantification process of the drugs. The purpose of any sample preparation is the sample clean-up and/or the extraction, enrichment or pre-concentration of the analytes to improve the analytical results. However, it should to be considered that any sample treatment will depend on both the nature of sample and the analytical technique to be employed, requiring an almost case-by-case development. Therefore, no universal sample preparation is available. The choice and optimization of a suitable sample pre-treatment is not easy, especially with highly complex sample matrices like biological fluids: plasma, serum, whole blood, urine, etc. Ideally, sample preparation should be as simple as possible, not only because it will reduce the time required, but also because the greater the number of steps, the higher the probability of introducing errors. If possible, sample preparation should be carried out without or minimum loss of the analytes while eliminating as many interferences as possible from the matrix. Finally, it should also include, when necessary, a suitable dilution or concentration of the analytes in order to obtain an adequate concentration for the subsequent analysis. Sometimes, it may also include the transformation of the analytes into different chemical forms that can facilitate their separation or detection.

Several sample preparation techniques have been developed; usually these involve the most common liquid–liquid extraction (LLE) and solid-phase extraction (SPE) techniques. SPE offers the potential for specific and accurate sample preparation, which when used with new and advanced technology, can be easily automated. Methods based on solid-phase extraction, on-line extraction, column-switching techniques, SPME or direct injection of samples into a HPLC column with back flushing have also been successfully described. In addition a new developed HPLC polymer stationary phase column consisting of a highly cross-linked hard gel of polyvinyl alcohol is applied for direct injection of human plasma and urine samples so that neither extraction nor column-switching is required [13].

A comparative study on LLE and SPE extraction processes for the isolation of 1,4-benzodiazepines is given in Fig. **1**, during the period 1996-2009. The linear curve with the positive slope indicates that extraction relating to SPE or to its synonyms is increasing.

Sample preparation techniques on the chromatographic analysis of 1,4-benzodiazepines in biological samples have been extensively discussed herein. Focus has been given to the classification as well as to the physico-chemical properties of these drugs that can be useful in the development of suitable analytical methods and their physiological action and metabolism.

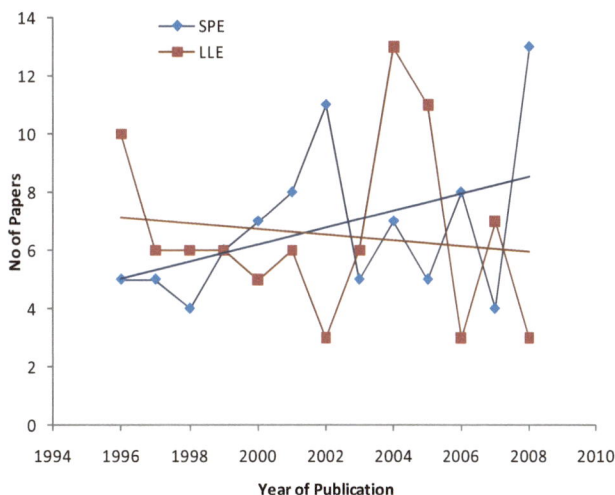

Figure 1: Comparison on SPE and LLE extractions applied for determination of 1,4-benzodiazepines since 1996-2009.

BENZODIAZEPINES-STRUCTURE, SYNTHESIS, PROPERTIES

Structure-Classification

The most important types of 1,4-benzodiazepines according to their chemical structure are discussed in this paragraph.

I Classically almost all active benzodiazepines are based on the 5-aryl-1,4-benzodiazepine structure having a carbonyl group (ketone) at position 2, abbreviated "one" except for those possessing a fused heterocyclic ring or a thionyl group. The aryl substituent at the 5-position is usually phenyl (e.g., oxazepam) or 2-halophenyl (lorazepam, flurazepam). Chemical structures of 1,4-diazepine, 1,4-benzodiazepine, 1,4-benzodiazepin-2-one and diazepam-a representative benzodiazepine are shown in Fig. **2**.

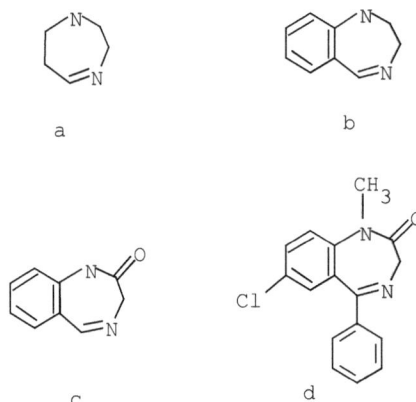

Figure 2: Chemical structures of a. 1,4-diazepine, b. 1,4-benzodiazepine, c. 1,4-benzodiazepin-2-one d. Diazepam-a representative benzodiazepine.

Other variations of 1,4-benzodiazepine structure include, **II** with an additional five membered ring annulation on the 1, 2-position; imidazo or diazolo derivatives (1, 3-diazole) such as midazolam, clinazolam, and triazolo derivatives (1,3,4-triazole) such as alprazolam, adinazolam, estazolam, triazolam and **III** Oxazolo benzodiazepines, cloxazolam, flutazolam, haloxazolam, mexazolam, oxazolam. Structures of some 1,4-benzodiazepines belonging to group I are given in Fig. **3**, and those from group II and III are given in Fig. **4**.

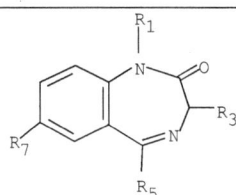

I 1,4-benzodiazepine

Benzodiazepines	R_1	R_3	R_5	R_7
Bromazepam	H	H	2'-pyridyl	Br
Camazepam	CH$_3$	(CH$_3$)$_2$-N-COO-	Phenyl	Cl
Clonazepam	H	H	2-Cl-phenyl	NO$_2$
Clorazepate	H	COOH	phenyl	Cl
Chlordiazepoxide*	-	H	phenyl	Cl
Delorazepam	H	H	2-Cl-phenyl	Cl
Diazepam	CH$_3$	H	phenyl	Cl
Ethyl loflazepate	H	CH$_3$CH$_2$COO	2-F-phenyl	Cl
Fludiazepam	CH$_3$	H	2-F-phenyl	Cl
Flunitrazepam	CH$_3$	H	2-F-phenyl	NO$_2$
Flurazepam	(C$_2$H$_5$)$_2$-N-CH=CH-	H	2-F-phenyl	Cl
Halazepam	CF$_3$CH$_2$	H	phenyl	Cl
Lorazepam	H	OH	phenyl	Cl
Lormetazepam	CH$_3$	OH	phenyl	Cl
Nordazepam	H	H	phenyl	Cl
Norfludiazepam	H	H	F-phenyl	Cl
Nimetazepam	CH$_3$	H	phenyl	NO$_2$
Nitrazepam	H	H	phenyl	NO$_2$
Oxazepam	H	OH	phenyl	Cl
Phenazepam	H	H	Cl-phenyl	Br
Pinazepam	CH=C-CH$_2$	H	phenyl	Cl
Prazepam	Cyclopropyl methylene	H	phenyl	Cl
Temazepam	CH$_3$	OH	phenyl	Cl
Tetrazepam	CH$_3$	H	1,2-dihydro cyclohexyl	Cl

*R_4= N-oxide and double bond at C_1-C_2, CH$_3$NH instead of O at position 2

Figure 3: Structures of selected 1,4-benzodiazepines.

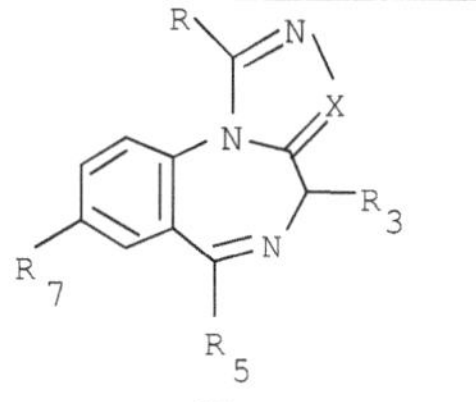

Benzodiazepines	Group	R	R_5	R_7
Adinazolam	Diazolo/ Triazolo II	-CH₂N=	phenyl	Cl
Alprazolam		CH₃	phenyl	Cl
Clinazolam		CH₃	Cl-phenyl	Cl
Estazolam		H	phenyl	Cl
Flumazenil		H	=O	F
Imidazenil		H	Br-phenyl	F
Midazolam		CH₃	F-phenyl	Cl
Triazolam		CH₃	Cl-phenyl	Cl

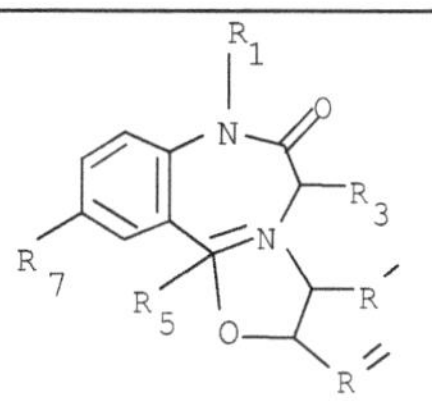

Benzodizepines		R_1	R_5	R_7
Cloxazolam	Oxazolo III	H	Cl-phenyl	Cl
Flutazolam		(CH₂)₂OH	F-phenyl	Cl
Haloxazolam		H	F-phenyl	Br
Mexazolam		H	Cl-phenyl	Cl
Oxazolam		H	phenyl	Cl

Figure 4: Other members of 1,4-benzodiazepines.

Miscellaneous benzodiazepines include thionyl triazolo benzodiazepines, brotizolam, cyclotiazepam, etizolam, Chlordiazepoxide and demoxepam which are N-oxide derivatives, with N-oxide group at position 4 and a methylamino group at position 2, and 1,5-benzodiazepine, clobazam, and 2,3-benzodiazepine.

Properties

Solubility

Benzodiazepines are weak basic drugs and as free bases are lipid-soluble and water-insoluble. 1, 4-benzodiazepines are soluble in organic solvents such as methanol, ethanol, dimethyl formamide and chloroform, but only slightly soluble in n-hexane or n-heptane and practically insoluble in water. In contrast, the salt forms (chlordiazepoxide and flurazepam hydrochlorides, loprazolam methanesulphonate, dipotassium clorazepate) are water soluble [14].

Stability

Stock solutions of 1, 4-benzodiazepines in methanol, ethanol or acetonitrile are stable for 3-6 months, when they were kept at -4°C in the dark. Benzodiazepines are stable in biological media when stored at -20°C for several weeks or months.[15-17] No significant degradation was revealed after at least four freeze/thaw cycles of plasma samples stored at 20°C.[18,19] Attention should be given to the time elapsed between sampling, centrifuging, if necessary and storing at -20 °C. Decomposition of several benzodiazepines was found both pH and formaldehyde concentration dependent, when they were exposed to various concentrations of formaldehyde and various pH.[20] Influence of hair bleaching on benzodiazepines showed that the concentrations of benzodiazepines decreased in bleached hair in comparison to non treated hair.[21]

Photo-decomposition/Thermal Decomposition

Studies showed that some of benzodiazepines are photolabile, and the photo-instability of alprazolam increases as the pH decreases.[22] Accelerated thermal, hydrolytic, and photochemical (UV radiation) degradations of alprazolam was found under several reaction conditions when the main photo-degradation products were triazolaminoquinoleine; 5-chloro-(5-methyl-4H-1,2,4-triazol-4-yl) benzophenone, and 1-methyl-6-phenyl-4H-s-triazo-(4,3-a)(1,4)benzodiazepinone.[23] N-oxide metabolites of benzodiazepines e.g. demoxepam readily undergo thermal decomposition giving the product, nordiazepam, common to many benzodiazepines.[24,25]

Hydrolysis

The acid-base characteristics of 1, 4-benzodiazepines are due to the nitrogen atom in position 4 which can be protonated, except in 4-N-oxide-derivatives. Other nitrogen atoms, as in the 7-amino derivatives, can also be protonated. The hydroxyl group in the 3-hydroxy derivatives can be deprotonated at high pH values, whilst the N-oxide group in 4-N-oxide-derivatives is protonated at low pH values. In aqueous or aqueous-alcoholic solution, most 1,4-benzodiazepines undergo hydrolysis, particularly under acidic or alkaline conditions. Depending upon the different conditions and the 1,4-benzodiazepine type, hydrolysis can affect the 4,5-azomethine group, the 1,2-amidic bonds, or both, producing the corresponding benzophenone as shown in Fig. **5**. In some cases, benzodiazepines are hydrolysed to give their corresponding benzophenones in strong aqueous acid media (e.g. HCl or H_2SO_4) at high temperature (60-120°C) over various periods of time. The disadvantage of benzodiazepine analysis through their benzophenones is that hydrolysis of different benzodiazepines can produce the same benzophenone, with the corresponding lack of selectivity. Acid hydrolysis of bromazepam undergoes reversible 4,5-azomethine bond cleavage.[26,27] The spectrophotometric methods used for determining individual benzodiazepines are generally based on their acid hydrolysis and the subsequent determination of the obtained benzophenone.[28] The native fluorescence of 1,4-benzodiazepines is very low but their fluorescence emission can be enhanced after acidic hydrolysis in presence of an alcohol (methanol or ethanol) enabling their fluorimetric detection.[29]

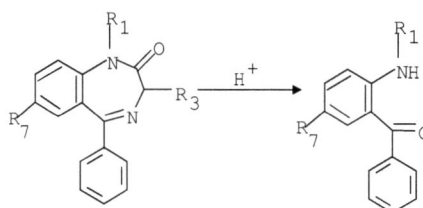

Figure 5: Acid Hydrolysis of 1,4-benzodiazepine.

Metabolism

Major metabolic pathway of 1,4-benzodiazepines involves hepatic hydroxylation via cytochrome P_{450} thus forming hydroxy- metabolites. 1,4-benzodiazepines can undergo metabolic reactions by two phases; phase I predominantly dealkylation, aliphatic and aromatic hydroxylation, reduction, and acetylation, and phase II conjugation reactions consisting largely of glucuronides as presented in Fig. **6**. In most cases the phase I metabolites have some biological activity, which may be greater or less than that of the parent, whereas the conjugates possess no significant activity.

Metabolism: Phase I Oxidation-Hydroxylation

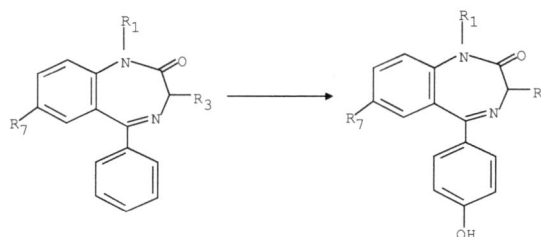

Phase II Conjugation-Glucuronidation

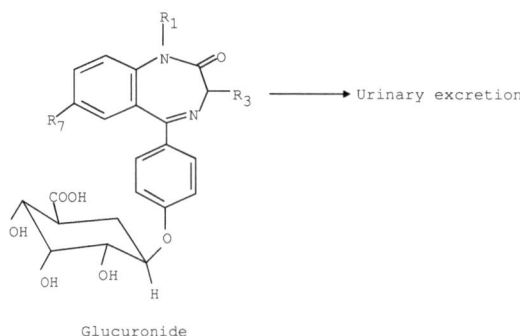

Glucuronide

Figure 6: Metabolism of 1,4-benzodiazepine (diazepam) through oxidation-reduction followed by subsequent glucuronidation.

Phase I

[a] *Oxidation*: Bio-transformation of many 1,4-benzodiazepines occurs by oxidative reactions, primarily in position 1 taking place in the liver; by dealkylation or demethylation of the nitrogen and in position 3; hydroxylation of the carbon.

[b] *Reduction:* A sub-group of 1,4-Benzodiazepines having 7-nitro substituents, such as nitrazepam, flunitrazepam, nimetazepam, and clonazepam is metabolized by reduction of the nitro group to form corresponding biologically inactive 7-amino and 7-acetamido derivatives or by demethylation to N-desmethyl metabolites. Metabolism of nitro-benzodiazepine [flunitrazepam] through reduction of nitro group and subsequent acidamido-benzodiazepine formation is shown in Fig. **7**.

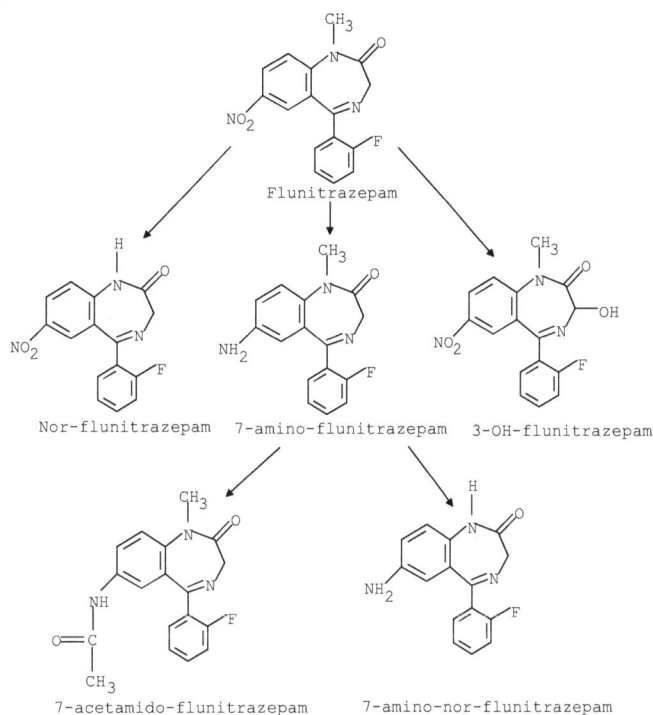

Figure 7: Metabolism of nitro-benzodiazepine (flunitrazepam) through reduction of nitro group and subsequent acidamido-benzodiazepine formation.

Phase II

Conjugation: Benzodiazepines are predominantly excreted as glucuronide conjugate, which is the target metabolite in urine or post-mortem blood samples. The 1 or 3-hydroxy substitution of 1,4-benzodiazepines allows direct conjugation to glucuronic acid, yielding pharmacologically inactive, water-soluble glucuronide conjugates that are excreted in urine. The 7-amino metabolites are subsequently converted to N-glucuronides, whereas the N-desmethyl metabolites are further hydroxylated and then glucuronidated. Usually, in order to liberate 1,4-benzodiazepines from their conjugates an enzymatic hydrolysis is required.[30-34] Metabolic pathways of benzodiazepines leading to the formation of oxazepam are given in Fig. **8**.

Physiological Action

Benzodiazepines produce their variety of effects by depressing the central nervous system and by modulating the $GABA_A$ [γ-aminobutyric acid$_A$] receptor, the most prolific inhibitory receptor within the brain. As far as is currently known, 1,4-benzodiazepines act by a single mechanism, interacting at the specific $GABA_A$ receptor in the brain to enhance the ability of the neurotransmitter γ-amino butyric acid, GABA to open a chloride ion channel and thereby hyperpolarize the neuronal membrane. Benzodiazepines effectively increase the chloride transport through ion channels and ultimately reduce the arousal of the cortical and limbic systems in the CNS [22].

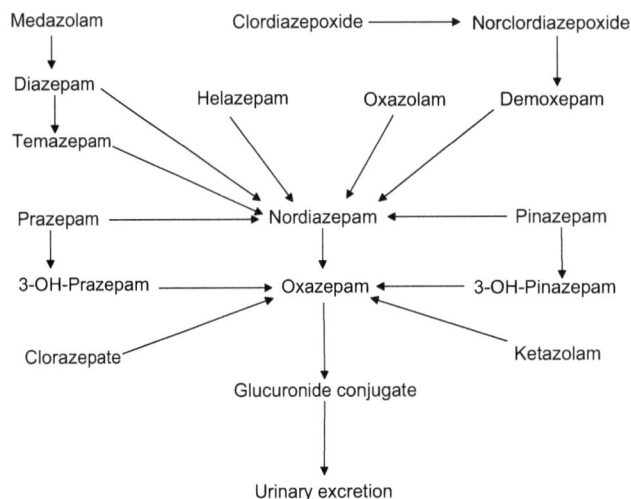

Figure 8: Biotransformation pathways of some 1,4-benzodiazepines leading to oxazepam formation and their urinary excretion as glucuronide conjugate.

Choice of sample, collection and storage

Blood, like serum or plasma, and urine are the biological samples usually analysed. Other biological samples such as saliva, liver tissues also have been analysed. Serum is the supernatant liquid collected by centrifugation after coagulation (about 30 min at room temperature) of a blood sample. Plasma is the supernatant liquid obtained by centrifugation of a blood sample collected in a tube containing an anticoagulant (e.g. heparin, EDTA, citrate or oxalate). [18,35] Plasma or serum samples can be kept for 6 h at room temperature, or for 1-2 days at 4°C. For longer storage period samples should be frozen at -20°C. The way in which samples are collected and stored can affect the final results of the analysis. Urine samples must be stored by freezing at -20°C or by the addition of a preservative agent such as toluene, boric acid or concentrated hydrochloric acid [14].

Sample preparation

Liquid-liquid extraction and solid-phase extraction methods as the sample pre-treatment prior to chromatographic analysis are summarized in Tables 1 and 2, respectively. In some occasions a single or simple treatment procedure is not enough to ensure safe and accurate results, thus the use of several consecutive sample treatments. In most cases, a single extraction or pre-concentration procedure is enough to reduce the sample complexity or to improve the LODs of the methods. 1,4-Benzodiazepines are usually present at trace levels (μgmL^{-1} or $ngmL^{-1}$) in a complex biological matrix and the potentially interfering compounds need to be removed before analysis. Therefore sample pretreatment including protein removal followed by extraction should be capable of concentrating the sample and reducing the amount of interfering substances. The most common samples used for the bio-analysis of 1,4-benzodiazepines requiring concentration are serum/plasma, blood, urine, hair and saliva. Blood, plasma and serum are often be deproteinized and hair needs incubation, while urine may require hydrolysis prior to the isolation procedure. Saliva needs no deproteination as it contains protein negligible.

Hydrolysis for Urine: Enzymatic Hydrolysis

Strongly acidic or basic media such as hydrochloric acid or sodium hydroxide for chemical hydrolysis of conjugates is not recommended because 1,4-benzodiazepines can be hydrolysed to the corresponding benzophenones.[26] Enzymatic hydrolysis generally causes no degradation of the parent molecule to the corresponding benzophenone. Most investigators prefer enzymatic digestion [hydrolysis] of plasma, urine, hair, tissue samples of benzodiazepines before extraction to liberate the conjugated fraction of the drug, especially for old stains strongly bound to the material.[36] With slight variations including temperatures [50-60 °C], the amount and source of enzyme used, pH of buffer [4-5] and time of incubation [2-4h] the same procedure was used for the hydrolysis of urine by different authors. β-Glucuronidase has been used as an appropriate enzyme to release benzodiazepines from their conjugates with the glucuronic acid.[30,31,37-42] Gluculase [β-Glucuronidase + Sulphatase] has also been used to release benzodiazepines from any type of conjugate.[43] Urine was incubated at 37 °C for 1-4 h using 0.2 M sodium acetate (pH 4.5-5.2) and *Helix pomatia* β-glucuronidase for hydrolysis.[44,45]

Protein Removal: Blood, Plasma Treatment

Various methods such as ultramicro-filtration [7] and equilibrium dialysis [46] remove proteins from blood samples. Usually plasma or serum protein precipitation consists of mixing one volume of plasma or serum with three volumes of acid [6% m/v $HClO_4$ [47,48,49], con. H_3PO_4 [50,51], 10% m/v trichloroacetic acid [52]] or organic solvent [methanol [53-55], isopropanol [56,57], acetonitrile [58,59], chloroform [60], acetone [57]] followed by vortex-mixing and centrifugation, which releases the 1,4-benzodiazepines from protein-binding sites removing 99% of the proteins.[61] Recoveries of the bound portion of drug are dependent upon the nature of the 1,4-benzodiazepine and the precipitation agent.[14] The addition of fatty acids that compete with 1,4-benzodiazepines for binding sites of proteins [7,40,62,63] or the addition of alkyl sulphates such as sodium octylsulphate, sodium dodecylsulphate [9] that disrupt the structure of proteins is an alternative way to release 1,4-benzodiazepines from proteins without precipitation.

Treatment of Hair: Buffer Incubation

Acid or alkaline hydrolysis of benzodiazepines, leading to decomposition into corresponding benzophenones were found to be unsuitable to extract the target drugs from the hair matrix. Methanol [32,64,65] or ammoniacal methanol [66,67] can be used in incubation, but the chromatograms obtained were often poor. Reports are available to avoid this problem using buffer incubation, like Soerensen buffer pH 7.5 [68], 0.5M Na_2HPO_4, pH 8.5 [39,69,70], 1 M NaOH [10], proteinase K [71,72]. Mixture of β-glucuronidase/ arylsulfatase at pH 4.0 [43], both methanol and 0.1 M hydrochloric acid [73] or 8M urea–0.2M thioglycolate solution [pH 3] are also reported to be used for hair incubation.[21] Hair samples are washed before incubation by hot water [21], isopropanol [72], and methanol [65,66] or consecutively by one or more solvents. Dichloromethane [74,75] or 0.1% sodium dodecylsulfate [9,67] are preferably used for washing.

Extraction Techniques

Chromatographic techniques, with few exceptions, require some form of isolation procedures to separate the benzodiazepines from biological matrices. These procedures can be separated into following distinct types:

1. Liquid–liquid extraction

2. Supercritical fluid extraction

3. Solid-phase extraction

4. Molecularly imprinted solid-phase extraction

5. On-line solid-phase extraction

6. Micro-extraction (SPME, LPME)

7. Non-extraction procedures

Liquid–liquid Extraction (LLE)

Liquid-liquid extraction is the most widely used method for the pre-treatment of biological samples. The selectivity and efficiency of this classical sample treatment depend mainly on the selection of the immiscible solvents, but other factors may also affect the distribution of the solute into both phases like the pH, the addition of a complexation agent, the addition of salts [salting out effect], etc. Although the use of LLE alone provides good results in terms of extraction efficiency and clean-up of the samples, it is often carried out in combination with other pre-concentration procedures.

Benzodiazepines and their metabolites are usually extracted as the neutral molecules from bio-fluids with a range of organic solvents under weakly alkaline conditions, with recoveries in excess of 90%. Some workers find it unnecessary to alkalize samples since the pK values of benzodiazepines are considerably below the physiological pH.[33,76,77] Some others recommend the use of a back extraction with aqueous acid solutions and basifying followed by extraction with organic solvent.[8,78] Solvent polarity and pH of the aqueous phases are the major factors to be considered. pH should be adjusted to a value at which the drug is in the neutral form but is not hydrolysed. A single step extraction involving 1 mL of sample and 5 mL of organic solvent is sufficient for all benzodiazepines except midazolam, oxazepam and lorazepam. However, some 1, 4-benzodiazepines were subjected to a double extraction owing to their lower lipid solubility and the combined extracts are evaporated to dryness prior to chromatographic analysis.[79]

The organic solvents usually chosen are diethyl ether, chloroform, ethyl acetate, butyl acetate and others such as neutral toluene, benzene, heptane or hexane to which a small amount of a more polar solvent such as methylene chloride, isoamyl alcohol or isopropanol is added. But there are no particular references for any solvent or combination of solvents to be employed for their extraction. One advantage of low boiling solvents is that they can be readily evaporated. Diethyl ether, however, is disadvantaged by its volatility and inherently high danger from fire. Usually a solvent evaporation step is required after extraction. The possible adsorption of the drug onto the glassware can be prevented by silanization of glassware or by the inclusion of l-2% of alcohol (ethanol or 3-methylbutan-l-ol) in a non-polar extractant such as hexane or heptane.

Solvents used for the extraction of 1,4-benzodiazepines include methanol [80], ethanol[81], toluene [33,82], benzene [83], diethylether [15,79,84-86] cyclohexane [87], ethylacetate [42,88-90] chloroform [16,56,60,91,92], dichloromethane [76,93-95], *n*-butylchloride [2,17,65,96-100], butylacetate [69,77] or *tert.*-butylmethyl ether.[101] Again extraction was performed using mixture of bases such as diethylether-chloroform (80:20, v/v) [74,75], 10% (v/v) isopropyl alcohol/ dichloromethane [78], chloroform-diethylether (95:5 v/v) [102], toluene-isoamyl alcohol [95:5 v/v] [103], toluene-hexane-isoamyl alcohol (78:20:2, v/v)[24], hexane/diethylether (20:80, v/v) [18,104], chloroform/isopropyl alcohol (9:1 v/v) [105], methylenechloride/diethylether (80/20, v/v) [39,68,70,104], ethylacetate:hexane (75:25 v/v) [53,106], n-hexane–dichloromethane (70:30 v/v) [9,58], *n*-hexane-chloroform (70:30, v/v) [107,108], hexane–ethylacetate (90:10, v/v) [109], n-hexane:ethylacetate (7:3 v/v) [71], diethylether–ethylacetate (1:1, v/v)[110], heptane-isoamyl alcohol (98:2 v/v) [111], dichloromethane:*n*-pentane (4:6 v:v) [112], CH$_2$Cl$_2$:MeOH (20:80 v/v)[32,113], 1-chlorobutane–dichloromethane (96:4, v/v) [114], cyclohexane–diethylether (31:69 v/v) [115], ethylacetate:heptane (4:1 v/v) [116], toluene:methylenechloride (7:3 v/v) [10,117] or n-pentane/ethylacetate (3:1, v/v) [118].

Liquid-extractions are most commonly conducted under slightly alkaline conditions by the use of variety of base solutions such as saturated NH$_4$Cl (pH 9.5) [65], NaOH [78,82,84,106, 107-109,111,115,119], 0.5 M KOH [87], sodium borate (pH 8.0-11.0) [15,24,56,60, 92,101,102,112,120], 1 M K$_2$CO$_3$, pH 10.5 [16,71,91,121], Soerensen buffer (pH 7.6) [68,74,75,93], sodium heparin [18], 40% (v/v) K$_2$HPO$_4$ (pH 9) [39,70,105,122,123], 25% (v/v) ammonia [32,96,99,113], 0.5 M Na$_2$HPO$_4$ [69], Na$_2$CO$_3$/NaHCO$_3$ buffer (pH 9) [88,97], ammonium acetate-formic acid (pH 8.2) [39,53], 0.1 M ammonium carbonate (pH 9.3) [52,85,116], 0.2 M Na$_2$CO$_3$ [98], solid NaHCO$_3$ [89], sodium sulphate [110], 0.75 M glycine buffer (pH 9) [79,95] or 3-(Cyclohexylamino)-2-hydroxy-1-propanesulfonic Acid Sodium Salt (CAPSO) buffer (pH 10.0) [118]. Unlike 1 M HCl was used for the extraction of oxazepam from plasma [86]. Sample preparation method by LLE for the chromatographic determination of 1,4-benzodiazepines in biological samples are summarized in Table **1**.

Table 1: Biological Sample Preparation By LLE For The Chromatographic Determination Of 1,4-Benzodiazepines.

Analytes	Matrix	Sample pretreatment	Extraction solvent/derivatizing agent	% R	Ref.
DZ, deuterated Analogs (IS)	Urine, hair, oral fluid	Urine samples + IS solution + acetate buffer (3 M), pH 4.6 + β-glucuronidase (incubation for 1 h at 56°C), 20 mg hair + IS solution + methanol (incubation for 2 h at 45°C, Oral fluid + IS solution	1-chlorobutane + ammonium chloride buffer (pH 9.2)	80-97	144
MDL, 1-HMDL, DZ (IS)	Plasma	1 mL Plasma + 20 μL I.S. 2 μgmL^{-1} + 0.5 mL NaOH (0.5 N), vortex- mix	4 mL *n*-hexan–chloroform (70:30, v/v)	86-93	108
15 BDZs	Oral fluid	0.5 mL Oral fluid sample + 50 μL IS (deuterated; 0.015-0.89 μmolL^{-1}) + 250μL 0.2 molL^{-1} ammonium carbonate buffer	1.3 mL ethylacetate:heptane (4:1)	50	116
DZ, NDZ	Plasma	1.0 mL Plasma + 100 μL IS (alprazolam) + sodium borate buffer (pH 9.3)	5.0 mL dichloromethane: *n*-pentane (4:6 v:v)	82-92	112
CLZ, MDL, FNZ, OZ	Whole blood	Sample + Standard solutions in methanol (1 mgmL^{-1})	1-chlorobutane	>90	17
FNZ, CLZ (IS)	Serum	1 mL Serum + 1 mL borate buffer (pH 9) + 50 μL I.S.	3 mL *tert.*-butylmethyl ether	--	101
CLZ, DZ, FNZ, MDL, OZ, MCLZ (IS)	Whole blood	1.0 mL Blood + 30 μL IS (10 mgmL^{-1}) + 50 μL ammonia (25%)	5 mL 1-chlorobutane (n-butyl chloride)	>90	99
CLZ, DLZ, DZ, TZ	Brain membrane	Synaptosomal + 50 mM Tris/HCl buffer (pH 7.4) at 4°C, Tris/HCl buffer	100 μL methanol	---	3
FMZ, Lamotrigine (IS)	Plasma	1 mL Plasma + 30 μL IS (100 μgmL^{-1} lamotrigine)+ 1 mL solid sodium bicarbonate	5 mL ethyl acetate	92	89
CBZ, *N*-DCBZ, DZ (IS)	Serum, urine	0.2-mL Serum or 0.4 mL urine + 200 ng (20 mL) diazepam methanolic solution (I.S.) + 0.5 M glycine	Dichloromethane	96-103	95
DZ, CLZ (IS)	Plasma	100 μL Plasma spiked + 10 μL IS in methanol + 300 μL acetonitrile, supernatant + borate buffer (0.5 mL; pH 9.0)	ethyl acetate–*n*-hexane (30:70, v/v; 5 mL)	>87	120
CLZ, DZ, FNZ, LZ, NZ, NDZ, FZ, OZ, *N*-MCLZ (IS)	Plasma, Urine	500 μL Blood + 31.3 μL *n*-methylclonazepam (IS, 6 μgmL^{-1}) + 300 μL borate buffer (0.1 molL^{-1}, pH 8)	600 μL chloroform	82-92	60
TL, NZ	Plasma, Cerebrospinal Fluid	10 μL IS, Nitrazepam (2.14 μgmL^{-1}) + 100 μL propan-2-ol + 90 μL plasma + 200 μL borate buffer (50 mM, pH 11.0)	720 μL chloroform	94	92

Table 1: cont....

Analytes	Matrix	Sample pretreatment	Extraction solvent/derivatizing agent	% R	Ref.
OZ, LZ, TZ	Rabbit Plasma	200 μL Plasma + 200 μL 1 N HCl + 50 μL IS (4 μgmL^{-1}) in acetonitrile	3 mL diethyl ether, vortex-mix, 1 min (0–4 ^{0}C), centrifuge	85-99	86
CLZ, DFZ, DZ, FNZ, LZ, MDL, NDZ, OZ, MCLZ (IS)	Blood	1 mL Blood + 30 μL IS (10 mgL^{-1}) + 50 μL ammonia solution at 25% (pH 11.5)	5 mL n-butylchloride	---	2
CDO, ESZ, FNZ, TL, PRZ (IS)	Rat Hair, Plasma	100 μL Plasma + 100 μL K$_2$CO$_3$ + 50 μL prazepam(IS) Hair incubation: Proteinase K or [MeOH:NH$_4$OH (25%)(20:1)]/ [MeOH:trifluoroacetic acid (TFA) (50:1)]/ Soerensen buffer/ 1 M NaOH digestion	2 mL n-hexane : ethyl acetate (7:3 v/v)	98	71
MDL, 1-HMDL, PRZ (IS)	Plasma	1 mL Plasma + 20 μL IS (prazepam, 68 ngmL^{-1}) + 1 mL 0.75 M glycine buffer (pH 9)	4 mL diethyl ether, centrifuge, organic phase + 1-mL 0.1 M acetate buffer (pH 4.7)	101-124	79
TL, NZ (IS)	Rat Plasma, Brain micro-dialysate	90 μL Plasma + nitrazepam (I.S.; 10 μL, 7.5 μM) + isopropanol (100 μL), centrifuge, supernatant + borate buffer	chloroform	--	56
AL, ESZ, MDL, 4-HESZ, P-ESZ, Cl-ESZ, 1-HAL, 4-HAL, 1-HMDL, 4-HMDL, 1-HA-d5, EST-d5(IS)	Rat Hair, Hair Root, Plasma	Wash: 1 mL 0.1% sodium dodecylsulfate, 0.2 mL methanol containing 100 ng of an IS (EST-d$_5$, or 1-HA-d$_5$), ultra sonication	Hair: 2 mL CH$_2$Cl$_2$:MeOH:/28% NH$_4$OH (20:80:2), Plasma: 1 mL n-hexane-CH$_2$Cl$_2$ (1:1)	88	9
MDL, α-HMDL Desmethyl clomipramine (IS)	Plasma	2.00 mL Plasma + 400 μL 1 M NaOH + 50 mL I.S.	5 mL heptane-isoamylalcohol (98:2 v/v)	85-111	111
CBZ, DCBZ, 4-HCBZ, 4- DCBZ, PRZ (IS)	Blood	1 mL Whole blood or 1 g tissue (forensic test; stomach, liver, kidney) + 20 μL IS (prazepam, 100 mgmL^{-1}) + 0.5 mL carbonate buffer (50 mM, pH 10.5)	6 mL n-hexane:ethylacetate (7:3, v/v) + 6 mL n-hexane:2-propanol (99:1, v/v)	76-99	121
23 BDZs, deuterated analogs	Plasma	Plasma (0.5 mL) + 0.05 mL of IS, mix + 5 mL sodium sulfate	5 mL diethylether–ethylacetate (1:1, v/v), centrifuge	87-113	110
CLZ, 3-MCLZ (IS)	Plasma	1 mL Plasma + 75 μL 1 μgmL^{-1} 3-methylclonazepam (IS) + 0.2 mL NaOH (0.1 M)	8 mL hexane–ethyl acetate (90:10, v/v), shake, centrifuge	88	134
MDL, 1-HMDL, CLZ (IS)	Infants plasma	Plasma (100 μL) + sodium hydroxide (1 M, 100 μL)	10% v/v isopropyl alcohol in dichloromethane + centrifuge + back extraction -phosphoric acid (0.02 M)	>70	78
MDL, α-HMDL, 4-HMDL	Rat serum	50 μL Serum + borate buffer (1 M, pH 9.0, 100 μL)	1-mL chloroform-diethyl ether (95:5) + vortex-mixe, centrifuge	90	102
BRZ, CLZ, CDO, EL, ESZ, FTZ, HLZ, LZ, NZ, OXL, TL, DZ (IS)	Serum	0.5 mL Serum + 200 μL 1 M potassium carbonate + 20 μL internal standard (diazepam)	3 mL chloroform + mix, centrifuge	>95	16
CLZ, DZ (IS)	Plasma	500 μL Plasma + IS (50 μL diazepam at 100 ngmL^{-1}) + mix	4 mL mixture hexane/diethylether (20:80, v/v)	86-89	18
MDL, 1-HMDL, 4-HMDL, 1,4-HMDL, DZ(IS)	Rat liver, plasma	Liver + saline (38ºC) + perfusion medium (120 mL Williams medium E, equilibrated with 95% O$_2$ and 5% CO$_2$)	plasma/perfuse medium + 200 μL of 0.1 M sodium hydroxide + 50 μL (IS) + 4 mL diethyl ether vortex mix, centrifuge	92	84
22 benzodiazepines	Urine	1.0 mL Urine + 20 μL internal standard (deuterated) (5 μgmL^{-1}), + stock solutions (2 μgmL^{-1}) + 1 mL of a 40% potassium phosphate buffer (pH 9)	Chloroform/isopropyl alcohol (9:1) (4 mL) + shake	84-98	105
LZ, NDZ (IS)	Serum	Serum sample + IS solution (300 ngmL^{-1} nordazepam, 10 μL), sonication	4 mL dichloromethane	94-98	76
CLZ, DZ, FNZ, FZ, LZ, MDL, N-DFZ, NDZ, OZ	Whole blood	1 mL Blood sample + 50 μL IS (5 mgL^{-1} methylclonazepam) + 50 μL 25% ammonia, Shake	5 mL n-butylchloride	83-99	96
AL, DZ-d5 (IS)	Hair	Decontamination: Hair + methylene chloride (5 mL, 2 min), dried, segmented (1 or 2 cm), incubation: 20 mg hair + 1 mL phosphate buffer at pH 8.4 + 1 ng IS diazepam-d$_5$	5 mL methylenechloride/ diethylether (80/20, v/v)	76	123
AL, BRZ, DZ, FNZ, LZ, LRZ, MDL, TTZ, TL, deuterated analogues (IS)	Plasma, saliva	0.5 mL Sample + 50 μL 1 mgL^{-1} IS (deuterated) + 0.5 mL pH 9.0 borate buffer or 0.5 mL 0.1 M ammonic carbonate buffer, pH 9.3, centrifuge	6-8 mL diethyl ether	70-87	15
AL, 7-ACLZ ,7-AFNZ, BRZ, CBZ, DZ, LZ, LRZ, MDL, NDZ, OZ, TZ, TTZ, TL, DZ-d$_5$(IS)	Hair	Decontamination: methylene chloride (2 × 5 mL for 2 min), segmentation, incubation: 1 mL of phosphate buffer pH 8.4 + 1 ng IS diazepam-d$_5$, overnight	5 mL methylene chloride/diethylether (90/10, v/v	32-76	70
CBZ, N-MCBZ, AL (IS)	Plasma/ LLE	0.5 mL Plasma + 20 μL alprazolam as IS (4μg/mL)	1.5 mL toluene	100	33
26 BDZs and metabolites, deuterated analogs (IS)	Blood, urine, hair	Blood: 250 μL samples + 50 μL IS, Urine: incubation: 250 μL samples + 50 μL IS solution + 200 μL acetate buffer, pH 4.6 + 25 μL β-glucuronidase, 1 h at 56°C Hair: decontamination (twice): dichloromethane, water, methanol, ultrasonication, Incubation: 1 mL methanol at 45°C for 2 h, shaking, centrifuge.	4 mL 1-chlorobutane + saturated ammonium chloride buffer (pH 9.2)	25-104	65
MDL, 1-HMDL, 4-HMDL, midazolam-d$_5$ (IS	Plasma	1 mL Plasma + 20 μL d$_5$ midazolam (I.S.) (100 ng), 1 mL buffer (pH 11)	5 mL 1-chlorobutane–dichloromethane (96:4, v/v)		114
MDL, 1-HMDL, 1-CDO(IS)	Plasma	200 μL Plasma + 50 μL IS (1 μgmL^{-1}) + 0.5 mL sodium carbonate–sodium hydrogen carbonate buffer (pH 9.5), vortex-mixing	5 mL 1-chlorobutane	79-90	97
MDL, α-HMDL	Plasma	50 μL methanol:ammonium acetate (80:20, v/v) + formic acid, pH 8.2 + 50 μL IS (flurazepam)	3 mL ethylacetate:hexane (75:25)	--	53
LZ, OZ (IS)	Plasma/LLE	0.5 mL Plasma + 20 μL oxazepam (IS) (10 ngμL^{-1}) + 1 mL acetonitrile, vortex mix, centrifuge + 0.5 mL sodium carbonate–sodium hydrogen carbonate buffer (pH 9.5)	5 mL n-hexane–dichloromethane (70:30 v/v)	72-84	58
23 benzodiazepines	Hair	Washing (twice): dichloromethane, incubation: 2 ng clonazepam-d$_4$ (IS) + 20 mg cut hair + Soerensen buffer, 14 h at 56ºC	2 mL dichloromethane/ether (80:20, v/v)	---	68
LZ, DZ-d5 (IS)	Urine, saliva, Hair	1 mL Urine (hydrolyze overnight) + β-glucuronidase at pH 5.2 500 μL saliva + 500 mL phosphate buffer pH 8.4 Hair decontamination: dichloromethane, segmentation, incubation: 1 mL phosphate buffer pH 8.4 (overnight)	2-5 mL dichloro-methane/diethylether (80/20, v/v), centrifugation	90-111	39

Table 1: cont….

Analytes	Matrix	Sample pretreatment	Extraction solvent/derivatizing agent	% R	Ref.
BRZ, LZ (IS)	Plasma	0.5 mL Spiked samples (in methanol) + 0.05 mL 0.06 M NaOH + 0.05 mL internal standard (lorazepam)	4 mL hexane–ethylcetate (7 : 3, v/v) centrifuge	73.7	106
BRZ, DZ (IS)	Plasma	Human plasma + 50 μL IS solution (500 ng ml-1diazepam), vortex-mix	diethylether–hexane (80:20, v/v) (4.0 mL) vortex-mix, centrifuge	80-99	104
BRZ, 3-HBRZ, CLZ, 7-ACLZ, CLZ-d4, 7-ACLZ-d4 (IS)	Urine, Hair	Hair washing: dichloromethane, incubation: 20 mg powdered hair + 1 mL Sorensen buffer (pH 7.6), 14 h at 56°C or 0.1 M NaOH 15 min at 95°C	2 mL dichloromethane, centrifug	---	119
MDL, DZ-d_5(IS)	Saliva, Plasma	0.5 mL Plasma or saliva + 50 μL IS (1 mgL^{-1}) + 0.5 mL borate buffer (pH 9.5) for plasma or 0.5 mL 0.1 M ammonium carbonate buffer (pH 9.3) for saliva	6-8 mL diethyl ether	65	85
MDL, 1-HMDL, AL (IS)	Plasma	2 mL Plasma + 200 μL IS (alprazolam; 250 ngmL^{-1}) + 0.5 mL NaOH (0.5 M), vortex-mix	5 mL n-hexane-chloroform (70:30, v/v), shaking, centrifuge	78-85	107
MDL, 1'-HMDL	Plasma	1-mL spiked plasma (1-100 ngmL^{-1}) + 25 μL IS (diazepam, 40 μgmL^{-1}, methanol) + 40 μL of 2% NaOH	3.5 mL cyclohexane–diethyl ether (31:69, v/v)	90	115
MDL	Plasma	Plasma samples (0.5 mL) + 1 μg flurazepam (IS+ 2.5 N NaOH	toluene	--	82
AL, α-HAL, deuterated analogues (IS)	Plasma	Plasma samples were buffered to alkaline pH	toluene/methylene chloride (7:3, v/v)	--	117
TL, 1-HTL, 1-HT-d4 (IS)	Rat hair	Rat hair + IS, 1-hydroxymethyltriazolam (1-HT-d$_4$)	CH$_2$Cl$_2$/MeOH/28% NH$_4$OH (20:80:2, v/v/v)	---	113
18 benzodiazepines	Blood	Samples were buffered to alkaline pH	butyl chloride	---	100
TL, 1-HTL, 4-HTL,1 ,4-DHTL, 1-HMTL-d4 (IS)	Rat, human hair	10 mg Hair (1 mm pieces) + 0.2-mL of 1-HT-d$_4$ (IS)	2 mL CH$_2$Cl$_2$:MeOH:28% NH$_4$OH (20:80:2 v/v/v)	88-92	32
BRZ, ESZ, NFDZ, AL, TL, deuterated BDZs(IS)	Urine	500 μL Urine + 50 μL 0.1 M sodium carbonate	Chloroform (400 μL)	96-107	124
MDL, HZ (IS)	Urine	1 mL Urine sample + 0.5 mL sodium borate	2 mL toluene-hexane-isoamyl alcohol (78:20:2, v/v)	--	24
MDL, FZ (IS)	Plasma	1 mL Plasma + 100 μL IS + 200 μL 0.5 M KOH, vortex-mix	3 mL cyclohexane	100	87
DZ, CZA, AL, FZ, MFZ, NDZ, OZ, LRZ, TZ, LZ	Plasma	500 μL Plasma + 200 μL Borax buffer (Sörensen, pH 9.0) + 100 μL IS (5 mgL^{-1} bromazepam)	5 mL dichloromethane	---	93
FNZ, 7-AFNZ, diaz-d5 (IS)	Hair	Decontamination: methylenechloride, 50 mg powdered hair + diazepam-d$_5$ (IS) + Soerensen buffer (pH 7.6), incubated for 2 h at 40°C.	5 mL diethylether-chloroform (80:20, v/v)	45, 90	74
AL, TL (d4) (IS)	Rat hair	10-25 mg cut rat hair + 50 μL triazolam-d$_4$ (5 ng) + 2 mL 1 N NaOH, digested overnight at 40°C, cooled,+ 6 N HCl (pH 9.0) + 1 mL sodium borate buffer	7 mL toluene:methylene chloride (7:3) BSTFA + 1% TMCS (N,O-bis(trimethylsilyl) trifluoroacetamide containing 1% Trimethyl chlorosilane)	92-99	10
NDZ, OZ, BRZ, DZ, LZ, FNZ, AL, TL, PRZ-d5 (IS)	Hair	Decontamination: 5 mL of methylenechloride, 50 mg powdered hair + 25 ng prazepm-d$_5$ (IS) + 1 mL Soerensen buffer (pH 7.6), incubated for 2 h at 40°C	5 mL diethyl ether-chloroform (80:20, v/v), 35 μL BSTFA-TMCS, for 20 min at 70°C.	48-90	75
FNZ, DFNZ, 7-AFNZ	Blood	Samples buffered at alkaline pH	Diethyl ether, diisopropyl ether, toluene-isoamyl alcohol mixture (95:5, v/v)	80	103
DZ, DDZ, OZ, TZ, LZ, DXP, NCBZ, NZ, 1-HMDL, AL, 1-HAL, 1-HTL, CLZ, FNZ, DFNZ, PNZ, LRZ	Blood	Whole blood + 100 μL 1 M acetate buffer pH 4.8	Ethyl acetate (500 mL) at pH 7.4, MTBSTFA + 1% TBDMSCl	---	42
FNZ, 7-AFNZ, DFNZ, 7-ADFNZ, 7-AFNZ-d3 (IS)	Serum, Urine	1 mL Serum + 10 μL 7-amino-FNZ-d$_3$ (IS) (1 μgmL^{-1}) + 4 mL methanol, centrifuge, supernatant +1 mL of 40% phosphate buffer (pH 9.0)	4 mL ethyl acetate, 0.5 mL CHCl$_3$ + 20 μgmL^{-1} 4-pyrolidinopyridine + 100 μL heptafluorobutyric anhydride	96	122
DZ, NDZ, BRZ, MZ (IS)	Whole blood	1 mL Whole blood + stock methanolic solution (2 - 20 ngmL^{-1}) + IS (medazepam, 1000 ngmL^{-1})	2 mL n-butyl acetate	--	77
MZ, NDZ, DZ, OZ, BRZ, CDO, PNZ, NZ, LZ, TZ, AL, MDL, 1-HAL, 1-HMDL, FZ (IS)	Whole blood	0.5 mL Whole blood + 0.5 mL of 0.5 M Na$_2$HPO$_4$ + flurazepam (200 ngmL^{-1} IS)	5 mL butyl acetate Derivatize-acetonitrile-MTBSTFA (80:20, v/v)	88-109	69
CDO, ESZ, FNZ, TL, PRZ (IS)	Rat hair, Plasma	Plasma: 100 μL Plasma + 50 μL prazepam (IS) + 100 μL K$_2$CO$_3$ Hair: Incubation: Proteinase K, [MeOH:NH$_4$OH (25%)(20:1)], [MeOH:trifluoroacetic acid (TFA) (50:1)], Soerensen buffer, 1 M NaOH digestion, or glucuronidase	2 mL n-hexane: ethyl acetate (7:3 (v/v) 50 μL ethyl acetate:BSTFA (2:1) or 50 μL BSTFA.	98	71
7-AFNZ	Urine	Urine samples + pH 9.0 (Na$_2$CO$_3$/ Na$_2$HCO$_3$ buffer)	Ethyl acetate N-methyl-N-(trimethysilyl)trifluoro + acetamide	---	88
BRZ, DZ, FNZ, HZ, MZ, NZ, OZ, TTZ	serum	0.5 mL Serum + 200 μL 1 M dipotassium carbonate	3 mL chloroform	---	91
NZ, CLZ, CBZ, DZ, DCBZ, DDZ, FNZ (IS)	serum	500 μL Serum + I.S, flunitrazepam (300 ngmL^{-1}) + 250 μL 0.8 M CAPSO buffer (pH 10.0)	4.0 mL n-pentane/ethyl acetate (3:1, v/v)	90-96	118

Supercritical Fluid Extraction (SFE)

Supercritical fluid technology as sample preparation is a rapidly expanding analytical technique known to be important in forensic sciences. Selected benzodiazepines were extracted from serum using a supercritical CO_2 mobile phase. The precision of the SFE method in one paper found in literature was shown to be comparable to the precision obtained with other classical preparation techniques of liquid-liquid and solid-phase extraction.[124]

Solid-phase Extraction

Sample preparation using solid-phase extraction was firstly introduced in the mid-1970s, replacing LLE due to its simplicity, selectivity and the better LODs that it provides. The approach, in which compounds of interest are retained on solid-phase adsorbents, followed by selective elution, has been intensively applied. Since then, SPE has gained a wide acceptance due to its several advantages such as the ease of automation, less organic solvent usage, no foaming or emulsion problems, shorter sample preparation or minimal handling time, high analyte recovery even at low concentrations, clean sample extracts and little or no need for concentration of the extract, extraction reproducibility, ability to increase selectively analyte concentration and commercial availability of many SPE devices and sorbents, including the use of molecular imprinted polymers (MIPs).

In solid-phase extraction the sample is poured directly onto a column packed with solid adsorbents like alumina, silica, chemically bonded silica, florisil or non-ionic or ionic exchange resins. However, disposable columns of various sizes and with a wide range of adsorbents are commercially available. These columns selectively adsorb benzodiazepines and their metabolites from bio-fluids at a pH of 9.0-11.0. Drugs are retained on the adsorbent surface. Undesirable compounds also adsorbed may be removed by washing with an appropriate solvent or buffer. Drugs and related compounds are then eluted by passing an appropriate elution organic solvent through the column making sample cleanup simpler, quicker and less laborious than the traditional solvent extraction procedure.

Octyl or Octadecylsilane-bonded cartridges (C_8, C_{18}) were commonly successfully used. Drummer [8] reviewed that C_2 column provided the best combination of high recovery and clean extracts from urine, compared to C_8, C_{18}, phenyl and cyclohexyl phases, whilst CN provided little retention on the cartridge due to its polar nature. Cyanopropyl (CN) cartridges can interact with the analytes by means of hydrophilic, lipophilic and hydrogen bond interactions, however low extraction yields are obtained. Lipophilic sorbents such as C_{18} and C_8 strongly retained the analytes giving low extraction yields, while the weakly lipophilic C_1 sorbent gave promising results in terms of both extraction yields and selectivity.[35]

Commercially available C_{18} bonded phase *Bond-Elut* [Varian] extraction columns have been used for the rapid preparation of blood, plasma, serum or urine samples to determine diazepam, N-desmethyldiazepam, nitrazepam, medazepam, flunitrazepam or midazolam and their metabolites.[125,114] Mixed mode *Bond-Elut certify* has been used for the determination of benzodiazepines and their metabolites in human plasma and urine [126], oral fluid [127], blood [128], hair [72]. The polymeric cartridges (*Oasis HLB and Abselut Nexus*) have advantages over classical C_{18}-bonded [Bond Elut C_{18}] silica cartridges for the extraction of some of benzodiazepines (diazepam, flunitrazepam, nitrazepam, oxazepam) in serum and urine.[129]

SPE cartridge DSC-18 (Supelco) (500 mg/3 mL) was applied for alprazolam, bromazepam, diazepam and flunitrazepam which were eluted with methanol in human biological fluids by HPLC-UV.[19] World Wide Monitoring *Clean Screen®* columns (ZSDAU 020) were applied in an LC–MS–MS method for the extraction of nine benzodiazepines in hair followed by analytes elution using 1.5 mL 2% ammoniated ethylacetate and 1.5 mL dichloromethane/isopropanol /ammonium hydroxide (78:20:2).[64]

A number of benzodiazepines in urine or plasma samples were extracted using an *Oasis MCX* and retained drugs were eluted with a mixture of methylenechloride:isopropanol:ammonia (78:20:2 v/v). [31,51] *Oasis HLB* SPE cartridges were used for the extraction of a large number of benzodiazepines either in whole blood or urine or plasma when drugs were eluted using CH_2Cl_2 [55] or dichloromethane–isopropanol (75:25) [130] or acetonitrile–tetrahydrofuran–water–formic acid (80:1:17:2, v/v) [34], respectively. In HPLC chiral separation lorazepam in human plasma was extracted with *Oasis HLB* SPE cartridge and elution was done with methanol.[131]

Flunitrazepam and its metabolite (7-aminoflunitrazepam) in whole blood and urine or estazolam in plasma or flurazepam, flunitrazepam, clobazam and clorazepate in serum were extracted onto CN SPE cartridges (cyanopropyl or butyl) followed by elution with ethylacetate–methanol or 20% ACN or 10% ACN in water, respectively.[132,133]

Commercially available C_{18} SPE-cartridges, *Sep-Pak, Chromabond, Empore disk, Ultra-clean, Strata,* were used for the extraction of various benzodiazepines in blood, plasma, urine, hair samples, and retained drugs were eluted with either chloroform or methanol or methylene chloride or acetone/dichloromethane (3:l). [43,134-136]

A number of benzodiazepines and their metabolites such as clonazepam, its major metabolite 7-aminoclonazepam in blood, urine or hair samples were extracted on *Isolute HCX* mixed-mode cartridges and drugs were eluted with dichloromethane/isopropanol/ammonium hydroxide (78:20:2, *v/v*) [40,73] or ethyl acetate.[42]

A large number of benzodiazepines including metabolites in blood or plasma samples were extracted under basic conditions onto *Chem Elut CE* cartridge packed with celite, and retained analytes were subsequently eluted by dichloroethane:isopropyl alcohol 97.5:2.5(v/v) [137] or *t*-butylmethylether.[138] Hydrolyzed derivatives of some benzodiazepines in human urine were extracted by ZSDAU020 SPE columns followed by elution with ethyl acetate/ammonium hydroxide [98:2, v/v].[38] *Toxitube A* cartridges were used for the extraction of 22 benzodiazepines in blood and urine.[68,139] 96-well microtiter plate was used for the isolation of midazolam in plasma. Concentrated to dryness under nitrogen, samples were reconstituted in water (20 mL) containing 0.1% TFA.[140]

A special double column device with X-5 resin solid-phase was designed for the extraction of 1,4-benzodiazepines in human plasma; drugs can be extracted at different pH in two different columns.[141,142] SPE sample preparation for the chromatographic determination of 1,4-benzodiazepines in biological samples are summarized in Table **2**.

Table 2: Sample Preparation Of Biological Samples By SPE For The Chromatographic Determination Of 1,4-Benzodiazepines.

Analytes	Matrix/	Sample pretreatment	Extraction sorbents	Elution	% R	Ref.
BRL, CLZ, MFNZ, DZ, FNZ, KTZ, LRL, LRZ, NZ, TL	Whole blood	1 mL Blood + 6 mL 0.1 M phosphate buffer, pH 6.0, vortex-mixing, sonification, centrifuge	Bond Elut Certify (3-mL; Varian) column	2 mL acetone-chloroform (50:50); 3 mL ethylacetate-ammonia solution (98:2),	50-90	128
7-AFNZ, FNZ, OZ, LZ, CDO, TZ, DZ, NDZ, NZ	Hair	Washing- hair sample + methanol-25% NH₄OH (1.5 mL, 20:1), sonication.	ZSDAU 020 Screen column	1.5 mL 2% ammoniated ethylacetate + 1.5 mL dichloromethane/isopropanol/ammonium hydroxide (78:20:2)	72-98	64
7-ANZ, 7-ACLZ, 7-AFNZ, AL, α-HAL, OZ, 3-HDZ, *N*-MDZ	Urine	Urine samples (0.5 mL) + 50 μL IS (deuterated) + 0.5 mL 0.1 M acetate buffer (pH 4.0) + β-glucuronidase at 60°C for 2h + 0.5 mL 0.1 M phosphate buffer (pH 7.5)	Oasis MCX column	2 mL chloromethane/isopropanol/NH₄OH (80:20:2, v/v/v)	56-83	31
FNZ, 7-AFNZ, NZ(IS)	Blood, urine	1 mL Blood/urine + 100 μL nitrazepam (IS) (1000 ngmL⁻¹) + sodium fluoride, vortex mix + 10 mL DI water, centrifugation	CECN₄ butyl end capped (10 mL) column	ethylacetate/methanol (80:20, v/v)	83-87	132
21 BDZs, trimipramine-*d₃* *(IS)*	Urine	1 mL Sorensen buffer + 50 μL IS (deuterated) solution + 100 μL working solutions (0.0004-1.0 mgL⁻¹), vortex-mixed for 10s	Oasis HLB (3 mL/60 mg) column	3 mL dichloromethane–propanol-2 (75:25 v/v)	77-110	130
DZ, FNZ, NZ, OZ, FNZ (IS)	Serum, urine	Serum + 100 μL flunitrazepam (IS, 2 μgmL⁻¹)	Oasis HLB, 1 mL Abselut Nexus	diethylether (2 × 1 mL)	95-103	129
23 benzodiazepines	Blood, urine	1 mL Urine and blood + 1 ng clonazepam-d₄ (IS)	Toxitube A1 (Varian)	50 mL acetonitrile/methanol (50/50, v/v)	---	68
PRZ, OZ, NDZ, DZ (IS)	Plasma	Spiked plasma + 500 μL 0.05 M NaH₂PO₄, vortex	Oasis HLB cartridge	0.5 mL acetonitrile–tetrahydrofuran–water–formic acid (80:1:17:2, v/v/v)	69-101	34
MDL, FNZ (IS)	Dog Plasma	200 μL Plasma + 8 μL flunitrazepam (IS) + 20 μL conc. orthophosphoric acid, vortex mix	Oasis 96-well 30 mg MCX (2 mL)	1 mL dicloromethane:isopropanol: ammonia (78:20:2, v/v/v)	77-95	50
23 BDZs, MBRZ (IS)	Blood	spiked blood sample + 50 μL methylbromazepam (IS)	Varian ChemElut cartridges	4 mL *t*-butylmethylether	60-91	138
MDL, AL (IS)	Plasma	30 μL Plasma + acetonitrile (100 μL) + IS (1220 ngmL⁻¹) + formic acid (0.2%), vortex-mix, centrifuge	96-well microtiter plate	water (20 μL) containing 0.1% TFA	---	141
FNZ, 7-AFNZ, N-DFNZ, 3-HFNZ, FNZ-d₇ (IS)	Urine, Plasma	1 mL Plasma + 3 mL 0.1 M phosphate buffer + 10 ng IS (deuterated), Urine samples + acetic acid buffered (pH 4.5) + 125 μL β-glucuronidase incubated for 1 h at 56⁰C	mixed-mode HCX cartridges, (3 mL/130 mg)	2 mL ammonia 25% (v/v) + 98 mL dichloromethane–isopropanol (70:30, v/v)	81-100	30
ESZ, AL (IS)	Plasma	1 mL Plasma + alprazolam (5 ng) in methanol(10 μL) + 5 mL 1 M sodium chloride	Sap-pak CN cartridge	5 mL 20% ACN in water	96	133
FNZ, FNZ-d (IS)	Plasma	1 mL Plasma sample+ 20 μL orthophosphoric acid + 40 μL IS (1 mgL⁻¹), vortex mix	Oasis MCX cartridge	1 mL methylenechloride : isopropanol : ammonia (78 : 20 : 2, v/v/v)	93-101	51

Table 2: cont….

Analytes	Matrix/	Sample pretreatment	Extraction sorbents	Elution	% R	Ref.
DZ, *N*-DDZ, TZ	Rat plasma, urine	1-mL Spiked urine or plasma + 1 N acetic acid (pH 5)	Sep-Pak Vac 3cc (500 mg)	2 × 2 mL methanol, 2 × 1 mL acetonitrile	71-84	63
CLZ, methoxycarbam azepine (IS)	Plasma	600 µL Sample + 500 µL of tris-buffer 0.8 M, pH 10.9, + 100 µL IS, vortex mix	Chem Elut CE cartridge	3 mL dichloroethane : isopropyl alcohol 97.5:2.5(v/v)	75	137
FNZ, 7-AFNZ, *N*-DFNZ, 3-HFNZ, FNZ-d₃, 7-AFNZ-d₃ (IS)	Blood, urine	0.5–1-mL Supernatant + 2 mL 0.01 M ammonium carbonate buffer (pH 9.3) + IS mixture	C₁₈ Bond Elut cartridge (200 mg)	2 × 0.5 mL methanol–0.5M acetic acid	92-99	125
DZ, NDZ, OX, TZ, LZ, CDO, NZ, FNZ, 7-AFNZ , d₅ BDZs (IS)	Hair	Washing: 0.1% sodium dodecylsulfate + deionized water + dichloromethane, 30 mg hair samples + 1.5 mL methanol / 25% aqueous ammonium hydroxide (20/1, v/v)	MISPE	methanol	93	67
FNZ, 7-AFNZ	Urine	Urine sample (2 mL) + 2 mL acetonitrile, vortex-mix, centrifuge	HLB extraction column	acetonitrile/water (5/95, v/v)	95-101	59
11 benzodiazepines and metabolites	Hair	Washing: isopropanol + phosphate buffer, digestion: Hair samples (10–30 mg) + proteinase K	BondElut Certify columns		---	72
LZ	Plasma	200-µL Plasma	Oasis HLB cartridge	5 mL of methanol	---	131
AL, BRZ, DZ, FNZ, mefenamic acid (IS)	Plasma, urine	Spiked plasma or Urine + ACN (200 µL)	DSC-18 (Supelco)	2 mL of methanol	81-115	19
MZ, DZ, CBZ, MDL	Plasma	Plasma samples + stock solutions (1 mgmL⁻¹ in methanol) + ethylacetate	DEC cartridge in the ASPEC	ethyl acetate	70	145
OZ, LZ, NDZ, DZ, FNZ, LRZ	Hair	Decontamination: warm water + acetone, 30-50 mg of the pulverised hair + 2 mL actetate buffer (pH=4) + 70 µL β-glucuronidase/arylsulfatase incubated for 2h, 40°C	Chromabond C₁₈ column	2 mL acetone/dichloromethane (3:l)	50-95	43
DZ, DDZ, OZ, TZ, LZ, DXP, NCBZ, NZ, 1-HMDL, AL, 1-HAL, 1-HTL, CLZ, FNZ, DFNZ, PNZ, LRZ	Urine	1 mL Urine + 10 µL of β-glucuronidase + 100 µL 1 M acetate buffer pH 4.8, heating for 40 min at 40⁰C	Isolute Confirm HCX columns	ethyl acetate (3 mL) MTBSTFA with 1% TBDMSCl	---	42
20 benzodiazepines, FDZ(IS)	Whole blood	1 mL Spiked blood + 3 mL 0.05 M phosphate buffer, pH 7.0	Oasis HLB cartridge	5 mL CH₂Cl₂	44-138	55
DZ, NDZ, 7-AFNZ, deuterated analytes(IS)	Hair	Wash: warm water + acetone, 30 mg pulverised hair + 8 M urea–0.2 M thioglycolate solution (pH 3) (incubated for 2 h at 60⁰C)	Chromabond C₁₈ columns	acetone/dichloromethane (3:l) heptafluorobutyric anhydride (HFBA)	---	21
FNZ, 7-AFNZ, d₇ (IS)	Oral fluid	1 mL Supernatant + internal standards (1 µgL⁻¹) + 4 mL of phosphate buffer (0.02 N, pH 6.0)	BondElut Certify cartridge 130 mg	2 × 1 mL dichloromethane/ isopropanol/ ammonia (78:20:2) (v/v/v), HFBA	83-90	127
CLZ and 7-ACLZ	Hair	Hair + 30 µL DZP-d₅ 1 µgmL⁻¹ + 3 mL methanol + 0.1 M hydrochloric acid (3 mL), incubated over night at 55°C, vortex-mixed + 1.93 M glacial acetic acid	Mixed-mode Isolute HCX	3 mL methylene chloride/ isopropanol/ ammonium hydroxide (78:20:2, v/v) HFBA at 60°C	---	73
CLZ,7-ACLZ, DZ D₅ (IS)	Urine	1 mL Urine sample + 1 mL DI water + IS (30 ngmL⁻¹) + 1 mL acetic aced buffer (pH 4.5) + 100 µL β-glucuronidase	HCX isolute (10 mL/200 mg) column	3 mL dichloromethen/ isopropanol/ ammonium hydroxide (78:20:2;v/v/v)	---	40
AL, FZ, OZ, LZ, DZ, TZ, MDL, NDZ, PRZ, DA-FZ, α-HAL, α-HMDL, α-HTL, 2-HEFZ, 7-AFNZ, 7-ACLZ, 7-ANZ	Urine	5 mL Urine samples + 100 µL phosphate buffer + β-glucuronidase, incubated at 50°C for 1 h	Clean-Screen ZSDAU020 columns	2 mL hexane 50 µL of N,O-bis(trimethylsilyl) trifluoro-acetamide + 1% trimethylchloro silane + 50 µL of ethyl acetate	---	38
FNZ, 7-AFNZ, NZ (IS)	Whole blood, urine	1 mL Whole blood/urine + Nitrazepam (IS) (100 mL of 1000 ngmL⁻¹)	(CECN4) butyl endcapped (10 mL, 200 mg sorbent)	2 × 3 mL ethyl acetate/methanol (80:20)25 µL ethyl acetate + 25 µL PFPA	83-77	132
CTZ	Plasma	Plasma samples + 20 µL IS (diazepam, 5 µgmL⁻¹ in methanol)	Strata C₁₈-E cartridge	1 mL of methanol	93-101	136
22 benzodiazepines	Plasma, urine	1 mL Whole blood/urine + BDZ drugs (50 ngmL⁻¹)	Toxi-tube A	100 µL acetonitrile 40 µL BSTFA, heated at 80⁰C for 20 min.	68-93	140
MDL, 1-HMDL, 4-HMDL, midazolam-d₅ (IS)	Plasma	250 µL Plasma + 10 µL I.S (100 ng), supernatant + 750 µL 0.5 M Na₂HPO₄ (pH 7)	Bond Elute C₁₈ (Varian-100) cartridge	1 mL acetonitrile–5 mM ammonium acetate–acetic acid (60:40:1,v/v/v)	---	114
BRZ, FNZ, CLZ, DZ, LZ, AL	Plasma, urine	Spiked samples + acetonitrile, centrifuge	LC-18	methanol:acetonitrile (50:50, v/v)	88-113	159
BRZ, FNZ, CLZ, DZ, LZ, AL1-HAL, 1-HTL	Plasma, urine, saliva	Spiked samples + acetonitrile, centrifuge	Nexus varian	methanol:acetonitrile (50:50, v/v)	95-107	160

Molecularly Imprinted Solid-Phase Extraction

The molecularly imprinted polymers (MIPs) of highly cross-linked polymers are synthesised in the presence of template molecules. After synthesis, the template is removed, leaving behind imprinted binding sites ([cavities) within the polymer network which are complementary in size, shape, and chemical functionality to the template. These binding sites are able to rebind the template molecule, or other molecules that have close structural similarity to the template molecule, in a strong and selective manner. The hair extraction method with MIPs is expected to remove matrix interferences, thus providing cleaner extracts than the corresponding SPE method, leading to a more sensitive and reliable analytical protocol. MIP procedure produced extracts with less matrix interferences than the classical SPE method. Molecularly imprinted solid-phase extraction (MISPE) protocols for diazepam extraction could be used as complementary methods to classical SPE for the analysis of benzodiazepines in hair samples. As MIP possesses group-selective binding nature it was successfully applied for some other benzodiazepine drugs such as nordiazepam, nitrazepam, chlordiazepoxide, temazepam, oxazepam, lorazepam, flunitrazepam, 7-aminoflunitrazepam. [67,143]

On-line Solid-phase Extraction

ASPEC (Automated Sample Preparation with Extraction Columns) System

ASPEC is a system that enables the fully automated extraction and determination of analytes in biological matrices. It includes extraction, clean-up, drying and transfer of the analytes (elution) to the chromatographic system. Benzodiazepines such as clobazam, medazepam, midazolam, diazepam in plasma, flunitrazepam and its metabolites 7-aminoflunitrazepam, *N*-desmethylflunitrazepam and 3-hydroxy flunitrazepam in urine or plasma, alprazolam, clonazepam, and nitrazepam in serum were isolated by automated ASPEC system using *Isolute HCX* mixed-mode cartridges or *Bond-Elut* C_{18} *(Varian) or Supelclean LC*-18 disposable cartridges. Methanol or ethyl acetate was used for elution step.[30,144,145]

Column Switching (CS)

Column switching technique was employed to elute the extracted analytes from the pre-column into a HPLC analytical column. It was applied to the simultaneous determination of five frequently prescribed benzodiazepines; clonazepam, diazepam, midazolam, oxazepam, flunitrazepam and main metabolites (norflunitrazepam, 7-amino- and 7-acetamido-flunitrazepam). The use of biocompatible extraction column as pre-column offered repeated direct injection of serum, plasma, urine supernatant or other complex matrices without any clean-up procedure.[66,130,146-149] Application of monolithic supports to online extraction of 1,4-benzodiazepines is the first published work dealing with online extraction by column switching on whole blood.[150] The protein component of the biological samples (serum, urine) was flushed through alkyl-diol-silica (ADS, LiChrospher RP-18) [146,147] or hydrophobic polymer (BioTrap 500 MS) [66,147,148] pre-column used in column switching technique for the direct and on-line extraction of benzodiazepines in the pores of the stationary phases. Online solid-phase extraction column with the combination of an *N*-vinylacetamide-containing hydrophilic polymer enhances the sensitivity, and eliminates tedious sample pretreatment of a number of benzodiazepines and metabolites [41] in urine and plasma [59,149] where sample extraction, clean-up and elution were performed automatically. Basic advantage of column switching over other techniques is potential time savings, though these techniques avoid an extraction step they do require more instrumentation.

Micro-extraction Methods

Solid-phase Micro Extraction (SPME)

The demand for reduction in extraction instrumentation size, decreased solvent use, and need for rapid and convenient sample preparation has led to the development of micro-extraction in analytical chemistry. SPME was firstly developed by Pawliszyn and co-workers in 1989 and became commercially available in 1993. Since its development, SPME has been increasingly used since its setup is small and convenient, and it can be used to extract analytes from very small samples. It provides a rapid extraction and transfer to analytical instrument and can be easily combined with other extraction and/or analytical procedures improving to a large extent the sensitivity and selectivity of the whole method. SPME has two processes which are equilibrium between analytes and the fiber coating, and desorption to the mobile phase. After the extraction the SPME fiber is withdrawn and inserted into the desorption devices interfaced with an HPLC/LC system and mobile phase is used to desorbs the analytes.[151]

SPME devices have been prepared using highly bio-compatible SPME capillary coated with restricted access material (RAM), alkyl-diol-silica (ADS) for the simultaneous fractionation of the protein component from a biological sample.[152] SPME can be automated, and direct in-tube extraction is performed for several benzodiazepines like clonazepam, diazepam, oxazepam, temazepam, nordiazepam and 7-aminoflunitrazepam,

N-desmethylflunitrazepam in human serum and urine. The extracted benzodiazepines can be desorbed from the capillary coating by means of the mobile phase flow and transported to the LC column for the separation. Between two silica fibers examined, polydimethylsiloxane/divinylbenzene (PDMS/DVB) proved to be most suitable than carbowax/templated resin (Carbowax/TPR-100) for extraction of benzodiazepines in urine.[1] In gas chromatographic analysis after analyte extraction with simultaneous *in situ* derivatization (acetylation or silylation) the SPME (polyacrylate fiber) device is transferred to GC injector for thermal desorption of midazolam and diazepam in human plasma [47,52] or urine [48]. Carbowax-divinylbenzene (CAX/DVB) coated fiber was used for the extraction of five 1,4-benzodiazepines such as diazepam, nordiazepam, oxazepam, temazepam, and lorazepam from urine in the highest amounts compared to the other tested fiber coatings examined: poly-arcylate (PA), 100 μm polydimethylsiloxane (PDMS), and poly(dimethyl-siloxane/divinylbenzene) PDMS/DVB.[153]

Liquid-phase Micro-Extraction (LPME)

Liquid-phase micro-extraction (LPME) was demonstrated independently by Dasgupta and Cantwell in 1996. This was based on a small drop of an organic solvent (called single-drop LPME or Solvent Dynamic Microextraction (SDME). Later, to improve the stability and reliability of single-drop LPME, Pedersen-Bjergaard and Rasmussen introduced hollow fiber liquid-phase micro-extraction (HF-LPME) in 1999, which have been widely applied in the field of drug analysis in recent years. The micro-extraction device consisted of a porous hollow fiber of polypropylene filled with extraction solvent (25 μL) was immersed in bio-samples with continuously vibration at 600 rpm for 50 min. An aliquot of the extraction solvent [butyl acetate: 1-octanol (1:1, v/v) for urine or hexylether:1-octanol (1:3, v/v) for plasma, with pre-concentrated analytes, diazepam and *N*-desmethyldiazepam, was injected directly into the capillary gas chromatograph.[54], A new polyvinylidene difluoride (PVDF) hollow fiber with higher porosity and better solvent compatibility showed advantages with faster extraction efficiency and operational accuracy compared to other polypropylene (PP) hollow fibers in the automated liquid-phase micro-extraction (HF-LPME) for flunitrazepam in biological samples.[154]

Non-Extraction Methodologies

Dialysis

In the last few years, dialysis has gained popularity as a sample preparation technique in determination of traces of analytes in protein-containing matrices, because the use of a semi-permeable membrane offers the possibility of removing macromolecular sample constituents as the dialysis membranes are designed to allow only small molecules to be sampled. Therefore, no sample pre-treatment is required as clean chromatograms can often be obtained. It has been successfully applied to a variety of biomedical samples prior to LC or GC analysis. Nakashima K. *et al.* [92] applied this method to determine plasma and brain microdialysate concentrations of triazolam, when microdialysates were directly injected onto the HPLC and no interference from compounds codialyzed with triazolam was observed.

In addition, if a trace-enrichment pre-column is incorporated in the set-up to overcome the dilution of the sample caused by the dialysis step, efficient sample clean up and analyte enrichment can be combined in the system in fully automated way. Clean up for the determination of benzodiazepines in plasma was based on performing the dialysis of samples using water as acceptor phase and trapping the diffused analytes on a PLRP-S copolymer pre-column, when desorption was made with ethyl acetate.[155]

Direct Injection (DI)

Bio-fluids either crude samples or after protein precipitation, and solution of pharmaceutical formulations in organic solvents have been directly subjected to analysis by chromatographic methods. Etizolam, brotizolam, lorazepam and triazolam in human plasma were analyzed injecting the filtered supernatant after protein precipitaion with 0.13 % formic acid onto *MSpakGF* polymer column, which enabled direct injection of crude biological samples. [13,62] Plasma spiked with nitrazepam, clobazam, oxazepam and lorazepam in acetonitrile at certain concentration was directly injected to the HPLC system set on an analytical hydrophobic shielded phase (Hisep) column equipped with a Hisep guard column.[156] Isolated supernatants of plasma samples were simply treated with acetonitrile to precipitate and remove proteins and were directly injected into the HPLC system equipped with *Zorbax Eclipse XDB* C_{18} column for lorazepam determination.[157]

When Micellar liquid chromatography (MLC) is applied to drug determination in biological fluids, such as serum and urine, direct injection of the sample without any pre-treatment is possible. The surfactant sodium dodecylsulphate (SDS) micelles tend to bind proteins competitively by releasing protein-bound drugs, so the substances are free to partition into the stationary phase, whereas the proteins, rather than precipitating into the column, are solubilised and eluted with or shortly after the solvent front. MLC procedures are reported for the determination of several 1,4-benzodiazepines injecting serum samples directly, without any pre-treatment giving LOD 2-6 ngmL^{-1}.[11,91,158]

However, direct injection of complex samples leads to the contamination of columns impairing their performance. Contamination often persists even when a pre-column is used to protect the analytical column. To avoid these problems, sample clean-up including enzymatic digestion, protein precipitation and solvent or solid-phase extractions is required.

CONCLUSION

Since sample preparation is often an inevitable step in biological fluid analysis to extract and concentrate drugs of interest from the matrix, several sample preparation techniques capable of reducing the amount of interfering substances have been developed. Although, extraction strategies such as solid-phase extraction and liquid-liquid extraction have been extensively used with success in extracting benzodiazepines from biological fluids, it is recognized that these methods are time consuming, tedious and often require complicated procedures, pre-concentration of the extract prior to instrumental analysis, these sample preparation approaches can suffer from poor automation capabilities and excess use of solvents except for modern techniques with less or no solvent consumption. The large amount of organic solvent used in the LLE extraction procedure causes problems with regards to health and the environment. Apart from these classical extraction techniques, liquid extraction and solid-phase extraction using reversed-phase silica sorbents have problems when there are differences in chemical nature (polarity, affinity, pH, etc.,) between extracted compounds and extraction solvents or solid-phase extraction sorbents. The classical sorbents commonly used are porous silica particles surface-bonded with C_{18} or other hydrophobic alkyl groups. Analysts have to watch carefully and control closely the extraction procedure. Therefore, it is difficult to achieve high, reproducible recoveries for analysis of numerous drugs, especially regarding a mixture of apolar compounds and polar ones, such as parent drugs and their polar metabolites.

To omit conventional sample pre-treatment, HPLC columns such as internal-surface reversed-phase silica support have been developed, which enables direct injection of biological samples into HPLC; these columns are usually used in a column-switching arrangement. Solid-phase micro-extraction (SPME) is a relatively new approach to sample preparation. It requires less organic solvents, which is important from an ecological and analytical view point. It is also a fast, solvent-free and an excellent performance technique. A new HPLC polymer stationary phase, *MSpak GF* column, which consists of a highly cross-linked hard gel of polyvinyl alcohol, has established rapid and simple chromatographic methods for analyzing benzodiazepines by direct injection of human plasma and urine samples avoiding any extraction procedure or any column-switching technique. Molecularly imprinted solid-phase extraction is a recent advancement in extraction which could be used as complementary methods to classical SPE for the analysis of benzodiazepines.

These moderate techniques need more sophisticated instrumentation, for ease of operation and thus in economic consideration, conventional LLE or SPE have been widely used still now by most of the researchers. Recently, some polymeric extraction cartridges are commercialized and can be used within the whole pH range and with many different polar and apolar organic solvents (methanol, chloroform, diethylether, etc.) contrary to classical reversed-phase silica extraction columns. Therefore, it is easier to find an appropriate extraction condition for a specific compound and especially for a mixture of analytes with different chemical properties.

REFERENCES

[1] Aresta A, Monaci L, Zambonin CG. Determination of delorazepam in urine by solid-phase microextraction coupled to high performance liquid chromatography. J Pharm Biomed Anal 2002; 28: 965–972.

[2] Bugey A, Staub C. Rapid analysis of benzodiazepines in whole blood by high-performance liquid chromatography: use of a monolithic column. J Pharm Biomed Anal 2004; 35: 555–562.

[3] Kavvadias D, Abou-Mandour AA, Czygan F, Beckmann H, Sand P, Riederer P, Schreier P. Identification of Benzodiazepines in *Artemisia dracunculus* and *Solanum tuberosum* Rationalizing Their Endogenous Formation in Plant Tissue. Biochem Biophys Res Commun 2000; 269: 290–295.

[4] Drummer OH. Benzodiazepines-Effects on human performance and behaviour. Forensic Sci Rev 2002; 14(1): 1-14.

[5]   Barbone F, McMahon AD, Davey PG, Morris AD, Reid IC, McDevitt DG, MacDonald TM. Association of road-traffic accidents with benzodiazepine use. The Lancet 1998; 352(24): 1331-1336.

[6]   Drummer OH, Ranson DL. Sudden Death and Benzodiazepines. Am J of Forensic Med & Patholo 1996; 17(4): 336-342.

[7]   Yuan H, Mester Z, Lord H, Pawliszyn J. Automated in-tube solid-phase microextraction coupled with liquid chromatography–electrospray ionization mass spectrometry for the determination of selected benzodiazepines. J Anal Toxicol 2000; 24(8): 718-725.

[8]   Drummer OH. Methods for the measurement of benzodiazepines in biological samples. J Chromatogr B 1998; 713: 201–225.

[9]   Toyóoka T, Kumaki Y, Kanbori M, Kato M, Nakahara Y. Determination of hypnotic benzodiazepines (alprazolam, estazolam, and midazolam) and their metabolites in rat hair and plasma by reversed-phase liquid-chromatography with electrospray ionization mass spectrometry. J Pharm Biomed Anal 2003; 30: 1773-1787.

[10]   Hiild KM, Crouch DJ, Wilkins DG, Rollins DE, Maes RA. Detection of alprazolam in hair by negative ion chemical ionization mass spectrometry. Foren Sci Int 1997; 84: 201-209.

[11]   Esteve-Romero J, Carda-Broch S, Gil-Agust M, Capella-Peiro M, Bose D. Micellar liquid chromatography for the determination of drug materials in pharmaceutical preparations and biological samples. Trends Anal Chem 2005; 24(2): 75-91.

[12]   Samanidou V, Uddin MN, Papadoyannis I. Benzodiazepines: Sample Preparation and HPLC Methods for Their Determination in Biological Samples-A Review Article. Bioanalysis 2009, in Press.

[13]   Lee XP, Kumazawa T, Sato J, Shoji Y, Hasegawa C, Karibe C, Arinobu T, Seno H, Sato K. Simple method for the determination of benzodiazepines in human body fluids by high-performance liquid chromatography–mass spectrometry. Anal Chim Acta 2003; 492: 223–231.

[14]   Berrueta LA, Gallo B, Vicente F. Biopharmacological data and high-performance liquid chromatographic analysis of 1,4-benzodiazepines in biological fluids: a review. J Pharm Biomed Ana. 1992; 10(2/3): 109-136.

[15]   Quintela O, Cruz A, de Castro A, Concheiro M, Lopez-Rivadulla M. LC-ESI-MS for the determination of nine selected benzodiazepines in human plasma and oral fluid. J Chromatogr B 2005; 825: 63-71.

[16]   Tanaka E, Terada M, Misawa S, Wakasugi C. Simultaneous determination of twelve benzodiazepines in human serum using a new reversed-phase chromatographic column on a 2 μm porous microspherical silica gel. J Chromatogr B 1996; 682: 173-178.

[17]   El Mahjoub A, Staub C. Stability of benzodiazepines in whole blood samples stored at varying temperatures. J Pharm Biomed Anal 2000; 23: 1057-1063.

[18]   Cavedal LE, Mendes FD, Domingues CC, Patni AK, Monif T, Reyar S, Pereira ADS, Mendes GD, De Nucci G. Clonazepam quantification in human plasma by high-performance liquid chromatography coupled with electrospray tandem mass spectrometry in a bioequivalence study. J Mass Spectrom 2007; 42: 81-88.

[19]   Samanidou VF, Pechlivanidou AP, Papadoyannis IN. Development of a validated HPLC method for the determination of four 1, 4-benzodiazepines in human biological fluids. J Sep Sci 2007; 30: 679-687.

[20]   Tracy TS, Rybeck BF, James DG, Knopp JB, Gannett PM. Stability of benzodiazepines in formaldehyde solutions. J Anal Toxicol 2001; 25(3): 166-173.

[21]   Yegles M, Marson Y, Wennig R. Influence of bleaching on stability of benzodiazepines in hair. Foren Sci Int 2000; 107: 87-92.

[22]   Huidobro AL, Ruperez FJ, Barbas C. Isolation, identification and determination of the major degradation product in alprazolam tablets during their stability assay. J Pharm Biomed Anal 2007; 44: 404-413.

[23]   Nudelman NS, Cabrera CG. Isolation and Structural Elucidation of Degradation Products of Alprazolam: Photostability Studies of Alprazolam Tablets. J Pharm Sci 2002; 91(5): 1274-1286.

[24]   Essien H, Lai SJ, Binder SR, King DL. Use of direct-probe mass spectrometry as a toxicology confirmation method for demoxepam in urine following high-performance liquid chromatography. J Chromatogr B 1996; 683: 199-208.

[25]   Valentine JL, Middleton R, Sparks C. Identification of Urinary Benzodiazepines and their Metabolites: Comparison of Automated HPLC and GC-MS after Immunoassay Screening of Clinical Samples. J Anal Toxicol 1996; 20(6): 416-424.

[26]   Panderi I, Archontaki H, Gikas E, Parissi-Poulou M. Acidic hydrolysis of bromazepam studied by high performance liquid chromatography: Isolation and identification of its degradation products. J Pharm Biomed Anal 1998; 17: 327-335.

[27]   Damjanović T, Popović G, Verbić S, Pfendt L. Study of acid hydrolysis of bromazepam. Can J Chem 2004; *82*(8): 1260-1265.

[28]   Gambart D, Cardenas S, Gallego M, Valcarce M. An automated screening system for benzodiazepines in human urine. Anal Chim Acta 1998; 366: 93-102.

[29]   Dolejsova J, Solich P, Polydorou ChK, Koupparis MA, Efstathiou CE. Flow-injection fluorimetric determination of 1,4-benzodiazepines in pharmaceutical formulations after acid hydrolysis. J Pharm Biomed Anal 1999; 20: 357–362.

[30] Jourdil N, Bessard J, Vincent F, Eysseric H, Bessard G. Automated solid-phase extraction and liquid chromatography–electrospray ionization-mass spectrometry for the determination of flunitrazepam and its metabolites in human urine and plasma samples. J Chromatogr B 2003; 788: 207–219.

[31] Hegstad S, Øiestad EL, Johansen U, Christophersen AS. Determination of Benzodiazepines in Human Urine using SPE and HPLC-ESI-Tandem Mass Spectrometry. J Anal Toxicol 2006; 30(1): 31-37.

[32] Toyóoka T, Kanbori M, Kumaki Y, Nakahara Y. Determination of Triazolam Involving Its Hydroxy Metabolites in Hair Shaft and Hair Root by Reversed-Phase Liquid Chromatography with Electrospray Ionization Mass Spectrometry and Application to Human Hair Analysis. Anal Biochem 2001; 295: 172–179.

[33] Rouini M, Ardakani YH, Hakemi L, Mokhberi M, Badri G. Simultaneous determination of clobazam and its major metabolite in human plasma by a rapid HPLC method. J Chromatogr B 2005; 823: 167–171.

[34] Valavani P, Atta-Politou J, Panderi I. Development and validation of a liquid chromatographic/ electrospray ionization mass spectrometric method for the quantitation of prazepam and its main metabolites in human plasma. J Mass Spectrom 2005; 40: 516–526.

[35] Mercolini L, Mandrioli R, Amore M, Raggi MA. Separation and HPLC analysis of 15 benzodiazepines in human plasma. J Sep Sci 2008; 31: 2619-2626.

[36] Mehta AC. High-pressure liquid chromatographic determination of some 1,4-benzodiazepines and their metabolites in biological fluids: a review. Talanta 1984; 31(1): 1-8.

[37] Lozano-Chaves ME, Palacios-Santander JM, Cubillana-Aguilera LM, Naranjo-Rodriguez I, Hidalgo-Hidalgo-de-Cisneros JL. Modified carbon-paste electrodes as sensors for the determination of 1,4-benzodiazepines: Application to the determination of diazepam and oxazepam in biological fluids. Sensors and Actuators B 2006; 115: 575–583.

[38] Klette KL, Wiegand RF, Horn CK, Stout PR, Magluilo JJr. Urine Benzodiazepine Screening using Roche Online KIMS Immunoassay with β-Glucuronidase Hydrolysis and Confirmation by Gas Chromatography–Mass Spectrometry. J Anal Toxicol 2005; 29: 193-200.

[39] Kintz P, Villain M, Cirimele V, Pepin G, Ludes B. Windows of detection of lorazepam in urine, oral fluid and hair, with a special focus on drug-facilitated crimes. Forensic Sci Int, 2004; 145: 131-135.

[40] Negrusz A, Bowen AM, Moore CM, Dowd SM, Strong MJ, Janicak PG. Elimination of 7-aminoclonazepam in urine after a single dose of clonazepam. Anal Bioanal Chem 2003; 376: 1198–1204.

[41] Miki A, Tatsuno M, Katagi M, Nishikawa M, Tsuchihashi H. Simultaneous Determination of Eleven Benzodiazepine Hypnotics and Eleven Relevant Metabolites in Urine by Column-Switching Liquid Chromatography–Mass Spectrometry. J Anal Toxicol 2002; 26(2): 87-93.

[42] Rasanen I, Neuvonen M, Ojanpera I, Vuori E. Benzodiazepine findings in blood and urine by gas chromatography and immunoassay. Forensic Sci Int 2000; 112: 191-200.

[43] Yegles M, Mersch F, Wennig R. Detection of benzodiazepines and other psychotropic drugs in human hair by GC/MS. Forensic Sci Int 1997; 84: 211-218.

[44] Lennestål R, Lakso H, Nilsson M, Mjörndal T. Urine monitoring of diazepam abuse-new intake or not? J Anal Toxicol 2008; 32: 402-407.

[45] El-Sohly MA, Gul W, Avula B, Murphy TP, Khan IA. Simultaneous analysis of thirty-five benzodiazepines in urine using liquid chromatography–mass spectrometry-time of flight. J Anal Toxicol 2008; 32: 547-561.

[46] Herraez-Hernandez R, Louter AJH, van de Merbel, Brinkman UATh. Automated on-line dialysis for sample preparation for gas chromatoggaphy: determination of benzodiazepines in human plasma. J Pharm Biomed Anal 1996; 14: 1077-1087.

[47] Frison G, Tedeschi L, Maietti S Ferrara SD. Determination of midazolam in human plasma by solid phase microextraction and gas chromatography/mass spectrometry. Rapid Commun. Mass Spectrom 2001; 15: 2497-2501.

[48] Staerk U, Kulpmann WR. High-temperature solid-phase microextraction procedure for the detection of drugs by gas chromatography–mass spectrometry. J Chromatogr B 2000; 745: 399–411.

[49] Zarghi A, Jenabi M. Assay of oxazepam in human plasma by reversed-phase high-performance liquid chromatography. Boll Chim. Farm 2001; 140(6): 455-457.

[50] Harris SR, Gedge JI, Nedderman ANR, Roffey SJ, Savage M. A sensitive HPLC-MS-MS assay for quantitative determination of midazolam in dog plasma. J Pharm Biomed Anal 2004; 35: 127–134.

[51] Kollroser M, Schober C. Simultaneous analysis of flunitrazepam and its major metabolites in human plasma by high performance liquid chromatography tandem mass spectrometry. J Pharm Biomed Anal 2002; 28: 1173–1182.

[52] Krogh M, Grefslie H, Rasmussen KE. Solvent-modified solid-phase micro-extraction for the determination of diazepam in human plasma samples by capillary gas chromatography. J Chromatogr B 1997; 689: 357-364.

[53] Frerichs VA, Zaranek C, Haas CE. Analysis of omeprazole, midazolam and hydroxy-metabolites in plasma using liquid chromatography coupled to tandem mass spectrometry. J Chromatogr B 2005; 824: 71–80.

[54] Ugland HG, Krogh M, Rasmussen KE. Liquid-phase microextraction as a sample preparation technique prior to capillary gas chromatographic-determination of benzodiazepines in biological matrices. J Chromatogr B 2000: 749: 85–92.

[55] Inoue H, Maeno Y, Iwasa M, Matoba R, Nagao M. Screening and determination of benzodiazepines in whole blood using solid-phase extraction and gas chromatography/mass spectrometry. Forensic Sci Int 2000; 113: 367–373.

[56] Nakashima K, Yamamoto KAl-Dirbashi OY, Kaddoumi A, Nakashim MN. Semi-micro column HPLC of triazolam in rat plasma and brain microdialysate and its application to drug interaction study with itraconazole. J Pharm Biomed Anal 2003; 30: 1809-1816.

[57] Honeychurch KC, Smith GC, Hart JP. Voltammetric behavior of nitrazepam and its determination in serum using liquid chromatography with redox mode dual-electrode detection. Anal Chem 2006; 78: 416-423.

[58] Muchohi SN, Obiero K, Kokwaro GO, Ogutu BR, Githiga IM, Edwards G, Newton CRJC. Determination of lorazepam in plasma from children by high-performance liquid chromatography with UV detection. J Chromatogr B 2005; 824: 333–340.

[59] Fuh M, Lin S, Chen L, Lin T. Determination of flunitrazepam and 7-aminoflunitrazepam in human urine by on-line solid phase extraction liquid chromatography–electrospray-tandem mass spectrometry. Talanta 2007; 72: 1329–1335.

[60] Wilhelm M, Battista H, Obendorf D. HPLC with Simultaneous UV and Reductive Electrochemical Detection at the Hanging Mercury Drop Electrode: A Highly Sensitive and Selective Tool for the Determination of Benzodiazepines in Forensic Samples. J Anal Toxicol 2001; 25(4): 250-257.

[61] Blanchard J. Evaluation of the relative efficacy of various techniques for deproteinizing plasma samples prior to high-performance liquid chromatographic analysis. J Chromatogr 1981; 226: 455-460.

[62] Lee X, Kumazawa T, Fujishiro M, Hasegawa C, Marumo A, Shoji Y, Arinobu T, Seno H, Sato K. Simple method for determination of triazolam in human plasma by high-performance liquid chromatography/tandem mass spectrometry. J Pharm Biomed Anal 2006; 41: 64–69.

[63] Abu-Qare AW, Abou-Donia MB. Chromatographic method for the determination of diazepam, pyridostigmine bromide, and their metabolites in rat plasma and urine. J Chromatogr B 2001; 754: 503–509.

[64] Miller EI, Wylie FM, Oliver JS. Detection of Benzodiazepines in Hair Using ELISA and LC–ESI-MS–MS. J Anal Toxicol 2006; 30: 441-448.

[65] Laloupl M, Fernandez MMR, De Boeck G, Wood M, Maes V, Samyn N. Validation of a Liquid Chromatography–Tandem Mass Spectrometry Method for the Simultaneous Determination of 26 Benzodiazepines and Metabolites, Zolpidem and Zopiclone, in Blood, Urine, and Hair. J Anal Toxicol 2005; 29(7): 616-626.

[66] El Mahjoub A, Staub C. Determination of benzodiazepines in human hair by on-line high-performance liquid chromatography using a restricted access extraction column, Forensic Sci Int 2001; 123: 17–25.

[67] Ariffin MM, Miller EI. Cormack PAG, Anderson RA. Molecularly Imprinted Solid-Phase Extraction of Diazepam and Its Metabolites from Hair Samples. Anal Chem 2007; 79: 256-262.

[68] Cheze M, Duffort G, Deveaux M, Pepin G. Hair analysis by liquid chromatography–tandem mass spectrometry in toxicological investigation of drug-facilitated crimes: Report of 128 cases over the period June 2003–May 2004 in metropolitan Paris. Forensic Sci Int 2005; 153: 3–10.

[69] Gunnar T, Ariniemi K, Lillsunde P. Determination of 14 benzodiazepines and hydroxy metabolites, zaleplon and zolpidem as *tert*-butyldimethylsilyl derivatives compared with other common silylating reagents in whole blood by gas chromatography–mass spectrometry. J Chromatogr B 2005; 818: 175–189.

[70] Villain M, Concheiro M, Cirimele V, Kintz P. Screening method for benzodiazepines and hypnotics in hair at pg/mg level by liquid chromatography–mass spectrometry/mass spectrometry. J Chromatogr B 2005; 825: 72–78.

[71] Scott KS, Nakahara Y. A study into the rate of incorporation of eight benzodiazepines into rat hair (HPLC, GC-MS). Forensic Sci Int 2003; 133: 47–56.

[72] Kronstrand R, Nyström I, Josefsson M, Hodgins S. Segmental ion spray LC–MS–MS analysis of benzodiazepines in hair of psychiatric patients. J Anal Toxicol 2002; 26(7): 479-484.

[73] Negrusz A, Bowen AM, Moore CM, Dowd SM, Strong MJ, Janicak PG. Deposition of 7-Aminoclonazepam and Clonazepam in Hair Following a Single Dose of Klonopin. J Anal Toxicol 2002; 26: 471-478.

[74] Cirimele V, Kintz P, Staub C, Mangin P. Testing human hair for flunitrazepam and 7-amino-flunitrazepam by GC/MS-NCI. Forensic Sci Int 1997; 84: 189-200.

[75] Cirimele V, Kintz P, Ludes B. Screening for forensically relevant benzodiazepines in human hair by gas chromatography-negative ion chemical ionization-mass Spectrometry. J Chromatogr B 1997; 700: 119-129.

[76] Kazemifard AG, Gholami K, Dabirsiaghi A. Optimized determination of lorazepam in human serum by extraction and high-performance liquid chromatographic analysis. Acta Pharm 2006; 56: 481–488.

[77] Aebi B, Sturny-Jungo R, Bernhard W, Blanke R, Hirsch R. Quantitation using GC–TOF-MS: example of bromazepam. Forensic Sci Int 2002; 128: 84–89.

[78] Lee TC, Charles B. Measurement by HPLC of Midazolam and its Major Metabolite, 1-Hydroxymidazolam in Plasma of Very Premature Neonates. Biomed Chromatogr 1996; 10: 65-68.

[79] Shiran MR, Gregory A, Rostami-Hodjegan A, Tucker GT, Lennard MS. Determination of midazolam and 19-hydroxymidazolam by liquid chromatography–mass spectrometry in plasma of patients undergoing methadone maintenance treatment. J Chromatogr B 2003; 783: 303–307.

[80]   Rizzo M, Sinopoli VA, Gitto R, Zappala M, De Sarro G, Chimirri A. High-performance liquid chromatographic determination of new 2,3-benzodiazepines. J Chromatogr B 1998; 705: 149–153.

[81]   Ghosh P, Reddy MM, Rao BS, Sarin RK. Determination of Diazepam in Cream Biscuits by Liquid Chromatography. J AOAC Int 2004; 87 (3): 569-572.

[82]   Van Brandt N, Philippe H, Paul M, Roger KV. A rapid high-performance liquid chromatographic method for the measurement of midazolam plasma concentrations during long-term infusion in ICU patients. Thera Drug Monit 1997; 19(3): 352-357.

[83]   Tomita M, Okuyama T. Application of capillary electrophoresis to the simultaneous screening and quantitation of benzodiazepines. J Chromatogr B 1996; 678: 331-337.

[84]   Jurica J, Dostalek M, Konecny J, Glatz Z, Hadasova E, Tomandl J. HPLC determination of midazolam and its three hydroxy metabolites in perfusion medium and plasma from rats. J Chromatogr B 2007; 852: 571-577.

[85]   Quintela O, Cruz A, Concheiro M, de Castro A, Lopez-Rivadulla M. A sensitive rapid and specific determination of midazolam in human plasma and saliva by liquid chromatography/electrospray mass spectrometry. Rapid Commun Mass Spectrom 2004; 18: 2976–2982.

[86]   Pham-Huy C, Villain-Pautet G, He Hua, Chikhi-Chorfi N, Galons H, Thevenin M, Jean-Roger C, Jean-Michel W. Separation of oxazepam, lorazepam, and temazepam enantiomers by HPLC on a derivatized cyclodextrin-bonded phase: application to the determination of oxazepam in plasma. J Biochem Biophys Methods 2002; 54: 287–299.

[87]   Portier EJG, de Blok K, Butter JJ, van Boxtel CJ. Simultaneous determination of fentanyl and midazolam using high performance liquid chromatography with ultraviolet detection. J Chromatogr B 1999; 723: 313–318.

[88]   Wang P, Liu C, Tsay W, Li J, Liu RH, Wu T, Cheng W, Lin D, Huang T., Chen C. Improved screen and confirmation test of 7-aminoflunitrazepam in urine samples for monitoring rlunitrazepam (Rohypnol) exposure. J Anal Toxicol (2002); 26(7): 411-418.

[89]   Zedkova L, Rauw GA, Baker GB, Coupland NJ. A rapid high-pressure liquid chromatographic procedure for determination of flumazenil in plasma. J Pharmacol Toxicol Methods. 2001; 46: 57– 60.

[90]   Bishop SC, Lerch M, McCord BR. Micellar electrokinetic chromatographic screening method for common sexual assault drugs administered in beverages. Forensic Sci Int 2004; 141: 7-15.

[91]   Capella-Peiro M, Bose D, Domınguez A, Gil-Agusti M, Esteve-Romero J. Direct injection micellar liquid chromatographic determination of benzodiazepines in serum. J Chromatogr B 2002; 780: 241–249.

[92]   Nakashima K, Yamamoto K, Al-Dirbashi OY, Nakashima MN. Deposition of Triazolam in rat by brain microdialysis and semi-micro column HPLCwith UV-detection. Biomed Chromatogr 2002; 16: 219–223.

[93]   de Jong LAA, Verwey B, Essink G, Muntendam A, Zitman FG, Ensing K. Determination of the benzodiazepine plasma concentrations in suicidal patients using a radioreceptor assay. J Anal Toxicol 2004; 28 (7): 587-592.

[94]   Cheze M, Villain M, Pepin G. Determination of bromazepam, clonazepam and metabolites after a single intake in urine and hair by LC–MS/MS Application to forensic cases of drug facilitated crimes. Forensic Sci Int 2004; 145: 123–130

[95]   Kunicki PK. Simple and sensitive high-performance liquid chromatographic method for the determination of 1, 5-benzodiazepine clobazam and its active metabolite *N*-desmethylclobazam in human serum and urine with application to 1,4-benzodiazepines analysis. J Chromatogr B 2001; 750: 41–49.

[96]   Bugey A, Rudaz S, Staub C. A fast LC-APCI/MS method for analyzing benzodiazepines in whole blood using monolithic support. J Chromatogr B 2006; 832: 249–255.

[97]   Muchohi SN, Ward SA, Preston L, Newton CRJC, Edwards G, Kokwaro GO. Determination of midazolam and its major metabolite 1-hydroxymidazolam by high-performance liquid chromatography-electrospray mass spectrometry in plasma from children. J Chromatogr B 2005; 821: 1–7.

[98]   Salem AA, Barsoum BN, Izake EL. Spectrophotometric and fluorimetric determination of diazepam, bromazepam and clonazepam in pharmaceutical and urine samples. Spectrochim Acta Part A 2004; 60: 771–780.

[99]   El Mahjoub A, Staub C. Simultaneous determination of benzodiazepines in whole blood or serum by HPLC/DAD with a semi-micro column. J Pharm Biomed Anal 2000; 23: 447–458.

[100] Rivera HM, Walker GS, Simsand DN, Stockham PC. Application of liquid chromatography-tandem mass spectrometry to the analysis of benzodiazepines in blood. Eur J Mass Spectrom 2003; 9(6): 599-607.

[101] Darius J, Banditt P. Validated method for the therapeutic drug monitoring of flunitrazepam in human serum using liquid chromatography–atmospheric pressure chemical ionization tandem mass spectrometry with an ion trap detector. J Chromatogr B 2000; 738: 437–441.

[102] Ma F, Lau CE. Determination of midazolam and its metabolites in serum microsamples by high-performance liquid chromatography and its application to pharmacokinetics in rats. J Chromatogr B 1996; 682: 109-113.

[103] Chłobowska Z, Świegoda C, Kościelniak P, Piekoszewski W. Identification and determination of flunitrazepam and its metabolites in blood by gas chromatography. Chemia Analityczna 2004; 49(1): 71-77.

[104] Laurito TL, Mendes GD, Santagada V, Caliendo G, de Moraes MEA, De Nucci G. Bromazepam determination in human plasma by high-performance liquid chromatography coupled to tandem mass spectrometry: a highly sensitive and specific tool for bioequivalence studies. J Mass Spectrom 2004; 39:168–176.

[105] El Sohly MA, Gul W, ElSohly KM, Avula B, Khan IA. LC-MS-(TOF) analysis method for benzodiazepines in urine samples from alleged drug-facilitated sexual assault victims. J Anal Toxicol 2006; 30(8): 524-538.

[106] Andraus MH, Wong A, Silva OA, Wada CY, Toffleto O, Azevedo CP, Salvadori MC. Determination of bromazepam in human plasma by high-performance liquid chromatography with electrospray ionization tandem mass spectrometric detection: application to a bioequivalence study. J Mass Spectrom 2004; 39: 1348–1355.

[107] Yasui-Furukoria N, Inoue Y, Tateishi T. Sensitive determination of midazolam and 1-hydroxymidazolam in plasma by liquid–liquid extraction and column-switching liquid chromatography with ultraviolet absorbance detection and its application for measuring CYP3A activity. J Chromatogr B 2004; 811: 153–157.

[108] Shimizu M, Uno T, Tamura H, Kanazawa H, Murakami I, Sugawara K, Tateishi T. A developed determination of midazolam and 1-hydroxymidazolam in plasma by liquid chromatography–mass spectrometry: Application of human pharmacokinetic study for measurement of CYP3A activity. J Chromatogr B 2007; 847: 275–281.

[109] Bares IF, Pehourcq F, Jarry C. Development of a rapid RP–HPLC method for the determination of clonazepam in human plasma. J Pharm Biomed Anal 2004; 36: 865–869.

[110] Kratzsch C, Tenberken O, Peters FT, Weber AA, Kraemer T, Maurer HH. Screening, library-assisted identification and validated quantification of 23 benzodiazepines, flumazenil, zaleplone, zolpidem and zopiclone in plasma by liquid chromatography/mass spectrometry with atmospheric pressure chemical ionization. J Mass Spectrom 2004; 39: 856–872.

[111] Ter Horst PGJ, Foudraine NA, Cuypers G, van Dijk EA, Oldenhof NJJ. Simultaneous determination of levomepromazine, midazolam and their major metabolites in human plasma by reversed-phase liquid chromatography. J Chromatogr B 2003; 791: 389–398.

[112] Atta-Politou J, Parissi-Poulou M, Dona A, Koutselinis A. A modified simple and rapid reversed phase liquid chromatographic method for quantification of diazepam and nordiazepam in plasma. J Pharm Biomed Anal 1999; 20: 389–396.

[113] Toyóoka T, Kanbori M, Kumaki Y, Oe T, Miyahara T, Nakahara Y. Detection of Triazolam and Its Hydroxy Metabolites in Rat Hair by Reversed-Phase Liquid Chromatography with Electrospray Ionization Mass Spectrometry. J Anal Toxicol 2000; 24(3): 194-201.

[114] Lausecker B, Hopfgartner G, Hesse M. Capillary electrophoresis–mass spectrometry coupling versus microhigh-performance liquid chromatography–mass spectrometry coupling: a case study. J Chromatogr B 1998; 718: 1–13.

[115] Eeckhoudt SL, Desager J, Horsmans Y, De Winne AJ, Verbeeck RK. Sensitive assay for midazolam and its metabolite 1′-hydroxymidazolam in human plasma by capillary high performance liquid chromatography. J Chromatogr B 1998; 710: 165–171.

[116] Øiestad EL, Johansen U, Christophersen AS. Drug Screening of Preserved Oral Fluid by Liquid Chromatography–Tandem Mass spectrometry. Clin Chem 2007; 53(2): 300-309.

[117] Crouch DJ, Rollins DE, Canfield DV, Andrenyak DM, Schulties JE. Quantitation of Alprazolam and α-Hydroxyalprazolam in Human Plasma Using Liquid Chromatography Electrospray Ionization MS–MS. J Anal Toxicol 1999; 23(6): 479-485.

[118. Imazawa M, Hatanaka Y. Micellar electrokinetic capillary chromatography of benzodiazepine antiepileptics and their desmethyl metabolites in blood. J Pharm Biomed Anal 1997; 15: 1503-1508.

[119] Cheze M, Villain M, Pepin G. Determination of bromazepam, clonazepam and metabolites after a single intake in urine and hair by LC–MS/MS application to forensic cases of drug facilitated crimes. Forensic Sci Int 2004; 145: 123–130.

[120] Muchohia SN, Ogutu BR, Newton CRJC, Kokwaro GO. High-performance liquid chromatographic determination of diazepam in plasma of children with severe malaria. J Chromatogr B 2001; 761: 255–259.

[121] Proenca P, Teixeira H, Pinheiro J, Marques EP, Vieira DN. Forensic intoxication with clobazam: HPLC/DAD/MSD analysis. Forensic Sci Int 2004; 143: 205-209.

[122] Snyder H, Schwenzer KS, Pearlman R, McNally AJ, Tsilimidos M, Salamone SJ, Brenneisen R, ElSohly MA, Feng S. Serum and Urine Concentrations of Flunitrazepam and Metabolites, after a Single Oral Dose, by Immunoassay and GC–MS. J Anal Toxicol 2001; 25(8): 699-704.

[123] Kintz P, Villain M, Cheze M, Pepin G. Identification of alprazolam in hair in two cases of drug-facilitated incidents. Forensic Sci Int 2005; 153: 222–226.

[124] Simmons BR, Stewart JT. Supercritical fluid extraction of selected pharmaceuticals from water and serum. J Chromatogr B 1997; 688(2): 291-302.

[125] Bogusz MJ, Maier R, Kruger K, Fruchtnicht W. Determination of flunitrazepam and its metabolites in blood by high-performance liquid chromatography–atmospheric pressure chemical ionization mass spectrometry. J Chromatogr B 1998; 713: 361–369.

[126] Kominami G, Nakamura M, Chomei N, Takada S. Radioimmunoassay for a novel benzodiazepine inverse agonist, S-8510, in human plasma and urine. J Pharm Biomed Anal 1999; 20: 145–153.

[127] Samyn N, De Boeck G, Cirimele V, Verstraete A, Kintz P. Detection of flunitrazepam and 7-aminoflunitrazepam in oral fluid after controlled administration of Rohypnol. J Anal Toxicol 2002; 26(4): 211-215.

[128] Verweij AMA, Hordijk ML, Lipman PJL. Liquid chromatographic-thermospray tandem mass spectrometric quantitative analysis of some drugs with hypnotic, sedative and tranquillising properties in whole blood. J Chromatogr B 1996; 686: 27-34.

[129] He H., Sun C., Wang X., Pham-Huy C., Chikhi-Chorfi N., Galons H., Thevenin M., Claude J, Warnet J. Solid-phase extraction of methadone enantiomers and benzodiazepines in biological fluids by two polymeric cartridges for liquid chromatographic analysis. J Chromatogr B 2005; 814: 385–391.

[130] Quintela O, Sauvage F, Charvier F, Gaulier J, Lachatre G, Marquet P. Liquid chromatography–tandem mass spectrometry for detection of low concentrations of 21 benzodiazepines, metabolites, and analogs in urine: method with forensic applications. Clin Chem 2006; 52(7): 1346–1355.

[131] Kanazawa H, Kunito Y, Matsushima Y, Okubo S, Mashige F. Stereospecific analysis of lorazepam in plasma by chiral column chromatography with a circular dichroism-based detector. J Chromatogr A 2000; 871: 181-188.

[132] Hackett J, Elian AA. Extraction and analysis of flunitrazepam/7-amino-flunitrazepam in blood and urine by LC–PDA and GC–MS using butyl SPE columns. Forensic Sci Int 2006; 157: 156–162.

[133] Miura M, Ohkubo T, Sugawara K, Okuyama N, Otani K. Determination of estazolam in plasma by HPLC with SPE. Anal Sci 2002; 18: 525-528.

[134] Kanazawa H, Okada A, Igarashi E, Higaki M, Miyabe T, Sano T, Nishimura R. Determination of midazolam and its metabolite as a probe for cytochrome P450 3A4 phenotype by liquid chromatography–mass spectrometry. J Chromatogr A 2004; 1031: 213–218.

[135] Vanhoenacker G, de l'Escaille F, De Keukeleire D, Sandra P. Analysis of Benzodiazepines in Dynamically Coated Capillaries by CE-DAD, CE-MS and CE-MS^2. J Pharm Biomed Anal 2004; 34: 595–606.

[136] Ahn S, Maeng H, Koo T, Kim D, Shim C, Chung S. Quantification of clotiazepam in human plasma by gas chromatography–mass spectrometry. J Chromatogr B 2006; 834: 128–133.

[137] Bolner A, Tagliaro F, Lomeo A. Optimised determination of clobazam in human plasma with extraction and high-performance liquid chromatography analysis. J Chromatogr B 2001, 750: 177–180.

[138] Sminka BE, Brandsma JE, Dijkhuizen A, Lusthof KJ, de Gier JJ, Egberts ACG, Uges DRA. Quantitative analysis of 33 benzodiazepines, metabolites and benzodiazepine-like substances in whole blood by liquid chromatography–(tandem) mass spectrometry. J Chromatogr B 2004; 811: 13–20.

[139] Pirnay S, Ricordel I, Libong D, Bouchonnet S. Sensitive method for the detection of 22 benzodiazepines by gas Chromatography-ion trap tandem mass spectrometry. J Chromatogr A 2002; 954: 235–245.

[140] Kapron JT, Pace E, Van Pelt CK, Henion J. Quantitation of midazolam in human plasma by automated chip-based infusion nanoelectrospray tandem mass spectrometry. Rapid Commun Mass Spectrom 2003; 17: 2019–2026.

[141] Qiu FH, Liu L, Luo Y, Liu F, Lu YQ. Systematic analysis of basic drugs in plasma using X-5 solid-phase extraction GC-FID and GC-MS. Yao Xue Xue Bao 1996; 31(4): 296-299.

[142] Qiu FH, Liu L, Luo Y, Lu YQ. A double column double pH solid phase extraction and capillary GC/FID method for rapid simultaneous determination of acidic and basic drugs in human plasma. Yao Xue Xue Bao 1996; 31(3): 205-208.

[143] Anderson RA, Ariffin MM, Cormack PAG, Miller EI. Comparison of molecularly imprinted solid-phase extraction (MISPE) with classical solid-phase extraction (SPE) for the detection of benzodiazepines in post-mortem hair samples. Forensic Sci Int 2008; 174: 40-46.

[144] Åkerman KK. Analysis of clobazam and its active metabolite norclobazam in plasma and serum using HPLC/DAD. Scand J Clin and Lab Invest 1996; 56(7): 609-614.

[145] Louter AJH, Bosma E, Schipperen JCA, Vreuls JJ, Brinkman UATh. Automated on-line solid-phase extraction-gas chromatography with nitrogen-phosphorus detection: determination of benzodiazepines in human plasma. J Chromatogr B 1997; 689: 35-43.

[146] Mullett WM, Pawliszyn J. Direct LC analysis of five benzodiazepines in human urine and plasma using an ADS restricted access extraction column. J Pharm Biomed Anal 2001; 26: 899–908.

[147] El Mahjoub A, Staub C. High-performance liquid chromatography determination of flunitrazepam and its metabolites in plasma by use of column switching technique: comparison of two extraction columns. J Chromatogr B 2001; 754: 271–283.

[148] El Mahjoub A, Staub C. High-performance liquid chromatographic method for the determination of benzodiazepines in plasma or serum using the column-switching technique. J Chromatogr B 2000; 742: 381–390.

[149] Goncalves JCS, Monteiro TM, Neves CSM, Gram KRS, Volpato NM, Silva VA, Caminha R, Goncalves MRB, Santos FM dos, Silveira GE da, Noel F. On-Line Solid-Phase Extraction Coupled With HPLC-MS-MS for Quantification of Bromazepam in Human Plasma: An Automated Method for Bioequivalence Studies. Therap Drug Monit 2005; 27(5): 601-607.

[150] Bugey A, Staub C. Application of monolithic supports to online extraction and LC-MS analysis of benzodiazepines in whole blood samples. J Sep Sci 2007; 30: 2967-2978.

[151] Mullett WM, Pawliszyn J. Direct Determination of Benzodiazepines in Biological Fluids by Restricted-Access Solid-Phase Microextraction. Anal Chem 2002; 74: 1081-1087.

[152] Walles M, Mullett WM, Pawliszyn J. Monitoring of drugs and metabolites in whole blood by restricted-access solid-phase microextraction coupled to liquid chromatography–mass spectrometry. J Chromatogr A 2004; 1025: 85–92.

[153] Luo Y, Pan L, Pawliszyn J. Determination of five benzodiazepines in aqueous solution and biological fluids using solid-phase microextraction with carbowax/DVB fiber coating. J Microcol Separ 1998; 10(2): 193-201.

[154] Cui S, Tan S, Ouyang G, Pawliszyn J. Automated polyvinylidene difluoride hollow fiber liquid-phase microextraction of flunitrazepam in plasma and urine samples for gas chromatography/tandem mass spectrometry. J Chromatogr A 2009; 1216(12): 2241-2247.

[155] Herraez-Hernandez R, Louter AJH, van de Merbel NC, Brinkman UATh. Automated on-line dialysis for sample preparation for gas chromatography: determination of benzodiazepines in human plasma. J Pharm Biomed Anal 1996; 14: 1077-1087.

[156] Pistos C, Stewart JT. Direct injection HPLC method for the determination of selected benzodiazepines in plasma using a Hisep column. J Pharm Biomed Anal 2003; 33: 1135-1142.

[157] Zhu H, Luo J. A fast and sensitive liquid chromatographic–tandem mass spectrometric method for assay of lorazepam and application to pharmacokinetic analysis. J Pharm Biomed Anal 2005; 39: 268–274.

[158] Ghosh P, Reddy MM, Rao BS, Sarin RK. Use of micellar mobile phases for the chromatographic determination of clorazepate, diazepam, and diltiazem in pharmaceuticals. J Chromatogr Sci 2000; 38(12): 521-527.

[159] Uddin MN, Samanidou VF, Papadoyannis IN. Development and validation of an HPLC method for the simultaneous determination of six 1,4-benzodiazepines from pharmaceutical and biological samples. J Liq Chromatogr Rel Technol 2008; 31(9): 1258-1282.

[160] Uddin MN, Samanidou VF, Papadoyannis IN. Validation of SPE-HPLC determination of 1,4-benzodiazepines and metabolites in blood plasma, urine, and saliva. J Sep Sci 2008; 31: 3704-3717.

CHAPTER 8

Control of the Level of Apoptosis by Different Analytical Techniques

Małgorzata Starek[1*], Monika Dąbrowska[1] and Jerzy Skuciński[2]

Jagiellonian University, Collegium Medicum, [1]Department of Inorganic and Analytical Chemistry, 9 Medyczna Str., 30-688 and [2]Institute of Emergency Medicine, 12 Michałowskiego Str., 31-126; Cracow, Poland

Abstract: The process of programmed cell death, or apoptosis is characterized by distinct morphological and biochemical mechanisms. Inappropriate apoptosis (too little or too much) is a factor in many human conditions including neurodegenerative diseases, autoimmune disorders and many types of cancer. Although many assays for apoptosis detection have been established so far, precise differentiating apoptosis and necrosis in single cells is still a hallenge. In this study we present the most common analytical techniques for the control of the level of apoptosis.

INTRODUCTION

Cell death may occur by two mechanisms: apoptosis, or programmed cell death, and necrosis, or cell death due to injury or trauma. Both types of cell death have their own specific and distinct morphological and biochemical hallmarks. Apoptotic cells share a number of common features, such as phosphatidylserine exposure, cell shrinkage, chromatin cleavage, nuclear condensation, and formation of pyknotic bodies of condensed chromatin. Necrotic cells exhibit nuclear swelling, chromatin flocculation, loss of nuclear basophilia, breakdown of cytoplasmic structure and organelle function, and cytolysis by swelling. Cell death can be induced by a wide variety of stimuli, such as growth factor withdrawal, heat shock, cold shock, radiation, heavy metals, genotoxic drugs, and a number of biological ligands such as Fas-L and tumor necrosis factor. Most if not all of these can induce both apoptosis and necrosis in a time- and dose- dependent manner [1, 2].

During apoptotic cell death the cell shrinks, thus activating proteolytic enzymes, the cell membrane losses its asymmetric structure due to phosphatydyloserine migration toward the external membrane layer. Chromatin in the nucleus undergoes condensation and marginalization. At the final stage of this process DNA fragmentation occurs. Chromatin undergoes multistage cleavage depending on activity of endonucleases. The first signal is fragmentation into long segments of 700, 300 kbp. The fragments of 300 kbp in length correspond to hexametric loops known as rosette structure. Afterwards, shorter fragments of 50 kbp corresponding to single DNA loops are formed. Next, further fragmentation into regular 200 bp intervals visible when using conventional agarose gel electrophoresis as characteristic apoptotic "ladder" pattern, may occur.

The next stage of apoptosis involves formation of apoptotic bodies due to cytoplasmic membrane blebbing and breaking. The apoptotic bodies contain well preserved cell and nucleus components and are quickly removed from the intercellular space due to phagocytosis. During the first apoptosis phase damages might be repaired and the programmed cell death process might be abandoned. After DNA fragmentation the cell losses its repairing ability.

The aim of this paper is not to discuss both processes thoroughly and differences between them, although this is necessary to some extent to understand the methods used for apoptosis detection. Recent papers imply its usefulness in cell death assessment not only by discrimination between dead and living cells, but also by allowing precise determination of the cell death type: apoptosis or necrosis. More specific techniques have been developed to determine cell death, and the combination of several methods is required to distinguish between apoptosis or necrosis. These techniques rely on specific morphological and molecular or biochemical changes associated with these two processes. This paper describes some of the techniques most commonly used to detect cell death.

TUNEL (TERMINAL DEOXYNULEOTIDYL TRANSFERASE-MEDIATED DEOXYURIDINE TRIPHOSPHATE BIOTYN NICK-LABELING)

TUNEL method based on the following principles: the oligonucleosomal DNA fragments in apoptotic cells contain 3'-hydroxyl groups (and 5'-phosphates) that arise from the cleavage of the phosphoribosyl backbone of the DNA helix. Breaks can be detected on the basis of the ability of the terminal deoxynuleotidyl transferase enzyme (TdT) to add nucleotide to the 3'-hydroxyl groups in genomic DNA. In TUNEL assay cells incubation with this enzyme along with labeled nucleotide triphosphate substrate (dNTPs) leads to incorporation of the labelled dNTP into nicked DNA strands. This allows visualization of DNA breaks – for example those in apoptotic degraded DNA strands [3, 4]. The detection of signals may be provide by light microscopy, fluorescence microscopy or flow cytometry.

Both apoptotic and necrotic cells are TUNEL-positive, while living cells are TUNEL-negative. Apoptotic cells are perfectly labeled compared to necrotic ones (stronger signal) due to larger number of DNA breaks in apoptotic cells.

However, the disadvantage of the TUNEL assay is the fact that it gives false positive results, especially in the case of cells treated with topoisomerase inhibitors, like etoposide, but not undergoing apoptosis [5].

Electrophoretic Methods

DNA degradation can be traced by using various electrophoresis techniques with native or denaturing gels. This is why apart from TUNEL method also such methods as pulsing electrophoresis enabling detection of long DNA fragments of the length ranging from few kbp to 10 mbp, and standard agarose electrophoresis to detect short DNA intervals (apoptotic ladders formed at the final stage of apoptosis) are also employed [6 – 8].

In the electrophoretic methods acrylamide gels can be used. These gels are formed by polymerization of acryl amide monomers into long chains with bis-acrylamide cross-linking. Crosslinking can be adjusted by changing the proportion of acrylamide to bis-acrylamide. The appropriate catalyst (PERS – potassium persulfate) and initiating agent (TEMED – N,N,N',N'-tetramethylethyldiamin) are necessary for this reaction. Acrylamide gels are used for separating fragments of size from several base pairs (20 % polyacrylamide), and also to separate single strand DNA molecules, for example for sequencing purposes.

Since migration rate in a single strand molecule gel depends not only on molecule size and charge, but also to large extent on its spatial structure, to determine its size precisely it is necessary to eliminate differences in migration rate caused by various molecule conformation. To do it agarose or polyacrylamide denaturing gels are used. Urea is the most common denaturing agent used in analysis o single strand DNA fragments in polyacrylamide gels.

To study potential apoptotic DNA changes agarose gel electrophoresis is commonly used. The resolution o this method depends on pore size in the gel. The pore size is determined with agarose concentration. More concentrated gels, i.e. 1.4-2.0 %, are used for separating smaller molecules (0.5-2.0 kbp). Less concentrated gels (0.5-0.7 %) are used to separate larger molecules (10-20 kbp and more). This technique allows identification of nucleosomic DNA fragments of 180-200 bp or multiples thereof - oligo- and polynucleosomic intervals. During electrophoretic analysis the characteristic DNA ladder is formed – one of markers of the apoptosis process.

Conventional agarose gel electrophoresis, which is used to direct internucleosomal DNA degradation in isolated nuclei, cells, or tissues, can be performed according to any of a variety of protocols. One commonly utilized protocol involves lysis of cells in SDS and EDTA, digestion with proteinase K, extraction with phenol to remove peptide fragments, application of the resulting DNA to a gel with suitable separation properties [e.g. 1-2 % (w/v) agarose], and staining with ethidium bromide. Variations on this method include: lysis of cells in a nondenaturing buffer [e.g. 20 mM Tris (pH 7-8) containing EDTA and a neutral detergent] to extract the chromatin fragments below 10 – 20 kb, which can then be quantitated and run on an agarose gel, thereby enhancing the sensitivity of the method for detection of low amounts of internucleosomal fragments [9, 10]. Direct end labeling of free 3' ends in the SDS/proteinase K-treated samples with a single α-^{32}P-labeled dNTP and DNA polymerase or terminal deoxynucleotidyl transferase (TdT) prior to electrophoresis, with autoradiographic detection after electrophoresis to enhance fragment detection [11, 12]. Transfer of the unlabeled DNA on the agarose gel to a nylon support followed by hybridization with a ^{32}P-labeled genomic DNA probe and autoradiography to detect nucleosomal fragments [13]. Use of DNA-binding dyes with enhanced quantum yield [e.g., SYBR Green from Molecular Probes (Eugene, OR)] to increase the ability to detect nucleosomal ladders [14].

Pulsing gel electrophoresis is used for separating large DNA molecules – from 2×10^4 to 10^7 bp, i.e. 20 kbp to 10 Mbp. This enables entire chromosomes, for instance yeast ones, to be separated. Electric field is switched on and off in short time intervals. When electric field is on, molecules migrate according to their size, and when the field is off, molecules have a tendency to relaxation and twisting in random loops. The time required for relaxation is directly proportional to molecule length. Afterwards, the direction of electric field is altered by 90 or 180 degrees compared to previous one. Longer molecules start to move slower than those of smaller length. The repeating field orientation changes lead gradually to separation.

The single-cell gel electrophoresis (SCGE; comet assay) allows detection of DNA fragmentation in single cells, and was initially used for DNA damage estimation.

During apoptosis cellular DNA is degraded by endonucleasis triggered with caspases. DNA is hydrolyzed preferably at internucleosomic spaces. This causes that fragments of 180 bp or multiples thereof are formed. The number of produced DNA fragments is larger for apoptotic than for necrotic cell. During single cell gel electrophoresis fragmented DNA leaves the cell and migrates toward the anode forming the pattern resembling tail of a comet. DNA remaining in the cell looks as comet head. Should no DNA breaking or cutting have occurred, comet head is only observed [15, 16].

Dna Visualization Methods

A coloration of DNA separated with gel electrophoresis can be executed by using ethidium bromide or another similar dyestuff, e.g. SYBR Gold. Multicolor fluorescent colorization is also used by applying laser induced fluorescence of DNA chain built-in fluorochromes. Detection with polyacrylamide gel silvering is less and less frequently used.

Dna Labeling Methods

Depending on particular application various DNA labeling methods are employed:

Radioactive isotope labeling

Numerous methods used in molecular biology are based on labeled molecules of nucleic acids. Good probe should be of high activity, appropriate purity and specificity. Nucleotides with one atom being replaced with a radioactive isotope are most frequently used for labeling purposes. Specific activity depends on the proportion of included radioactive isotopes to normal ones. Also shorter isotope half-life increases specific activity. Two isotopes are most commonly used: phosphorus ^{32}P (embodied in position a or g of nucleotide triphosphate) and sulfur ^{35}S (embodied in place of oxygen atom in relevant phosphate radical – a or g). Radioactive phosphorus is used when highly active probe is required to provide high sensitivity and short detection time. Sulfur is used primarily for DNA sequencing and to study protein metabolism.

Nick Translation

DNase I introduces single strand breaks (nick) into a molecule. Polymerase I due to activity of its 5' - 3' exonuclease removes nucleotides ahead of it, while building new ones behind, including radioactive nucleotides. Hence, nick is translated. The efficiency of incorporation is approx. 50 %. This method brings the best results when whole plasmids are labeled and it rather not recommended for linear fragments.

Random Priming

This technique, most commonly used, is based on hybridization of oligonucleotides (6 to 9 nucleotides) of random sequence to DNA to be labeled. Next, Klenow polymerase (fragment of DNA polymerase of no exonucleolitic 5' - 3' activity) synthesizes complementary DNA strand starting from the 3' OH end of incorporated starter. The reaction mixture contains also radioactive nucleotides gradually included into newly synthesized strand. The length of labeled fragments does not affect efficiency of the reaction. This allows a high activity probe to be obtained. This method is excellent in labeling both whole plasmids and its linear fragments.

DNA Ends Labeling

5' end labeling: in this method polynucleotide T4 kinase is used that moves phosphate group from position g ATP in DNA or RNA containing hydroxyl group at 5' end. Since DNA molecules normally contain phosphate group at 5'end it is necessary to use a alkaline phosphatase to remove it. The labeling 5'end is most frequently used for marking synthetic oligonucleotides.

3' end labeling: in this method terminal transferase adds deoxyribonucleotides at 3' end. No matrix is required. As substrate for this enzyme single strand as well as dual-strand DNA.

Filling sticky ends: this method uses Klenow fragment that builds missing nucleotides at 3' end in molecules etched with restriction enzymes forming sticky 5' ends. The incorporation efficiency is sometimes as high as 90 %.

After labeling it is necessary to separate not incorporated radioactive DTPs that could affect hybridization quality creating too high background, thus decreasing specificity of the reaction. Separation is achieved by column filtering with special filter bed that passes macromolecules, while stopping fine-molecule compounds.

Quantitative assay of labeling is achieved by measuring probe activity. This can be done with two devices: Geiger-Mëller counter and scintillation counter - both count radioactive decays in unit time. Autoradiogaphy allows labeled molecules to be detected visually in order to localize the fringe that hybridized with the probe. Older visualization method consists in exposing a plate coated with photographic emulsion sensitive to radiation and placed in special cassettes. The plate placed on the filter becomes blackened at the spot where hybridization with the probe occurred. The cassette is kept at -70^0 C for a period of time inversely proportional to the probe activity. Such low temperature enhances fringe sharpness. The plate is developed then. Novel method consists in filter exposure in other cassettes y using special device termed phosphoimager connected to a computer. This device enables cassette surface scanning and direct processing of obtained data.

Bioluminescence Method (the ADP:ATP Ratio)

Apoptosis is an active energy–requiring process. The cellular ATP level is an important determinant for cell death, either by apoptosis or necrosis. A cell stays alive as long as a certain ATP level is maintained. When ATP falls belowe this level apoptosis ensues provide enough ATP is still available for energy–requiring apoptotic processes such as enzymatic hydrolysis of macromolecules, nuclear condensation and bleb formation. Only when there is a severe drop in cellular ATP controlled cell death ceases and ushers in necrosis.

A decreased cellular ATP level is characteristic for cell death, but there is no systematic investigation whether the decrease is the cause or the consequence of cell death. An ADP/ATP ratio of about 0.2 was the critical discriminator between survival and apoptosis in all cell types [17].

ATP (Adenosine 5'-Triphosphate) plays a critical role in all living beings as an energy source for various enzymatic activities and as a direct prekursor in RNA synthesis. Adenosine diphosphate (ADP) or adenosine monophosphate (AMP) resulting from ATP-dependent reactions was rephosphorylated by cellular ATP synthetic activity. Preamble cells were reported to produce ATP continuously for a dozen of hours [18 – 22].

A conventional luciferin-luciferase method was established and many investigators have been using it to measure static ATP concentration [23 – 25]. Recently, the improved luciferin-luciferase assay was developed to measure ATP synthetic activity of purified enzymes in vitro or on a glass surface in real time [26, 27]. Furthermore, in whole mammalian cells, dynamic change of ATP concentration was measured by expression of the recombinant luciferase in vivo [28, 29]. The application of this reporter luciferase method to measure a dynamic change of ATP concentration in bacterial cells has been difficult because variation in luciferase activity caused by differential transcription, translation, or mRNA/enzyme stability would affect interpretation of the results [25, 30]. And difficulty of luciferin penetration through the bacterial membrane was another problem.

Cellular ATP content was measured employing the luciferin–luciferase method described by Stanley and Wiliams [31]. Briefly, cells were treated with 1.0 mL of boiling Tris buffer to extract the ATP, and the content was transferred to a scintillation vial along with 1.0 mL each of phosphate and arsenate buffers. Luciferin–luciferase (5 mg/mL) was added at 0.1 mL, and mixed thoroughly and placed in the Packard Trilab Liquid Scintillation 2500TRcounter, which was set in the single photon count mode. Light emission was recorded precisely at 30 sec as counts/min. Protein content was determined on a portion of the cell sample, and ATP was expressed as nanomoles per milligram of cell protein.

ATP was measured using bioluminescence based on luciferin–luciferase reaction [32]. The reaction which results in the generation of measurable light at a wavelength of 562 nm is given below:

$$ATP + Luciferin + O_2 \xrightarrow{Luciferase} Oxyluciferin + AMP + PP_i + CO_2 + Light (562 \text{ nm})$$

The experiments were either set up in triplicate in 96–well, luminometer plates (Wallac) or pipetted into 96–well luminometer plates (Berthold) from 12–well clear, tissue culture plates (Corning) or 96–well clear, tissue culture plates (Costar). The nucleotides were released from the cell suspensions by addition of an equal volume (in this case 100 μL) of cell nucleotide–releasing reagent (NRR). This releasing reagent also contained the luciferin–luciferase, nucleotide–monitoring reagent (NMR). The ATP levels were measured using a luminometer and expressed as the number of relative light units (RLU). Under these optimal conditions and at concentrations of ATP less than 10^{-6} M, the RLU were directly proportional to the amount of ATP present. The ATP signal was allowed to decay for 10 min to a steady state. After 10 min the ADP in the wells was converted to ATP by the addition of 20 μL of ADP converting reagent (ADPCR). An immediate reading was taken to determine the baseline ADP RLU (ADP 0). After 5 min incubation to allow for conversion of ADP to ATP, a third reading was taken (ADP 5). The ratio of ADP:ATP for each well was calculated from these three readings as follows:

$$(ADP\ 5\ RLU - ADP\ 0\ RLU) / ATP\ RLU$$

Necrosis induced by heating viable cells at 56^{0} C for 1 h was found to give mean ADP:ATP ratios of 20.2 (S.E.M. 4.5, n=10). This compared with a mean value of 0.07 (S.E.M. 0.02) for the control cells.

The role of ATP in apoptosis remains controversial. Recent evidence suggests that cellular ATP levels dictate whether cells die by apoptosis or necrosis. Apparently, low levels of ATP with an increase in reactive oxygen species (ROS) result in cell death by necrosis [33, 34]. Certainly, using heat or cold shock to include necrosis, the characteristic features of the necrotic cells is the dramatic reduction in ATP together with the relative increase in ADP compared with control cells.

Fragmentation of DNA is a late event in apoptosis, this would suggest that changes in the ADP:ATP ratio are representative of late apoptosis. This would seem logical because apoptosis is an energy–dependent process and oxidative phosphorylation would need to be maintained beyond the initiation phase.

CONCLUSIONS

Apoptosis is regarded as a carefully regulated energy-dependent process, characterized by specific morphological and biochemical features. The importance of understanding the mechanistic machinery of apoptosis is vital because programmed cell death is a component of both health and disease, being initiated by various physiologic and pathologic stimuli. The application of various analytical techniques allows different cell death pathways to be identified and the level of apoptosis to be determined that could have an impact on therapeutic strategy.

REFERENCES

[1]   Wyllie AH. Apoptosis and the regulation of cell numbers in normal and neoplastic tissues: an overview. Cancer Metastasis Rev 1992; 11: 95-103.

[2]   Gavrieli Y, Sherman Y, Ben-Sasson SA. Identification of programmed cell death in situ via specific labeling of nuclear DNA fragmentation. J Cell Biol 1992; 119: 493-501.

[3]   Reynolds JE, McBain JA, Zhou P, Eastman A, Craig RW. Characterization and detection of cells death by apoptosis. Encyclopedia of Cell Technology. Interscience Publication Book: Spier RE & Wiley A. Ed 2000: 536-550.

[4]   Searle J, Kerr JF, Bishop C. Necrosis and apoptosis: distinct modes of cell death with fundamentalny different significance. Pathol Annu 1982; 17: 229-259.

[5]   Hotz MA, Gong J, Traganos F, Darzynkiewicz Z. Flow cytometric detection of apoptosis: comparison of the assays of in situ DNA degradation and chromatin changes. Cytometry 1994; 15: 237-244.

[6]   Luo Y, Kassel D. Detection of early and late stage apoptosis with field inversion gel electrophoresis. Biotechniques 1996; 21: 812-816.

[7]   Hermann M, Lorenz HM, Voll R, et al. A rapie and simple metod for the isolation of apoptotic DNA fragments. Nucleid Acid Res 1994; 24: 5506-5507.

[8]   Simao TA, Andrada-Serpa MJ, Mendonca GAS, et al. Detection and analysis of apoptosis in peripheral blood cells from breast cancer patients. Braz J Med Biol Res 1999; 32: 403-406.

[9]   Igo-Kemenes T, Greil W, Zachau HG. Prepartation of soluble chromatin and specific chromatin fractions with restriction nucleases. Nucleic Acids Res 1977; 4: 3387-3400.

[10]   Kyprianou N, Issacs JT. Activation of Programmed Cell Death in the Rat Ventral Prostate after Castration. Endocrinology 1988; 122: 552-562.

[11]   Rosl F. A simple and rapid method for detection of apoptosis in human cells. Nucleic Acids Res 1992; 20: 5243.

[12]   Tilly J, Hsueh A. Microscale autoradiographic method for the qualitative and quantitative analysis of apoptotic DNA fragmentation. J Cell Physiol 1993; 154: 519-526.

[13] Mesner W, Winters TR, Green SH. Nerve growth factor withdrawal-induced cell death in neuronal PC12 cells resembles that in sympathetic neurons. J Cell Biol 1992; 119: 1669-1680.

[14] Mesner P, Kaufmann SH. Methods utilized in the study of apoptosis. Adv Pharmacol 1997; 41: 57-87.

[15] Yasuhara S, Zhu Y, Matsui T, et al. Comparison of comet assay electron microscopy, and flow cytometry for detection of apoptosis. J Histochem Cytochem 2003; 51: 873-885.

[16] Reiss RA, Rutz B. Silver-stained comet assay for detection of apoptosis. BioTechniques 1999; 27: 926-930.

[17] Richter Ch, Schweizer M, Cossarizza A, Franceschi C. Control of apoptosis by the cellular ATP level. FEBS Lett 1996; 378: 107-110.

[18] Fujio T, Furuya A. Effect of magnesium ion and chelating agents on enzymatic production of ATP from adenine. Appl Microbiol Biotechnol 1985; 21: 143-147.

[19] Mori H, Iida A, Fujio T, et al. A novel process of inosine 5'- monophosphate production using overproduction using overexpressed guanosine/inosine kinase. Appl Microbiol Biotechnol 1997; 48: 693-698.

[20] Fujio T, Maruyama A. Enzymatic production of pyrimidine nucleotides using Corynebacterium ammoniagenes cells and recombinant Escherichia coli cells: enzymatic production of CDP-choline from orotic acid and choline chloride (part 1). Biosci Biotechnol Biochem 1997; 61:956-959.

[21] Fujio T, Teshiba S, Maruyama A. Conctruction of a plasmid carrying both CTP synthetase and a fused gene formed from cholinephosphate cytidyltransferase and choline kinase genes and its application to industrial CDP-choline production: enymatic production of CDP-choline from orotic acid (part II). Biosci Biotechnol Biochem 1997; 61: 960-964.

[22] Fujio T, Nishi T, Ito S, et al. High level expression of XMP aminase in Escherichia coli and its application for the industrial production of 5'-guanylic acid. Biosci Biotechnol Biochem 1997; 61: 840-845.

[23] McElroy WD, Seliger HH, White EH. Mechanism of bioluminescence, chemiluminescence and enzyme function in the oxidation of firefly luciferin. Photochem Pchotobiol 1969; 10: 153-170.

[24] DeLuca M, McElroy WD. Kinetics of the firefly luciferase catalyzed reactions. Biochemistry 1974; 26: 921-925.

[25] Schneider DA, Gourse RL. Relationship between growth rate and ATP concentration in Escherichia coli: a bioassay for available cellular ATP. J Biol Chem 2004; 279: 8262-8268.

[26] Turina P, Samoray D, Graber DP. H^+/ATP ratio of proton transport-coupled ATP synthesis and hydrolysis catalyzed by CF_0F_1-liposomes. EMBO J 2003; 22: 418-426.

[27] Itoh H, Takahashi A, Adachi K, et. al. Mechanically driven ATP synthesis by F_1-ATPase. Nature 2004; 427: 465-468.

[28] Bowers KC, Allshire AP, Cobbold PH. Bioluminescent measurement in single cardiomyocytes of sudden cytosolic ATP depletion coincident with rigor. J Mol Cell Cardiol 1992; 24: 213-218.

[29] Koop A, Cobbold PH. Continuous bioluminescent monitoring of cytoplasmic ATP in single isolated rat hepatocytes during metabolic poisoning. Biochem J 1993; 295: 165-170.

[30] Di Tomaso G, Borghese R, Zannoni D. Assay of ATP in intact cells of the facultative phototroph Rhodobacter capsulatus expressing recombinant firefly luciferase. Arch Microbiol 2002; 26: 317-326.

[31] Stanley PE, Williams SG. Use of the liquid scintillation spectrometer for determining adenosine triphosphate by the luciferase enzyme. Anal Biochem 1969; 29: 381-392.

[32] Higashi T, Isomoto A, Tyuma E, Kakishita E, Uomoto M, Nagai K. Quantitative and continuous analysis of ATP release from blood platelets with firefly luciferase luminescence. Thromb Haemost 1985; 53: 65-69.

[33] Leist M, Single B, Castoldi AF, Kuhnle S, Nicotera P. Intacellular adenosine triphosphate (ATP) concentration: a switch in the decision between apoptosis and necrosis. J Exp Med 1997; 185: 1481-1486.

[34] Decaudin D, Marzo I, Brenner C, Kroemer G. Mitochondria in chemotherapy-inducted apoptosis: a prospective novel targret of cancer therapy. Int J Oncol 1998; 12: 141-152.

CHAPTER 9

Cytokinins: Progress and Developments in Analytical Methods

Jean Wan Hong Yong, Liya Ge and Swee Ngin Tan*

Natural Sciences and Science Education Academic Group, Nanyang Technological University, Singapore

Abstract: Since the discovery in the 1950s, cytokinins have been considered to play various important roles in life science. Many analytical procedures, therefore, have been developed for determination of the types, levels and metabolic profiling of endogenous cytokinins. The primary focus of this comprehensive review is on the various analytical methods designed to meet the requirements for cytokinin analyses in complex matrices with special emphasis on gas chromatography (GC), liquid chromatography (LC) and capillary electrophoresis (CE), mostly combined with mass spectrometry (MS). The advantages and drawbacks of the described analytical methods are discussed. As plant tissue contains cytokinins in trace amounts (usually at levels below 30 pmol per gram of fresh weight), the sample pre-treatment steps (extraction, pre-concentration and purification) for cytokinins are also reviewed. Finally, the present status and future trends of the analytical approaches are outlined.

INTRODUCTION

As a group of phytohormones, cytokinins exhibit a wide variety of bio-functions within plant growth and development, including cell division, cell differentiation, apical dominance, formation and activity of shoot meristem, induction of photosynthesis gene expression, inhibition of leaf senescence, nutrient mobilization, seed germination, root growth and stress response [1-2]. In addition, interesting therapeutic effects of cytokinins were reported in the relevant literature, such as the suppression of tumor growth [3-4], preventing blood clots [5], delaying the onset of human fibroblast ageing [6], and rescuing human mRNA splicing defect [7].

The majority of naturally occurring cytokinins are adenine derivatives (Table **1**), substituted at the N^6 position by either an isoprenoid side chain or an aromatic ring (designated isoprenoid and aromatic cytokinins, respectively) [1-2]. In each group, there are small variations in the side-chain structure between individual representatives, such as in the presence/absence of double bonds, additional hydroxyl or methyl groups, and their stereoisomeric positions [8]. In most cases, nucleosides, nucleotides, glucosides and other sugar-conjugates have also been found, implying that there is a metabolic network for their inter-conversion.

Table 1: Structures, names and abbreviations of cytokinins

R_1	R_2	R_3	Compound	Abbreviation
	H	H	*trans*-zeatin	Z
	R	H	*trans*-zeatin riboside	ZR
	G	H	*trans*-zeatin 9-glucoside	Z9G
	-	H	*trans*-zeatin 7-glucoside*	Z7G
	H	G	*trans*-zeatin *O*-glucoside	ZOG
	R	G	*trans*-zeatin riboside *O*-glucoside	ZROG
	RP	H	*trans*-zeatin riboside-5'-monophosphate	ZMP
	H	H	*cis*-zeatin	cZ
	R	H	*cis*-zeatin riboside	cZR
	G	H	*cis*-zeatin 9-glucoside	cZ9G
	H	G	*cis*-zeatin *O*-glucoside	ZOG
	R	G	*cis*-zeatin riboside *O*-glucoside	cZROG
	RP	H	*cis*-zeatin riboside-5'-monophosphate	cZMP

Table 1: cont....

Structure			Name	Abbreviation
	H	H	dihydrozeatin	DZ
	R	H	dihydrozeatin riboside	DZR
	G	H	dihydrozeatin 9-glucoside	DZ9G
	H	G	dihydrozeatin O-glucoside	DZOG
	R	G	dihydrozeatin riboside O-glucoside	DZROG
	RP	H	dihydrozeatin riboside-5'-monophosphate	DZMP
	H	-	isopentenyladenine	iP
	R	-	isopentenyladenine riboside	iPR
	G	-	Isopentenyladenine 9-glucoside	iP9G
	RP	-	isopentenyladenine riboside-5'-monophosphate	iPMP
	H	-	benzylaminopurine riboside	BA
	R	-	benzylaminopurine	BAR
	G	-	benzylaminopurine 9-glucoside	BA9G
	RP	-	benzylaminopurine-5'-monophosphate	BAMP
	H	-	*ortho*-topolin	oT
	R	-	*ortho*-topolin riboside	oTR
	G	-	*ortho*-topolin 9-glucoside	oT9G
	H	-	*meta*-topolin	mT
	R	-	*meta*-topolin riboside	mTR
	G	-	*meta*-topolin 9-glucoside	mT9G
	H	-	*para*-topolin	pT
	R		*para*-topolin riboside	pTR
	G	-	*para*-topolin 9-glucoside	pT9G
	H	-	kinetin	K
	R	-	kinetin riboside	KR
	RP	-	kinetin riboside-5'-monophosphate	KMP

*In Z7G, β-D-glucopyranosyl group is substituted at N^7.

H: hydrogen; R: β-D-ribofuranosyl; RP: β-D-ribofuranosyl-5'-monophosphate; G: β-D-glucopyranosyl.

Since plant tissue extracts represent rather complex multi-component mixtures, and typically cytokinins occur in trace quantities (usually at levels below 30 pmol g^{-1} of fresh weight [1-2]), it is necessary to apply operational and dependable extraction, pre-concentration and purification techniques, which prevent enzymes catalyzing metabolic conversions and the degradation of cytokinins, and also provide samples of sufficient purity for further analysis of the endogenous cytokinins [9-10].

Most importantly, studying cytokinins requires powerful analytical tools able to detect and identify trace amounts of these compounds in plant samples. For a brief review of analytical methods, please refer to [11]. Most early studies utilized radioimmunoassay (RIA) [12], enzyme-linked immunosorbent assays (ELISA) [13], gas chromatography (GC) [14-15] and high-performance liquid chromatography (HPLC) [15-16] for identification and quantification of cytokinins. However, unequivocal identification of cytokinin compounds using these methods presents difficulties such as cross-reactions of antibodies with structurally related substances and/or coelution of interfering compounds together with the analytes.

The development of user-friendly bench-up mass spectrometers has revolutionized analytical chemistry, enabling many laboratories to switch from fairly unspecific immunoassays and physicochemical techniques to a methodology with high sensitivity and selectivity [17-18]. Combined with MS detection, GC approaches of derivatized cytokinins specifically yield unambiguously identification [9]. Later on, liquid chromatography–electrospray ionization-tandem mass spectrometry (LC-ESI-MS/MS) under multiple-reaction monitoring (MRM) has attracted attention for the quantitative analysis of cytokinins, due to its selectivity, high sensitivity, and the fact that it does not require tedious derivatization steps [19-20]. Recently introduced ultra-performance liquid chromatography (UPLC) and nanoflow-LC combined with MS enable more sensitive analysis of cytokinins in plant materials [8, 21]. Moreover, capillary electrophoresis (CE), especially when couple to MS detector, has been approved to be a complementary method for cytokinin analyses [22-23]. This comprehensive review summarizes analytical methods currently available to investigators for the analysis of cytokinins. Section

2 gives an overview of current developments in sample preparation techniques, considered according to the extraction and different purification approaches; Section 3 describes various analytical methods including immunological, GC, LC and CE techniques in detail; and Section 4 outlines the concluding remarks and future prospects of cytokinin determination.

SAMPLE PREPARATION

Sample preparation is a key procedure in modern chemical analysis. It has been estimated that 60-80% of the work activity and operating cost in an analytical laboratory is spent on preparing samples for introduction into an analytical instrument. Because cytokinins are present in very low amounts in a wide range of structurally related compounds, the sample preparation methods are critical for the further quantification of endogenous cytokinins. The comprehensive procedures used for the extraction, pre-concentration and purification of cytokinins are outlined in Fig. **1**.

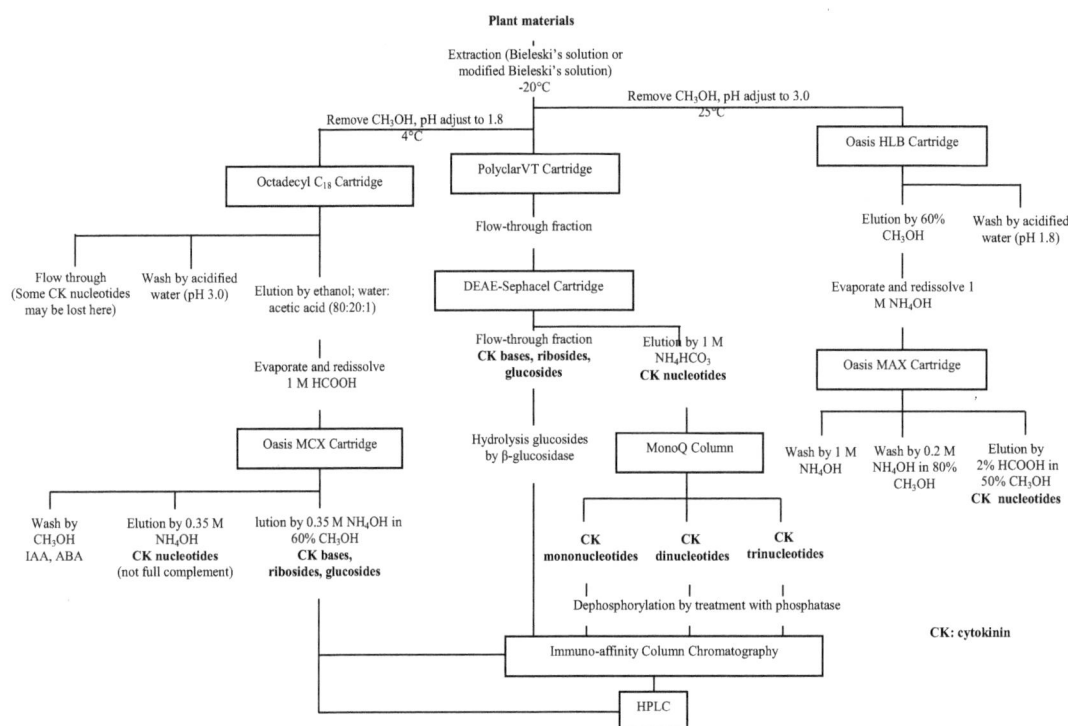

Figure 1: The comprehensive procedures used for extraction, pre-concentration and purification of cytokinins.

Extraction Techniques

The qualitative and quantitative aspects of cytokinins extracted from plant tissues vary with the types of extraction solvent and procedure employed [14]. In order to prevent any enzymatic degradation of cytokinins including conversion, e.g. conversion of cytokinin nucleotide to its riboside, plant material should be immediately frozen after harvesting. For further identifying and quantifying cytokinins and their metabolites in plant tissue, the samples must instantly be homogenized and extracted with a suitable solvent [14]. The isolation and identification of picomolar quantity of cytokinin (less than 30 pmol g^{-1} of fresh weight) are often hampered by the presence of an excess amount of polyphenols, carbohydrates, terpenoids, and other impurities in plants cells.

For a typical extraction, plant tissue is frozen in liquid nitrogen and dropped into a mixture of methanol/chloroform/water/formic acid (12/5/2/1, v/v/v/v), which is known as Bieleski's solvent, using 10 ml of solvent per g (fresh weight) of tissue [24]. The tissue is allowed to stand in the solvent for 18 h at -20°C. It is important that the tissue is allowed to stand long enough for the solvent to penetrate it before homogenized in methanol/90% formic acid/water (6:1:4, v/v/v) at 0°C using 5 ml of solvent per g (fresh weight) of tissue.

With the comparison of the extraction efficiency of three different extraction solvents: (a) 80% (v/v) methanol, (b) Bieleski's solvent and (c) modified Bieleski's solvent (methanol/water/formic acid; 15/4/1, v/v/v), it was found that the modified Bieleski's solvent significantly prevented dephosphorylation of cytokinin mononucleotides and gave high yields of cytokinins in plant extracts [25]. Therefore, it is considered to be the most suitable solvent for extraction of cytokinins.

Liquid-liquid Partition

During liquid-liquid partition, the plant material is normally homogenized by pestle and mortar in liquid nitrogen [26-27]. Cytokinins are then extracted with cold acetone (three times within 24 h). After centrifugation (8000 rpm, 10 min) the supernatant is evaporated in vacuo to dryness. The residue is dissolved in lukewarm (38°C) distilled water acidified to pH 3.5 with HCl. A triple extraction of an aqueous solution of cytokinins with butanol saturated with water acidified to pH 3.5 removes chlorophyll and other impurities, along with traces of cytokinins. These traces of cytokinins are then recovered by separating the butanol layer and back-extracting it three times with water acidified to the same pH. The acidified aqueous layers containing cytokinins are combined and then neutralized to pH 7 with KOH. A subsequent multiple extraction of this aqueous layer with alkaline butanol (butanol/ammonia [40%], 9:1) served to transfer the cytokinins into the butanol phase and free them of salts and other mixtures, which remain in the aqueous layer. The final butanol layer, containing the cytokinins, was isolated and evaporated to dryness for further purification and/or analysis.

Solid-phase Extraction

Solid-phase extraction (SPE) has been one of the main sample pretreatment techniques for extraction and clean-up of phytohormones from various plants [23, 28-30]. One of the distinct advantages of SPE method is that a high extraction recovery can usually be obtained for many compounds with a suitable sorbent and operating procedure, even under situations when other traditional extraction techniques, such as liquid extraction, may not be suitable.

Pre-concentration of the cytokinins has commonly been achieved using SPE with C_{18} cartridges [31]. A more effective but more complex approach is to purify plant extracts by passage through linked columns of polyvinylpolypyrrolidone power (PVPP), DEAE–cellulose or DEAE-Sephacel, and C_{18} SPE [10]. Fast and efficient separation of cytokinins could be achieved by using mixed-mode SPE bearing both the reverse-phase and ion-exchange characteristics [28]. In this type of SPE approach, the C_{18} SPE cartridges were successfully used as a pre-concentration tool, while further sample purification was carried out using mixed-mode cation exchanger (MCX) SPE cartridges [23]. High extraction recoveries were obtained for cytokinin bases, ribosides and glucosides with this SPE approach [23, 28]. Purification of cytokinins using mixed-mode-SPE, as compared to DEAE Sephadex and C_{18} SPE method, was suitable for the removal UV absorbing contaminants with higher extraction recoveries of cytokinins [25].

However, due to the relatively high polarity of cytokinin nucleotides and thus leading to poor retention by C_{18} cartridges during SPE, use of an anion-exchange sorbent is an efficient alternative step for the separation of cytokinin nucleotides from cytokinin bases and sugar conjugates. Therefore, an efficient dual-step SPE method was been developed for the pre-concentration and purification of cytokinin nucleotides using HLB and MAX cartridges [23]. A method through a series of anion-exchange column chromatography steps was also developed to discriminate the various nucleotides, including cytokinin mono-, di- and tri- nucleotides [29]. In the above paper, cytokinin nucleotides fraction gotten from anion-exchange SPE (PolyclarVT cartridge and DEAE-Sephacel cartridge) further separated into different mono-, di- and tri- nucleotide fractions using a MonoQ column with HPLC.

Immunoaffinity Purifications

Immunoaffinity purification approach has been shown as another feasible procedure for trace analysis of cytokinins in biological samples. It is known that immunoaffinity purification methods, which are based on antibody-antigen interactions, can provide selective sample enrichment, and thus greatly lower detection limits of trace analyses [32]. Therefore, when an immunoaffinity approach is used as purification step before final analysis, highly purified cytokinin preparations containing only traces of other UV-absorbing material could be obtained.

Immunoaffinity chromatography (IAC) based on generic polyclonal and monoclonal antibodies have been applied to get samples of high purity which contain unusual amount of very different cytokinin and their metabolites [16, 32-33]. The set-up a suitable affinity system requires an appropriate antibody (high affinity, rapid binding of cytokinin, resistance to harsh elution conditions, reusability), and consideration of matrix factors that are not specific to phytohormone analysis [33]. Generic cytokinin monoclonal antibodies are frequently used in this respect [32, 34]. Immunoaffinity columns purify according to structural similarities, and thus hold the promise to trap as yet unknown cytokinins. Examples for this have already been reported [16, 33], and some new physiologically active compounds may well turn up. The considerable degree of single-pass enrichment allows for purifying these metabolites for structure determination [14, 35]. IAC has higher selectivity than conventional SPE, but the throughputs offered by previous IAC procedures for cytokinins have generally been low.

An efficient off-line batch immunoextraction method was developed and optimized for the purification of new cytokinins and their corresponding ribosides [32]. The sensitivity of the assay could be significantly enhanced by including an immunoaffinity chromatography purification step. The combination of simple C_{18} SPE with batch IAE provides fast, easy to use and cost-effective technique for routine samples processing.

High-performance Liquid Chromatography

As plant extracts often contain multiple cytokinins, HPLC is commonly performed as a final step in sample preparation immediately preceding analysis [15]. Indeed, HPLC enables a rapid, high-resolution purification of cytokinins from plant extracts prior to analysis by MS, immunoassay, or bioassay. Reversed-phase (RP) columns (e.g. C_{18}) are the most commonly used for cytokinin fractionation. For preparative purification of cytokinins, a column size of 150×10 mm i.d. can be a good compromise between cost and sample loading capacity [14]. Due to the minimal sample volume, HPLC conventional analytical columns with 4.6 mm i.d were also widely used for separation and fractionation of cytokinins [36-37]. Acidic aqueous buffer and organic solvent (i.e. methanol or acetonitrile) are most widely used mobile phases for cytokinins. The gradient elution is most often used for HPLC analyses of cytokinins [37-38]. An alternative approach for cytokinin separation is to use the more economical isocratic eluation program [36]. Peaks were usually identified from their retention times by comparison with standards.

Furthermore, to increase selectivity and reliability, a comprehensive two-dimensional HPLC has been developed for fast and efficient purification of multiple phytohormones followed by MAX SPE [39]. It is certain that multidimensional HPLC can well separate cytokinins and provides clean enough fractions for further confirmation of identities.

ANALYTICAL TECHNIQUES FOR CYTOKININS

Immunological Techniques

Due to their low detection limits, immunological techniques used to be methods of choice for analysis of trace cytokinins before hyphenated techniques such as LC-MS or GC-MS have been widely applied [40]. When individual HPLC fractions of plant extracts are analyzed by immunological methods, the main disadvantage - cross-reactivity of the antibodies - could be partially overcome [12-13, 34]. Nevertheless, immunological still is a time-consuming method which involves many steps and several reactants.

Immunoassays

Over the last several decades several quantitative immunoassay methods have been developed to measure cytokinins. Immunoassay techniques including, RIA, ELISA and scintillation proximity assay (SPA) can be used as a sensitive and viable alternative for the determination of cytokinins [12-13, 41-42].

RIAs are very sensitive, and able to detect nano- or picomoles of molecules and have provided a large amount of information on the biochemical processes dealing with ligand-receptor systems. For a detailed RIA protocol, please refer to Weiler [43]. Due to the strong cross reactions, polyclonal antibodies for RIA can not be used directly on individual cytokinins in crude extracts from plant materials. Substances interfering with the RIA (e.g. phenols), therefore, should be removed from the crude plant extracts [44]. Moreover, the crude extract preferred to be further separated by HPLC and the resulting LC fractions can be used to determine the levels of individual cytokinins by RIA [45]. Because of the radioactive waste resulting from such assays, many molecules are now assayed with non-radioactive immunoassays and other techniques.

Compared with RIA, ELISA is less expensive and easier to set up, and moreover the problems associated with disposal of radioactive waste could be avoided. For the detection of cytokinins, avidin-biotin amplified ELISA, immunoaffinity purification and immunocytochemical techniques were developed [13, 46-48]. The use of an ELISA allowed the detection of these cytokinins over the range of 0.3 to 7 pmol for the isopentenyladenine-type and 1 to 1000 pmol for the zeatin-type [46]. Hapten-homologous and hapten-heterologous competitive ELISAs were developed for detecting endogenous cytokinin levels in crude plant extracts without intense purification, which allowed the use of minute amounts of plant extracts for cytokinin analysis [48].

SPA is a novel radioisotopic technique, applicable to assays involving ligand-antibody binding, which eliminates the need to separate free and bound ligand and to use scintillation fluid as required in conventional RIA [41-42]. SPA was first described by Wang *et al.* [41] as a sensitive assay for quantification of cytokinins as free bases. The lowest detectable amounts of iP, Z and S-DZ were < 0.01, 0.01 and 0.02 ng, respectively. Yong *et al.* used the above SPA method to measure the level of xylem-derived cytokinins entering a cotton leaf, and the cytokinin levels in the same leaf [42]. Unfortunately, until now SPA has rarely been used for analysis of cytokinins. However, as the precision of SPA is superior to the heterogeneous RIA assays due to fewer manipulative processes including lack of centrifugation, SPA may become useful for analysis of cytokinins in different plant materials in the future.

Due to the strong cross-reactivity, the different cytokinins normally have to be separated prior to immunoassays. A disadvantage of the cytokinin immunoassays, compared with LC-MS, is that they provide combined estimates of the contents of groups (free bases, ribosides, 9-glucosides and nucleotides) of cytokinins, due to their lower specificity, rather than estimates of specific cytokinins [12-13, 41-42]. On the other hand, the values (cytokinin equivalents) obtained from cytokinin immunoassays allow for several important cytokinins to be screened collectively by group-specific antibodies in simple plant materials (microalgae) [49]. The effective range and sensitivity of the immunoassays are still similar to those noted for LC-MS.

Immunolocalization of Cytokinins

Despite considerable problems which will still have to be overcome, a few reports have shown that antibodies may be useful tools to trace cytokinins in plant tissues and even cells at the ultrastructural level. Immunolocalization of endogenous cytokinins provides a complementary vision of their involvement in morphogenic processes [50-52].

Immobilisation by chemical fixation or freezing minimizes diffusion of the low molecular weight cytokinin compounds in plant tissues. Associated primary antibodies in sections or permeabilised cells can be detected by secondary antibodies linked to enzymes, fluorescent molecules or electron opaque markers, which allow detection by either light or electron microscopy [50-51]. Immunolocalization techniques have already found their application in various studies related to the physiology of these cytokinins. With immunolocalization studies, it is expected that compare the localization of cytokinins at the tissue and subcellular level could be compared [52].

Gas Chromatography

Since the early 1970's, GC-based methods have been used as a suitable analytical technique for the separation and quantification of cytokinins [14-15].

A derivatization step of the analytes is always needed, because underivatisated cytokinins are not volatile compounds, which are not suitable for the direct GC analysis. Derivatization methods have some inherent technical problems, such as hydrolysis of the derivatives, multiple-products formation, and limited volatility [53-54]. However, GC-MS was a reliable and specific means for the identification and quantification of cytokinins until the recent introduction of using a combination of HPLC with mass spectrometers. Structures of almost all naturally occurring cytokinins were elucidated by GC-MS before the 1990s [14-15].

Derivatization Methods

Appropriate and stable derivatization of cytokinins is critical for successful GC analysis. Generally, derivatization for GC-MS analyses is performed with N – methyl – N - (trimethylsilyl)trifluoroacetamide (MSTFA) and aims at increasing volatility and thermal stability of the cytokinin compounds. Briefly, derivatization procedure is silylation (trimethylsilyl) with MSTFA in 50% pyridine containing 1% trimethylcholorsilane for hours at room temperature, or a shorter time at 80°C [15]. The detection limit is about 10 μg of anhydrous cytokinins even if the sample is not completely derivatized.

Other complications associated with derivatisation, such as trimethylsilyl [14, 55], permethyl [56], *t*-butyldimethylsilyl [57], trifluoroacetyl [58] and acetyl [9] derivatives, were also reported. Relevant information on derivatization studies is briefly summarized in Table **2**. Pentafluorobenzyl derivatives of cytokinin free bases for negative ion MS were reported by Hocart *et al.* [59]. A novel method with selection of the N – methyl – N - (*tert.*-butyldimethylsilyl)trifluoroacetamide (MTBSTFA) reagent as the most comprehensive chemical derivatization protocol for the GC-MS analysis of cytokinins and other phytohormones [60].

Table 2: Typical GC approaches for the analysis of cytokinins.

Method	Derivatization	Cytokinins	LOD	Ref
GC-FID	Trimethylsilylation	iP, iPR, K, DZ, Z, KR, ZR	5 ng	[55]
GC-MS	Trimethylsilylation, permethylation	Glucosides of BA and Z	NS	[56]
GC-MS	*t*-Butyldimethylsilylation	cZ, Z, DZ, iP, BA	NS	[57]
GC-ECD	Trifluoroacetylation	iP, iPR, K, BA, Z, DZ, ZR	pg range	[58]
GC-MS	Acetylation	more than 30 cytokinins	ng range	[9]
GC-MS	Pentafluotobenzylation, *t*-Butyldimethylsil-ylation	cZ, Z, DZ, iP	NS	[59]
GC-MS	MTBSTFA	mT, Z	0.3, 0.9 pmol	[60]

However, GC-MS analysis is always associated with some inherent drawbacks for most of the commonly used derivatives. For example, trimethylsilyl [14, 55], trifluoroacetyl [58] and *t*-butyldimethylsilyl [59] derivatives were subjected to hydrolysis [61]. Acetylation provided stable and easy to prepare derivatives, but their volatility was not satisfactory [9].

Chromatographic Conditions

The separation conditions have not changed much since 1970s, although currently fused-silica capillary columns are used instead of packed glass columns. The non-polar stationary-phases such as BP1 (methylsilicone bonded phase) [61], DB-5 (5% phenylmethylpolysilaxane) [62], PTE-5 (Supelco) [63] are usually used. Common internal diameters of fused-silica capillary columns range between 0.2 and 3 mm coated with 0.1 to 2 μm films of stationary phases depending on amount of material to be injected into the column. Column oven heating rates are often 4-8°C/min between 150 and 310°C with total time of separation approximately 20 min.

Detection

The separation power of capillary GC and selectivity of MS detection make GC-MS a powerful technique for cytokinin analyses [15, 64]. The main advantage of GC-MS-based identification, when compared with soft-ionization techniques is differentiation of various sugar moieties. Moreover, GC-MS has been demonstrated as robust and well-established approach in plant metabolomics study [65]. The mass spectral library databases and deconvolution software are available for extracting meaningful information. Therefore, the GC-MS based techniques will enable us to identify unknown compounds (e.g. novel cytokinin metabolites) in plant tissues after derivatization.

Besides MS detection, the use of two other types, namely flame ionization detector (FID) and electron capture detector (ECD), were been reported [9, 56]. The main drawback of FID is the limited sensitivity which makes it inapplicable in trace analysis of cytokinins. Even though ECD is quite sensitive, but halogenated derivatives are required in order to obtain satisfactory detection limits.

Liquid Chromatography

In contrast to GC, LC usually does not require tedious derivatization steps. In addition to be the purification method for cytokinins, LC is a particularly suitable chromatographic technique for cytokinin analyses as they exhibit gradations in polarity and are readily detected by UV absorbance [66]. However, as absorbance at a single UV wavelength is inadequate for this purpose, the most widely used procedure for quantification of cytokinins is an isotope dilution MS, especially LC-ESI-MS [34, 67].

Chromatographic Conditions and Separation

Cytokinin free bases and their sugar conjugates are relatively hydrophobic compounds which behave as weak bases, therefore well separated on RP columns under acidic conditions [14]. However, the more ionic cytokinin nucleotides are not so well separated by RP-HPLC. Typically the nucleotides are converted to ribosides with alkaline phosphatase [67], or derivatized [29] to lower their polarity to improve retention and resolution on RP-type columns.

Like HPLC purification, analyses of cytokinins by LC are carried out using RP-C_{18} or C_8 columns [34, 37, 68-69]. The volatile eluent additives such as acetic or formic acid and their ammonium salts are usually added to solvents containing aqueous methanol/acetonitrile [37, 68-69]. To achieve good separation, gradient elution by increasing content of organic modifier is often used [68-70]. Analytical HPLC columns with i.d. ranging from conventional (4.6 mm), narrow-bore (2.1 mm), micro (1 mm) and capillary columns (0.3 mm) [37, 68-70] are used for cytokinin analyses.

It is well known that the sensitivity of LC–ESI-MS can be increased by lowering the flow rate of the mobile phase, which can be achieved by using narrow diameter LC columns [71]. Application of new analytical methods makes possible new direction in cytokinin research. UPLC is new technique using higher pressure (up to 1000 bar) and smaller particulate packing materials (1.7 μm). Therefore, it extends the limits of what has been, and is more achievable than conventional HPLC instrumentation. UPLC retains the practicality and principles associated with that conventional HPLC, whereas it could achieve higher resolutions, lower sensitivities, and rapid separations [72]. Representative UPLC chromatograms of cytokinins were shown in Fig. **2**. Compared to the HPLC analysis [34], UPLC analysis was 4-fold faster [8]. The highly sensitive nanoflow-LC has also been applied for comprehensive analysis of phytohormones [21].

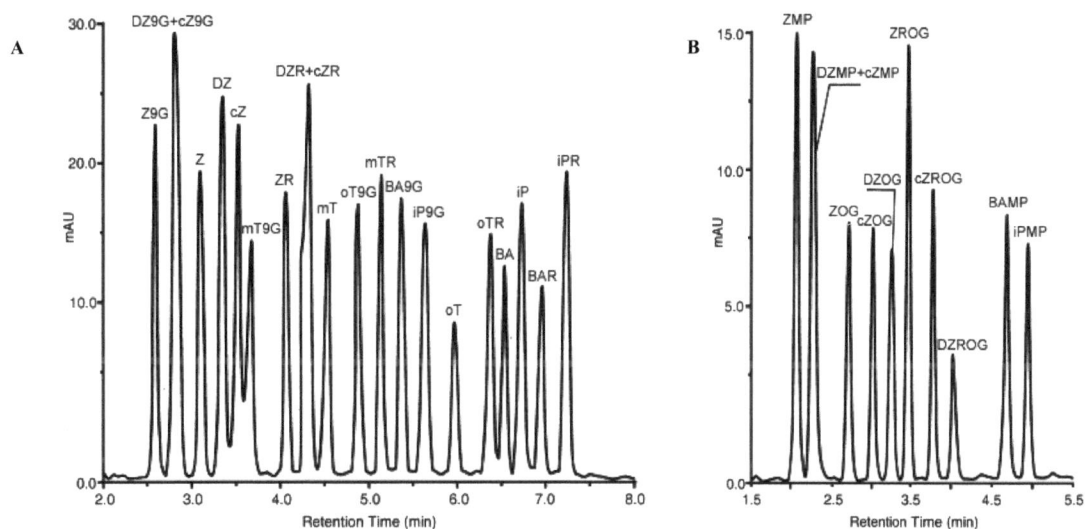

Figure 2: UPLC-UV chromatogram (268 nm) of (A) 21 cytokinin bases, ribosides, and 9-glucosides; and (B) 11 cytokinin O-glucosides and nucleotides containing 10 pmol of each derivative per injection. Adapted from [8], with permission.

Detection

As cytokinins exhibit strong UV absorbances between 220 and 300 nm, UV detection is suitable for their quantification [14]. Coincidently, UV-VIS absorbance detector is the most widely used detector for LC. Moreover, additional spectral information of cytokinins can be obtained on individual peaks or portions of individual peaks as they elute from a LC column. Therefore, LC-UV is widely used to separate and detect cytokinins, including fractionation for bioassays, immunoassays or volatile derivatives for GC. However, using this non-specific UV absorbance method of detection requires significantly higher amount of sample which needs extensive purification.

The LC-MS approach offers a new tool to detect, quantify and characterize cytokinins in plant tissue extracts at biologically meaningful levels. Table **3** presents a summary of the most representative LC-MS methods for cytokinin analyses. Further improvement of LC systems as well as mass analyzers (in the form of increasing ion transport efficiency) may overcome the low detect ability of cytokinins. Different ionization techniques were used for MS analyses of cytokinins in combination with RP-HPLC, including thermospray (TS) [73], fast atom bombardment (FAB) [74-77], atmospheric pressure ionization (API) [78-79], and electrospray ionization (ESI) [8, 19-20 34, 67, 77-78, 80-83].

Table 3: Typical LC-MS analytical methods for the determination of cytokinins

Analytes	Sample Matrix	Sample preparation	LC conditions				MS conditions		Ref
			column	Mobile phase	Flow rate	Elution	Source	Analyzer	
Z9G, DZ9G, cZ9G, Z, DZ, cZ, mT9G, ZR, DZR, cZR, mT, oT9G, mTR, BA9G, iP9G,oT, oTR, BA, iP, BAR and iPR ZMP, DZMP, cZMP, ZOG, cZOG, DZOG, ZROG, cZROG,DZROG, BAMP abd iPMP	*Arabidopsis thaliana,* ecotype Colombia, seedlings; Leaves of *Populus × canadensis* Moench, cv. *Robusta*	SCX or C18 SPE, and IAE.	BEH C18 (1.7 μm; 50 mm × 2.1 m m)	A: 15 mM ammonium formate (pH 4.0); B: MeOH	0.25 ml min⁻¹	gradient 0-8 min, 10-50% B	ESI	Triple quadrupole	[8]
6-(2-hydroxy-3-methoxybenzylamino)purine riboside, 6-(2,4-dimethoxybenzylamino) purine riboside and other 6-benzyladenosine derivatives	*Arabidopsis thaliana, A. tumefaciens*	C₁₈ SPE, IAC for *A. thaliana* octadecylsilica SPE, DEAE-Sephadex SPE, C₁₈ SPE, and immunoextraction for *A. tumefaciens*	C₁₈ column (1.7 μm; 150 mm × 2.1 m m) Symmetry C18; 5 μm; 0.3 × 150 mm	A: MeOH; B: 5 mM formic acid. A: 5 mM formic acid and 2% MeOH; B: MeOH with 0.05% formic acid.	0.25 mL min⁻¹ 5 μl min⁻¹	gradient 0-9 min, 30-64% A. gradient 0-15 min, 70-0% A.	ESI ESI	Triple quadrupole Q-TOF	[19]
ZMP, Z7G, Z9G, ZOG, DZ9G, Z, ZROG, ZR, DZ, DZR, iPMP, iP and iPR	Conifer Tree *Abies nordmanniana*	C₁₈ SPE	Zorbax XDB column (3.5 μm, 2.1 × 150 mm)	A: 0.1% acetic acid with pH adjusted to 8.0 with NH₃; B: MeOH	NS	gradient 0-15 min, 23-57% B.	ESI	Triple quadrupole	[20]
Z9G, DZ9G, Z, DZ, cZ, mT9G, ZR, DZ, cZR, mT, oT9G, mTR, BA9G, iP9G, oT, oTR, BA, iP, BAR and iPR	Cytokinin-autonomous tobacco BY-2 cell, fully expanded leaves of *Populus×canadensis* Moench., cv *Robusta*	two ion-exchange chromatography steps, SCX SPE, DEAE - Sephadex SPE, C₁₈ SPE and IAC.	Symmetry C₁₈ column (5 μm, 150 mm × 2.1 m m)	A: 15 mM formic acid adjusted to pH 4.0 by ammonium hydroxide; B: MeOH.	250 μl min⁻¹ (50% effluent was introduced into ESI source)	gradient 0-25 min, 10-50% B; 25-30 min, 50% B.	ESI	Single quadrupole	[34]
Propionylated iP, Z, DZ, ZR, iP, ZMP, iPMP, ZOG, Z7G, Z9G, ZROG and DZR	*Arabidopsis thaliana* wild-type variant Colombia plants	C₁₈ SPE and MCX SPE	10 × 1 mm BetaMax Neutral drop-in guard cartridge with 5-μm particles	A: 3% formic acid; B: 3% formic acid in ACN	0-13min 10 μl min⁻¹; 13.1-17 min 40 μl min⁻¹; 17.1-19 min 20 μl min⁻¹; 20 min 10 μl min⁻¹.	gradient 0-2 min, 5% B; 2-10 min 5-55% B; 10-12 min,55% B; 12.1-13 min, 80% B; 13.1-17 min, 100% B; 17.1 -20 min, 5% B.	ESI	Triple quadrupole	[67]
Z, DZ, ZR, DZR, Z9G, DZ9G, iP and iPR	transgenic homozygote and hemizygote as well as wild-type *Nicotiana tabacum* species, and cauliflower samples	DEAE–cellulose SPE and C₁₈ SPE or IAC.	LUNA C₈ column (5 μm, 15 cm× 1.0 mm).	A: 10 mM ammonium acetate; B: MeOH	60 μl min⁻¹	gradient 0-20 min, 10-80% B.	ESI	Triple quadrupole	[69]
Derivatized Z, iP, ZR, iP, ZMP, iPMP, ZOG, Z7G, Z9G and ZROG	*Arabidopsis thaliana*	strong cation-exchange cartridge (SCX) DEAE-Sephadex anion exchanger C18 cartridge IAC	Capillary LC column (150 × 0.3 mm packed with 4 μm Symmetry ODS packing material).	A: 98% water, 1% formic acid and 1% glycerol; B: 97% ACN, 1% water, 1% formic acid and 1% glycerol.	4 μl min⁻¹	gradient 10-12 min, 20-50% B; 12-40 min, 50% B.	frit-FAB	Double focusing magnetic sector	[74]
Propionylated oTOG, 2MeSoTOG and BA9G	*Chenopodium rubrum* cells	SCX SPE and C₁₈ SPE. Alkaline phosphatase treatment of nucleotides.	Capillary LC column (150 × 0.3 mm packed with 4 μm Symmetry ODS packing material).	50% aqueous ACN, 1% glycerol, 1% formic acid	0-5 min, 20 μl min⁻¹; 5-45 min, 4.5 μl min⁻¹.	isocratic	frit-FAB	Double focusing magnetic sector	[76]
pT, mT, pTR, oT, mTR, BA, MeoT, MemT, oTR, BAR, MeoTR and MemTR	*Arabidopsis thaliana* and *Populus × canadensis* leaves	octadecylsilica SPE, DEAE-Sephadex SPE, C₁₈ SPE, and IAC. Alkaline phosphatase treatment of nucleotides.	Symmetry C₁₈ column (3.5 μm, 150 mm × 2.1 m m).	A: MeOH; B: 0.1% formic acid adjusted to pH 2.9 with ammonium	0.25 ml min⁻¹ (25% effluent was introduced into ESI source)	gradient 0-5 min, 30-15% A; 5-25 min, 15-40% A; 25-30 min, 40% A.	ESI	Single quadrupole	[77]

Table 3: cont....

Analytes	Sample Matrix	Sample preparation	LC conditions				MS conditions		Ref
			column	Mobile phase	Flow rate	Elution	Source	Analyzer	
Propionylated pT, mT, pTR, oT, mTR, BA, MeoT, MemT, oTR, BAR and MeoTR			Capillary LC column (150 × 0.3 mm packed with 4 µm Symmetry ODS packing material)	55% aqueous ACN, 1% glycerol, 1% formic acid.	0-5 min, 20 µl min⁻¹; 5-65 min, 4.0 µl min⁻¹.	isocratic	frit-FAB	Double focusing magnetic sector	
Z, Z, Z9G, ZOG, ZROG, ZMP, DZ, DZR, DZ9G, DZOG, DZROG, DZMP, iP, iPR, iP9G, iPMP, BA, BAR, BA9G, BAMP, mT, and oT	*Physcomitrella patens*	DEAE-Sephadex SPE, octadecylsilica SPE and IAC	BEH C18 (1.7 µm; 150 mm × 2.1 m)	A: 15 mM ammonium formate (pH 4.0); B: MeOH.	0.25 ml min⁻¹	gradient 0-10 min, 10-50% B.	API	Triple quadrupole	[78]
Z, ZR, Z7G, Z9G, DZ, DZR, iP, iPR, iP9G, ZOG and ZROG	*Macadamia integrifolia*	C₁₈ SPE Alkaline phosphatase treatment of nucleotides.	C₁₈ column (3 µm, 20 mm × 2.1 m) Guard column (C₁₈, 5 µm, 4 mm × 2.0 mm)	A: 10 mM ammonium acetate; B: 350 ml MeOH, 100 ml ACN, 50 ml 10 mM ammonium acetate	0.1 ml min⁻¹	gradient 0-3 min, 5-10% B; 3-27 min, 10-43% B; 27-30 min, 43-80% B; 30-33 min, 80-100% B.	API	Q-TOF	[79]
Propionylated iPMP, iPR, ZMP, cZMP and cZR	pea roots	SCX SPE, DEAE-Sephadex SPE, C₁₈ SPE, and IAC. Alkaline phosphatase treatment of nucleotides.	Symmetry C₁₈ column (5 µm, 150 mm × 0.3 m m).	A: 15 mM ammonium formate (pH 4.0); B: MeOH.	5 µl min⁻¹	gradient 0-25 min, 10-50% A.	ESI	Q-TOF	[80]

NS, not state

Although the use of frit-FAB MS has been reported [51-54], ESI-MS is currently the most common for LC-MS method in cytokinin analyses. Compared to the one involving frit-FAB LC-MS system, the sensitivity of ESI is fairly high with lower background [8, 19-20 34, 67, 77-78, 80-83]. In 1997, the first application of LC-ESI-MS/MS with MRM for cytokinin determination was reported as a fast method for the quantification of 16 different cytokinins with a detection limit of 1 pmol [82]. Subsequent improved gradient elution together with capillary column provided a detection limit at the low femtomolar level [83].

It is obvious that the main advantage of the LC-MS approach over that of the GC-MS is the elimination of a derivatization step. In order to increase sensitivity, however, pre-column derivatization for LC-MS cytokinin analyses was used to give stronger quasi-molecular ion currents and to obtain more spectral information [67, 80, 74-77].

As expected, UPLC combined with MS, which provides significant advantages concerning selectivity, sensitivity and speed, is undoubtedly a suitable system for the study of cytokinins [8, 19, 78]. Schwartzenberg *et al.* successfully applied an efficient UPLC-MS/MS method to establish the cytokinin profile of the bryophyte *Physcomitrella patens*, which is possible to simultaneously analyze 40 cytokinins [78]. Doležal *et al.* applied UPLC-MS/MS method to isolate new cytotoxic members of the aromatic cytokinin family present endogenously in some living organisms [19]. Several hydroxymethoxybenzyladenosines as well as dimethoxybenzyladenosines were detected and tentatively identified from their UPLC retention times, antibody cross-reactivities and specific MRM diagnostic transitions in *Arabidopsis thaliana* as well as *Agrobacterium tumefaciens* extracts.

Capillary Electrophoresis

CE is one of the most powerful separation techniques particularly suitable for the analyses of cytokinins, due to its speed, high resolving power, and minimal sample and buffer requirement [22-23, 30]. Although some successful examples have been reported, the detection limit of CE is somewhat higher when compared with HPLC and GC, which is a consequence of lower amounts of sample injected during analysis and also in having a shorter optical path length. Therefore, most CE applications in cytokinin analyses require an enhancement of the detection sensitivity by using more specific detection systems, e.g. MS, and by on/off-line sample pre-concentration to increase sample solute concentration [22-23]. A review by Ge *et al.* [84] provides some comprehensive information pertaining to this aspect of cytokinin analyses.

Electrophoretic Conditions and Separation

Compared with HPLC or GC, the CE separation of cytokinins is not widely described, probably due to the very low concentrations of free cytokinins present [84]. As it is well known, without changes in the instrumental hardware, CE separations can be carried out using several different operation modes. The different CE modes used for the separation of cytokinins are capillary zone electrophoresis (CZE) and micellar electrokinetic chromatography (MEKC). Table **4** outlines the optimum electrophoretic conditions used for the analyses of cytokinins in various biological samples.

Table 4: The optimum separation conditions for cytokinins using different CE approaches [84-85]

Analytes	Sample Matrix	Sample preparation	Mode	Buffer	Capillary dimensions	Separation voltage	Injection	Detection	Reference assay
iP, iPR, Z, ZR, DZ,DZR, BA and BAR	STD sugar beet	SPE	CZE	150 mM phosphoric acid, pH 1.8.	77 cm (effective length 61 cm)×75 µm.	20 kV	10 mbar, 0.05 min.	UV 265 nm	NS
	STD, tobacco		MEKC	20 mM SDS, 50 mM borate, pH 9.2.			10 mbar, 0.1 min.		
BA, BA9G, BAR, mTR, oTR, KR, ZR, DZR, iP, and iPR	STD	NS	CD-modified CZE	100 mM phosphate-Tris (pH 2.5) buffer with 25 mM γ-CD.	47 cm (effective length 40 cm)×50 µm.	20 kV	NS	UV 200 nm	NS
BA, K, and other phytohormones including ABA, IAA, NAA, GA and 2,4-D	STD, transgenic tobacco flower	LLE	MEKC	50 mM borate containing 50 mM SDS, pH 8.0.	48.5 cm (40 cm effective length)×50 µm.	15 kV	5 s at 50 mbar.	UV 210 nm	NS
Z, DZ, ZOG, DZOG, mTR, iP and BA	STD, coconut water	dual-step SPE	MEKC	A combination of 10 mM phosphate and 10 mM borate buffer containing 50 mM SDS, pH 10.4.	57 cm (effective length 47 cm)×76 µm.	15 kV	5 s under a pressure of 0.5 psi.	UV 254 nm	HPLC, LC-MS
oT, mT, pT, oTR, mTR and pTR	STD, coconut water	dual-step SPE	MEKC	20 mM boric acid and 50 mM SDS, pH 8.0, with an extra 20% (v/v) MeOH added.	60 cm (effective length 50 cm)×76 µm.	15 kV	5 s under a pressure of 0.5 psi.	UV 269 nm	HPLC, LC-MS/MS
K and KR	STD, coconut water	dual-step SPE	CZE	100 mM ammonium phosphate buffer, pH 2.5.	40 cm (effective length 30 cm)×76 µm.	15 kV	5 s under a pressure of 0.5 psi.	UV 269 nm	HPLC, LC-MS/MS
iP, DZ, Z, BA, K, oT, DZOG, ZOG, DZR, ZR, oTR and KR	STD, coconut water	dual-step SPE	CZE	25 mM ammonium formate/formic acid buffer (pH 3.4) and 3% ACN (v/v).	65 cm ×50 µm.	25 kV	on-line sample stacking injections	MS, MS/MS	NS
DZMP, iPMP, cZMP, ZMP, BAMP and KMP	STD, coconut water	dual-step SPE	CZE	25 mM ammonium formate/formic acid buffer (pH 3.8) and 2% MeOH (v/v).	57 cm ×50 µm.	gradient separation voltage (25 kV for 32 min, and then a linear gradient to 30 kV in 5 min, finally 30 kV to the end of separation)	on-line sample stacking injections	MS/MS	NS
oT, mT, pT, oTR, mTR, pTR, oT9G, mT9G, pT9G, ZR, cZR, Z and cZ	STD, banana pulp	dual-step SPE	Partial filling-MEKC	50 mM ammonium formate/ammonium hydroxide at pH 9.0. Micellar solution with 70 mM ALS and 10% MeOH was injected for 90 s at 50 mbar before the sample and 120 s at 50 mbar after the sample.	100 cm × 50 µm i.d.	20 kV	on-line sample stacking injections	MS, MS/MS	NS

STD, Standard; NS, not stated; GA, gibberellic acid; ABA, abscisic acid; IAA, indole-3-acetic acid, NAA, α-naphthaleneacetic acid, 2,4-D, 2,4-dichlorophenoxyacetic acid.

Tacking into account of electrophoretic conditions, the electrolyte composition was extremely important to obtain a good CE separation. Please refer to Table **4** for the different types of buffers used [84-85].

Since nonvolatile buffer in ESI-MS at a relatively high concentration results in a significant loss of electrospray efficiency and ion source contamination, volatile buffer systems (e.g. ammonium formate buffer) are generally preferred for cytokinin analyses using CZE-MS [22-23]. CZE is particularly suited for the direct separation of highly charged cytokinin nucleotide compounds without any derivatization step, compared to HPLC analysis [23]. Furthermore, due to its experimental simplicity and less purified sample required, CZE was applied for the determination of the physicochemical constants for cytokinins and their analogs [86-87].

The separation selectivity of MEKC is improved compared to CZE, since surfactant is added to buffer solution to form micelles. The use of micelle-forming surfactant solutions can give rise to separations that resemble a typical RP-LC but with the known benefits of CE. For ionic cytokinins, MEKC separations are based on both the degree of ionization and hydrophobicity. The MEKC results were consistent with the results obtained using HPLC or LC-MS method for the determination of cytokinins in coconut water [30, 88]. The optimized MEKC method developed for topolin and topolin riboside isomers took less than half the time for a typical HPLC running [88]. The shortened analysis time is an advantage of CE yet to be exploited as a routine analytical tool for cytokinins.

Detection

Small sample volume injections make detection a significant challenge in CE. So far, UV and MS detectors are used for the determination of cytokinins. As the standard detector on many commercial CE instruments, UV detector is popular for cytokinins. In addition to UV detector, MS has been carried out online coupled with CE, which provides unsurpassed opportunities in the identification and structure elucidation of cytokinins. Currently, ESI is the most common interface between CE and MS, as it can produce ions directly from liquids at atmospheric pressure, and with high sensitivity and selectivity. The first CZE-MS method for the analysis of 12 cytokinins was developed and applied to screen for cytokinins in coconut water [22]. CZE-MS/MS method was also developed for the analysis of cytokinin nucleotides without sample derivatization [23]. The direct coupling of CE to MS has provides an additional sensitivity and most notably, high selectivity for the cytokinins. More recently, a new method based on MEKC directly coupled to ESI-MS was developed for the simultaneous separation and determination of 13 structurally similar cytokinins, including certain geometric and positional isomers [85]. In the MEKC-MS method, partial filling technique was used to prevent the micelles from reaching the MS as this is detrimental to its signal. The typical CE-MS electropherograms of cytokinins were shown in Fig. **3**.

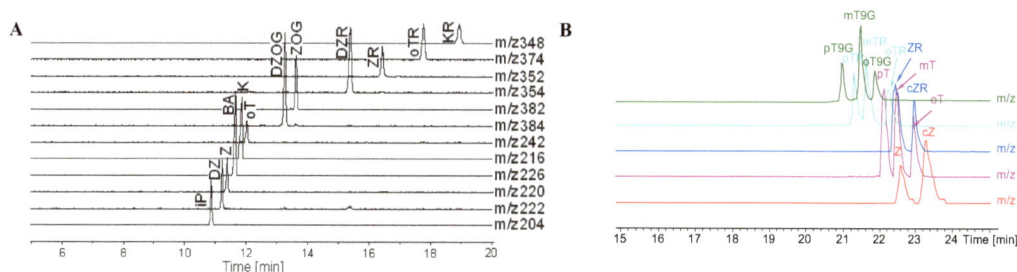

Figure 3: Selected ion mass electropherograms of (A) 12 cytokinin standards mixture obtained by CZE-MS; and (B) 13 cytokinin standard mixtures obtained by PF-MEKC-MS under optimized conditions. Adapted from [22, 85], with permission.

CONCLUSIONS AND PERSPECTIVES

This comprehensive review article has summarized the various methods used for separation and determination of cytokinins. Since free cytokinins present in plants are at extremely low levels, this implies the preliminary sample preparation steps are of primary importance influencing the reliability and reproducibility of the analysis. Therefore, the comprehensive sample preparation steps prior to cytokinin analyses are provided in this review.

It could be generally concluded the modified Bieleski's solvent (methanol/water/formic acid; 15/4/1, v/v/v) was considered as the most suitable extraction solvent; while novel SPE procedures were characterized by high recoveries and an acceptable degree of solute purity, and immunoaffinity and HPLC approaches allow further purification of most cytokinins.

As an established classical approach, immunoassays (especially ELISAs) are still commonly used as a sensitive and viable method for the determination of endogenous cytokinins [12-13, 41-42, 50]. Despite the widespread activity in the development of immunoanalytical methods for the detection of cytokinins, there is only a limited number of commercial immunoassay kits. Not only are these kits costly, but they also often rely on laborious purification of the plant extract samples. Whatever the limitations of this technique, so far immunological appears to be the only means of getting information about endogenous cytokinin distribution at the cellular and subcellular levels [50].

There is a variety of analytical methods in use for cytokinins, most offering high sensitivity and selectivity. Currently, most of attention is devoted to the separation topic, which represents most of the recently published papers and is covered with application of GC, LC, and CE. Simple combinations, such as HPLC-UV, CE-UV, GC-FID, might be useful and informative. However, the use of GC-MS (tandem-MS) [9], LC-MS (tandem-MS) [19-20, 34], and CE-MS (tandem-MS) [22-23] was proved to provide convincing and satisfactory results of cytokinin analysis, in all cases performed. From these results, we concluded that MS has rapidly become a high sensitive and selective tool for cytokinin analyses. This hyphenated combination ensures more reliable detection, identity confirmation and quantification of cytokinins as well as the identification of novel cytokinins.

There is, to our opinion, little doubt that, LC with tandem MS detection will continue to play a dominant role in cytokinin analysis in the near future. Next to excellent sensitivity, this technique can provide structural information based on fragmentation. Unlike the well established GC-MS (or tandem MS) method, LC-MS/MS could analysis cytokinins without derivatization.

The recently developed UPLC and nanoflow-LC technique could be novel approaches for the analyses of cytokinins. To date, however, limited work has reported to apply UPLC and nanoflow-LC as a separation technique for analysis of cytokinins [8, 21]. As the basic separation principles of CE are different from those of HPLC and other chromatographic techniques, it is an attractive complementary technique in the analyses of cytokinins. MS detection has been used in conjunction with CE to determine different cytokinins in biological matrices [22-23]. In order to screen numerous samples, on-line sample pretreatment (pre-concentration and interference removal) and CE separation in microchip formats require further development. It is anticipated that the development of various analytical methods will enable us to unravel some of the mysteries concerning about cytokinins in plants as well as their beneficial effects in medicinal fields [1-7].

REFERENCES

[1]   Mok DWS, Mok MC. Cytokinin metabolism and action. Annu Rev Plant Physiol Plant Molec Biol 2001; 52: 89-118.
[2]   Sakakibara H. Cytokinins: Activity, biosynthesis, and translocation. Annu Rev Plant Biol 2006; 57: 431-49.
[3]   Choi BH, KimW, Wang QC, Kim DC, Tan SN, Yong JWH, Kim KT, Yoon HS. Kinetin riboside preferentially induces apoptosis by modulating Bcl-2 family proteins and caspase-3 in cancer cells. Cancer Lett 2008; 261(1): 37-45.
[4]   Cheong J, Goh D, Yong JWH, Tan SN, Ong ES. Inhibitory effect of kinetin riboside in human heptamoa, HepG2. Mol. BioSyst. 2009; 5: 91-98.
[5]   Hsiao G, Shen MY, Lin KH, Chou CY, Tzu NH, Lin CH, Chou DS, Chen TF, Sheu JR. Inhibitory activity of kinetin on free radical formation of activated platelets in vitro and on thrombus formation in vivo. Eur J Pharm 2003; 465(3): 281-7.
[6]   Rattan SIS, Sodagam L. Gerontomodulatory and youth-preserving effects of zeatin on human skin fibroblasts undergoing aging in vitro. Rejuvenation Res 2005; 8(1): 46-57.
[7]   Slaugenhaupt, S. A.; Mull, J.; Leyne, M.; Cuajungco, M. P.; Gill, S. P.; Hims, M. M.; Quintero, F.; Axelrod, F. B.; Gusella, J. F. Rescue of a human mRNA splicing defect by the plant cytokinin kinetin. Hum Mol Genet 2004; 13: 429-436.
[8]   Novák O, Hauserová E, Amakorová P, Doležal K, Strnad M. Cytokinin profiling in plant tissues using ultra-performance liquid chromatography-electrospray tandem mass spectrometry. Phytochem 2008; 69(11): 2214-24.
[9]   Björkman PO, Tillberg E. Acetylation of cytokinins and modified adenine compounds: a simple and non-destructive derivatization method for gas chromatography-mass spectrometric analysis. Phytochem Anal 1996; 7(2): 57-68.
[10]   Tucker GA, Roberts JA, Eds. Plant hormone protocols, Totowa, NJ: Humana Press Inc. 2000.
[11]   Tarkowski P, Ge L, Yong JWH, Tan SN. Analytical methods for cytokinins. Trend Anal Chem 2009; 28(3): 323-35.
[12]   Grayling A, Hanke DE. Cytokinins in exudates from leaves and roots of red Perilla Phytochem 1992; 31(6): 1863-68.
[13]   Maldiney R, Leroux B, Sabbagh I, Sotta B, SossountzovL, Miginiac E. A biotin-avidine enzyme immunoassay to quantify three phytohormones: auxin, abscisic acid and zeatin-riboside. J Immunol Meth 1986; 90: 151-8.
[14]   Horgan R, Scott IM. In: Rivier L, Crozier A, Eds. Principles and practice of plant hormone analysis. London: Academic Press 1987; Chapter 5, pp.303.
[15]   Teller G.. In: Mok DWS, Mok MC, Eds. Cytokinins: Chemistry, Activity and Function. Boca Raton, FL: CRC Press 1994.
[16]   Nicander B, Stahl U, Bjorkman PO, Tillberg E. Immunoaffinity co-purification of cytokinins and analysis by high-performance liquid chromatography with ultraviolet-spectrum detection Planta 1993; 189 (3): 312-20.
[17]   De Hoffmann E, Charette J, Stroobant V, Mass spectrometry: principles and applications Paris, France, Wiley. 1996; pp.1-8.
[18]   Pan, X, Wang X. Profiling of plant hormones by mass spectrometry. J Chromatogr B 2009; 877(26): 2806-13.
[19]   Doležal K, Popa I, Hauserová E, Spíchal L, Chakrabarty K, Novák O, Kryštof V, Voller J, Holub J, Strnad M. Preparation, biological activity and endogenous occurrence of N6-benzyladenosines. Bioorg Med Chem 2007; 15(11): 3737-47.
[20]   Rasmussen HN, Veierskov B, Hansen-Møller J, Nørbæk R, Nielsen UB. Cytokinin Profiles in the Conifer Tree Abies nordmanniana: Whole-Plant Relations in Year-Round Perspective. J Plant Growth Regul 2009; 28(2):154-66.
[21]   Izumi Y, Okazawa A, Bamba T, Kobayashi A, Fukusaki E. Development of a method for comprehensive and quantitative analysis of plant hormones by highly sensitive nanoflow liquid chromatography–electrospray ionization-ion trap mass spectrometry Anal Chim Acta 2009; 648(2): 215-25.

[22] Ge L, Yong JWH, Tan SN, Ong ES. Determination of cytokinins in coconut (*Cocos nucifera* L.) water using capillary zone electrophoresis-tandem mass spectrometry. Electrophoresis 2006; 27(11): 2171-81.

[23] Ge L, Yong JWH, Tan SN, Yang XH, Ong ES. Analysis of cytokinin nucleotides in coconut (*Cocos nucifera* L.) water using capillary zone electrophoresis-tandem mass spectrometry after solid-phase extraction. J Chromatogr A 2006; 1133 (1-2) 322-31.

[24] Bieleski RL. The problem of halting enzyme action when extracting plant tissues. Anal Biochem 1964; 9: 431-42.

[25] Hoyerová K, Gaudinová A, Malbeck J, Dobrev PI, Kocábek T, Solcová B, Trávnícková A, Kamínek M. Efficiency of different methods of extraction and purification of cytokinins. Phytochem 2006; 67(11): 1151-9.

[26] Letham DS. Regulators of cell division in plant tissues. Planta 1974; 118(4): 361-4.

[27] Horgan R. In: Hillman JR, Ed. Isolation of plant growth substances. Society for experimental biology, seminar series 4. Cambridge, Cambridge University Press 1978; pp. 97-114.

[28] Dobrev PI, Kamínek M. Fast and efficient separation of cytokinins from auxin and abscisic acid and their purification using mixed-mode solid-phase extraction. J Chromatogr A 2002; 950(1-2): 21-9.

[29] Takei K, Yamaya T, Sakakibara H. A method for separation and determination of cytokinin nucleotides from plant tissues. J Plant Res 2003; 116(3): 265-9.

[30] Ge L, Yong JWH, Tan SN, Yang XH, Ong ES. Analysis of some cytokinins in coconut (Cocos nucifera L.) water by micellar electrokinetic capillary chromatography after solid-phase extraction. J Chromatogr A 2004; 1048(1): 119-26.

[31] Guinn G, Brummett DL. Solid-phase extraction of cytokinins from aqueous solutions with C18 cartridges and their use in a rapid purification procedure. Plant Growth Regul. 1990; 9(4): 305-14.

[32] Hauserová E, Swaczynová J, Doležal K, Lenobel R, Popa I, Hajdúch M, Vydra D, Fuksová K, Strnad M. Batch immunoextraction method for efficient purification of aromatic cytokinins. J Chromatogr A 2005; 1100(1): 116-25.

[33] MacDonald EMS, Morris RO. Isolation of cytokinins by immunoaffinity chromatography and analysis by high-performance liquid chromatography radioimmunoassay. Meth Enzymol 1985; 110: 347-58.

[34] Novák O, Tarkowski P, Tarkowská D, Doležal K, Lenobel R, Strnad M. Quantitative analysis of cytokinins in plants by liquid chromatography-single-quadrupole mass spectrometry. Anal Chim Acta 2003; 480(2): 207-18.

[35] Tarkowská D, Doležal K, Tarkowski P, Åstot C, Schmülling T, Holub J, Fuksová K, Sandberg G, Strnad M. Identification of new aromatic cytokinins in Arabidopsis thaliana and poplar leaves by LC-(+)ESI-MS and capillary liquid chromatography/frit - fast atom bombardment mass spectrometry. Physiol. Plant. 2003; 117 (4): 579-590.

[36] Genkov T, Ivanov I, Ivanova I. Analysis of cytokinins by immunoassay and high performance liquid chromatography of *in vitro* cultivated dianthus caryophyllus. Bulg J Plant Physiol. 1996; 22(3-4): 95-104.

[37] Ge L, Yong JWH, Goh NK, Chia LS, Tan SN, Yang XH, Ong ES. Identification of kinetin and kinetin riboside in coconut (*Cocos nucifera* L.) water using a combined approach of liquid chromatography-tandem mass spectrometry, high performance liquid chromatography and capillary electrophoresis. J Chromatogr B 2005; 829(1-2): 26-34.

[38] Fernandez B, Centeno M, Feito I, Sanchez-Tames R, Rodriguez A. Simultaneous analysis of cytokinins, auxins and abscisic acid by combined immunoaffinity chromatography, high performance liquid chromatography and immunoassay. Phytochem. Analysis 1995; 6(1): 49-54.

[39] Dobrev PI, Havlicek L, Vagner M, Malbeck J, Kaminek M. Purification and determination of plant hormones auxin and abscisic acid using solid phase extraction and two-dimensional high performance liquid chromatography. J Chromatogr A 2005;1075 (1-2):159-166.

[40] Weiler EW. In: Linskens HF, Jackson JF. Eds. Immunology in Plant Sciences, Berlin, Springer-Verlag, 1986; pp.1-17.

[41] Stevenson WG, Friedman PL. In: Hennekens CH, Ed. Clinical trials in cardiovascular disease. Philadelphia, WB Saunders Co. 1999; pp. 217-30.

[42] Wang J, Letham DS, Taverner E, Badenoch-Jones J, Hocart CH. A procedure for quantification of cytokinins as free bases involving scintillation proximity immunoassay. Physiol Plant 1995; 95(1): 91–8.

[43] Yong JWH, Wong SC, Letham DS, Hocart CH, Farquhar GD. Effects of elevated [CO2] and nitrogen nutrition on cytokinins in the xylem sap and leaves of cotton. Plant Physiol 2000; 124(2): 767-80.

[44] Weiler EW. Radioimmunoassay for trans-zeatin and related cytokinins. Planta 1980; 149(2): 155-162.

[45] Cook NC, Bellstedt DU, Jacobs G.. Endogenous cytokinin distribution patterns at budburst in Granny Smith and Braeburn apple shoots in relation to bud growth. Scientia Hort 2001; 287(1-2): 53-63.

[46] MacDonald EMS, Akiyoshi DE, Morris RO. Combined high performance liquid chromatog' raphy-radioimmunoassay for cytokinins . J Chromatogr 1981; 214(1): 101-9.

[47] Cahill DM, Weste GM, Grant BR. Changes in Cytokinin Concentrations in Xylem Extrudate following Infection of Eucalyptus marginata Donn ex Sm with Phytophthora cinnamomi Rands. 1986; 81(4):1103-9.

[48] Blintsov AN, Gusakovskaya MA, Ermakov IP. Differential analysis of major natural cytokinins by enzyme immunoassay. Appl. Biochem. Microbiol. 2000; 36(4): 398-403.

[49] Székács A, Hegedus G, Tóbiás I, Pogány M, Barna B. Immunoassays for plant cytokinins as tools for the assessment of environmental stress and disease resistance. Anal Chim Acta 2000; 421(2): 135-46.

[50] Jirásková D, Poulíčková A, Novák O, Sedláková K, Hradecká V, Strnad M. High throughput screening technology for monitoring phytohormone production in microalgae. J Phycol 2009; 45(1): 108-18.

[51] Casanova E, Valdés AE, Fernández B, Moysset L, Trillas MI. Levels and immunolocalization of endogenous cytokinins in thidiazuron-induced shoot organogenesis in carnation. J Plant Physiol 2004; 161(1): 95-104.

[52] Dewitte W, Chiappetta A, Azmi A, Witters E, Strnad M, Rembur J, Noin M, Chriqui D, Van Onckelen H. Dynamics of Cytokinins in Apical Shoot Meristems of a Day-Neutral Tobacco during Floral Transition and Flower Formation Plant Physiol 1999; 119(1): 111-22.

[53] Kärkönen A, Simola LK. Localization of cytokinins in somatic and zygotic embryos of Tilia cordata using immunochemistry. Physiol Plant 1999; 105(2): 355-65.

[54] Trione EJ, Banowetz GM, Krygier BB, Kathrein JM, Sayavedra-Soto L. A quantitative fluorescence enzyme immunoassay for plant cytokinins. Anal Biochem 1987; 162(1): 301-8.

[55] Palni LMS, Tay SAB, MacLeod JK. In: Linkens HF, Jackson JF, Eds. Modern methods of plant analysis. New series, Vol. 3. Berlin: Springer-Verlag 1987.

[56] Most BH, Williams JC, Parker KJ. Gas chromatography of cytokinins. J Chromatogr 1968; 38(1): 136-8.

[57] MacLeod JK, Summons RE, Letham DS. Mass spectrometry of cytokinin metabolites. Per(trimethylsilyl) and permethyl derivatives of glucosides of zeatin and 6-benzylaminopurine. J Org Chem 1976; 41(25): 3959-67.

[58] Hocart CH, Wong OC, Letham DS, Tay SAB, MacLeod JK. Mass spectrometry and chromatography of t-butyldimethylsilyl derivatives of cytokinin bases. Anal Biochem 1986; 153(1): 85-96.

[59] Ludewig M, Dörffling K, König WA. Electron-capture capillary gas chromatography and mass spectrometry of trifluoroacetylated cytokinins. J Chromatogr A 1982; 243(1): 93-8.

[60] Hocart CH, Wang J, Letham DS. Derivatives of cytokinins for negative ion mass spectrometry. J Chromatogr A 1998; 811(1-2): 246-9.

[61] Birkemeyer C, Kolasa A, Kopka J. Comprehensive chemical derivatization for gas chromatography-mass spectrometry-based multi-targeted profiling of the major phytohormones. J Chromatogr A 2003; 993(1-2): 89-102.

[62] Scott IM, Horgan R. Mass-spectrometric quantification of cytokinin nucleotides and glycosides in tobacco crown-gall tissue. Planta 1984; 161(4): 345-54.

[63] Chou CC, Chen WS, Huang KL, Yu HC, Liao LJ. Changes in cytokinin levels of Phalaenopsis leaves at high temperature. Plant Physiol Biochem 2000; 38(4): 309-14.

[64] Baroja-Fernandez E, Aguirreolea J, Martinkova H, Hanus J, Strnad M. Aromatic cytokinins in micropropagated potato plants. Plant Physiol Biochem 2002; 40(3): 217-24.

[65] Müller A, Düchting P, Weiler EW. A multiplex GC-MS/MS technique for the sensitive and quantitative single-run analysis of acidic phytohormones and related compounds, and its application to Arabidopsis thaliana, Planta 2002; 216(1): 44-56.

[66] Fiehn O, Kopka J, Trethewey RN, Willmitzer L. Identification of uncommon plant metabolites based on calculation of elemental compositions using gas chromatography and quadrupole mass spectrometry. Anal Chem. 2000; 72(15): 3573-80.

[67] Chen C. In: Linskens HF, Jackson JF, Eds. High performance liquid chromatography in plant sciences. Berlin: Springer 1987; pp.23.

[68] Nordström A, Tarkowski P, Tarkowska D, Dolezal K, Åstot C, Sandberg G, Moritz T. Derivatization for LC-electrospray ionization-MS: a tool for improving reversed-phase separation and ESI responses of bases, ribosides, and intact nucleotides. Anal Chem 2004; 76(10): 2869-77.

[69] Sjut V, Palmer MV. Reversed-phase high-performance liquid chromatography of plant hormones: some useful differences in stationary phase selectivity. J Chromatogr A 1983; 270: 309-12.

[70] Van Rhijn JA, Heskamp HH, Davelaar E, Jordi W, Leloux MS, Brinkman UAT. Quantitative determination of glycosylated and aglycon isoprenoid cytokinins at sub-picomolar levels by microcolumn liquid chromatography combined with electrospray tandem mass spectrometry. J Chromatogr A 2001; 929(1-2): 31-42.

[71] Fernández B, Centeno ML, Feito I, Sánchez-Tamés R, Rodriguez A. Simultaneous analysis of cytokinins, auxins and abscisic acid by combined immunoaffinity chromatography, high performance liquid chromatography and immunoassay. Phytochem Anal 1995; 6(1): 49-54.

[72] Abian J, Oosterkamp AJ, Gelpí E. Comparison of conventional, narrow-bore and capillary liquid chromatography/mass spectrometry for electrospray ionization mass spectrometry: practical considerations. J Mass Spectrom 1999; 34(4): 244-54.

[73] Li R, Dong L, Huang J. Ultra performance liquid chromatography–tandem mass spectrometry for the determination of epirubicin in human plasma. Anal Chim Acta 2005; 546(2): 167-73.

[74] Prinsen E, Redig P, Strnad M, Galís I, van Dongen W, van Onckelen H. In: Gartland KMA, Davey M, Eds. Methods in molecular biology, agrobacterium protocols. New Jersey: Humanae Press 1995; pp.245-62.

[75] Åstot C, Dolezal K, Moritz T, Sandberg G. Precolumn derivatization and capillary liquid chromatographic/frit-fast atom bombardment mass spectrometric analysis of cytokinins in Arabidopsis thaliana. J Mass Spectrom 1998; 33(9): 892-902.

[76] Åstot C, Dolezal K, Moritz T, Sandberg G. Deuterium *in vivo* labelling of cytokinins in Arabidopsis thaliana analysed by capillary liquid chromatography/frit-fast atom bombardment mass spectrometry. J Mass Spectrom 2000; 35(1): 13-22.

[77]   Doležal K, Åstot C, Hanuš J, Holub J, Peters W, Beck E, Strnad M, Sandberg G. Identification of aromatic cytokinins in suspension cultured photoautotrophic cells of Chenopodium rubrum by capillary liquid chromatography/frit - fast atom bombardment mass spectrometry. Plant Growth Regul 2002; 36(2): 181-9.

[78]   Tarkowská D, Doležal K, Tarkowski P, Åstot C, Holub J, Fuksová K, Schmülling T, Sandberg G, Strnad M. Identification of new aromatic cytokinins in Arabidopsis thaliana and Populus × canadensis leaves by LC-(+)ESI-MS and capillary liquid chromatography/frit–fast atom bombardment mass spectrometry. Physiol Plant 2003; 117(4): 579-90.

[79]   von Schwartzenberg K, Núñez MF, Blaschke H, Dobrev PI, Novák O, Motyka V, Strnad M. Cytokinins in the bryophyte Physcomitrella patens: analyses of activity, distribution, and cytokinin oxidase/dehydrogenase overexpression reveal the role of extracellular cytokinins. Plant Physiol 2007; 145: 786- 800.

[80]   Fletcher AT, Mader JC. Hormone profiling by LC-QToF-MS/MS in dormant Macadamia integrifolia: correlations with abnormal vertical growth. J Plant Growth Regul 2007; 26(4): 351-61.

[81]   Stirk WA, Novák O, Václavíková K, Tarkowski P, Strnad M, van Staden J. Spatial and temporal changes in endogenous cytokinins in developing pea roots. Planta 2008; 227(6): 1279-89.

[82]   Bredmose N, Kristiansen K, Nørbæk R, Christensen LP, Hansen-Møller J. Changes in concentrations of cytokinins (CKs) in root and axillary bud tissue of miniature rose suggest that local CK biosynthesis and zeatin-type CKs play important roles in axillary bud growth. J Plant Growth Regul 2005; 24(3): 238-50.

[83]   Prinsen E, van Dongen W, Esmans E, van Onckelen H. HPLC linked electrospray tandem mass spectrometry: a rapid and reliable method to analyse indole-3-acetic acid metabolism in bacteria. J Mass Spectrom 1997; 32(1): 12-22.

[84]   Prinsen E, van Dongen W, Esmans EL, Van Onckelen HA. Micro and capillary liquid chromatography-tandem mass spectrometry: a new dimension in phytohormone research. J Chromatogr A 1998; 826(1): 25-37.

[85]   Ge L, Tan S, Yong JWH, Tan SN, CE for cytokinin analyses: A review. Electrophoresis 2006; 27(23): 4779-91.

[86]   Ge L, Tan SN, Yong JWH, Hua L, Ong ES. Separation of cytokinin isomers with a partial filling-micellar electrokinetic chromatography-mass spectrometry approach. Electrophoresis 2008; 29(10): 2024-32.

[87]   Barták P, Bednář P, Kubáček L, Stránský Z. Advanced statistical evaluation of complex formation constant from electrophoretic data. Anal Chim Acta 2000; 407(1-2): 327-36.

[88]   Barták P, Bednář P, Stránský Z, Boček P, Vespalec R. Determination of dissociation constants of cytokinins by capillary zone electrophoresis. J Chromatogr A 2000; 878 (2): 249-59.

[89]   Ge L, Yong JWH, Tan SN, Yang XH, Ong ES. Analysis of positional isomers of hydroxylated aromatic cytokinins by micellar electrokinetic chromatography. Electrophoresis 2005; 26(9): 1768-77.

130   *Reviews in Pharmaceutical and Biomedical Analysis*, 2010, 130-148

CHAPTER 10

Pressurized Liquid Extraction in Phytochemical Analysis

Peng Li[1] and Shao-Ping Li*[2]

[1]Faculty of Life Science and Technology, Kunming University of Science and Technology, Kunming, Yunnan, P. R. China and [2]Institute of Chinese Medical Sciences, University of Macau, Macao SAR, P. R. China

Abstract: Pressurized liquid extraction (PLE) is an innovative sample preparation technique which has been developed as an alternative to conventional extraction methods in many areas, such as environmental, food and pharmaceutical analysis. The aim of the current review is to summarize the application of PLE technique in phytochemical analysis in last decade. The parameters, which may affect the extraction efficiency and selectivity, including the nature of solvent, temperature, pressure, extraction time and sample particle size and so on were explored. In addition, the procedure for method development and the parameters optimization strategies, univariate approach, orthogonal analysis and central composite design, were also discussed. Due to the obvious advantages of high extraction efficiency, short preparation time, little solvent consumption and good reproducibility, PLE undoubtedly have a broad application in phytochemical analysis.

Keywords: Pressurized liquid extraction (PLE); Phytochemical analysis; Sample preparation

INTRODUCTION

Herbal medicines and their derived products are widely used as health and/or therapeutic products in many countries. Unlike Western pharmaceuticals, herbal products are usually comprised of a complex mixture of different phytochemical constituents (mainly secondary metabolites such as flavonoids, saponins, alkaloids, anthraquinones, volatile oils, etc) from multiple herbs rather than a single chemical or a simple combination of several chemicals. Their chemical natures are considerably varied. Even if the different batches of the same herbal drug, the chemical constituents may be obviously changed due to the different harvest time, plant location and processing method and so on. Therefore, phytochemical analysis is necessary for quality control of herbal medicines, where the efficient and selective sample preparation method is very important.

Actually, the first and usually the most important step for phytochemical analysis is sample preparation. Since 70-80% of analysis time is spent on sample preparation and more than 60% of analysis error derives from nonstandard sample pretreatment, an appropriate approach for sample preparation is very important for qualitative and quantitative analysis of herbal medicine [1]. Individual steps used in sample preparation include sample collection, drying, comminution, homogenization, extraction and analyte enrichment if necessary [2]. Among of these procedures, extraction techniques play a unique and crucial role, which serve to selectively isolate analytes from potentially interfering sample components while getting these analytes into a form suitable for analysis [3, 4].

Although the development of instrumental techniques has grown rapidly nowadays, in most laboratories decades-old extraction procedures are still in common use [3]. Even in the monograph collected in the United States, Chinese and Japanese pharmacopeias, some conventional extraction methods such as distillation, reflux extraction and Soxhlet extraction are still commonly used. However, such methods suffer from a variety of disadvantages, including long extraction time, relatively large amount of organic solvents consumption and unsatisfactory extraction efficiency [5]. With the requirement of laboratory automation and fast quantitative assays for drug analysis, conventional extraction techniques are increasingly becoming the bottleneck of phytochemical analysis. In order to overcome these drawbacks, looking for a high efficient, fast and automatic extraction technique is crucial.

The advent of pressurized liquid extraction, PLE (also known as accelerated solvent extraction/ASE and pressurized fluid extraction/PFE) has complied with this tendency, which combines elevated temperature and high pressure to achieve fast and efficient extraction of the analytes from the (semi-)solid matrices. Initially, this extraction technique is mainly applied to sample preparation in environmental [6-9] and food analysis [10-13]. The distinct advantages of PLE are recently being exploited in phytochemical analysis. Especially, the application

*Address correspondence to this author Shao-Ping Li at: Institute of Chinese Medical Sciences, University of Macau, Macao SAR, P. R. China; Tel: +853-8397 4692; Fax: +853-2884 1358; E-mail: lishaoping@hotmail.com

of PLE on phytochemical analysis has shown a sharp increase since 2005 (Fig. **1**) due to its significant economies in time and solvents with high extraction efficiency [14, 15]. Since the basic set-up and principles of PLE have been described before [6], the aim of the current review is to summarize the application of PLE technique in phytochemical analysis in last decade. The parameters (solvent, temperature, pressure, time, sample particle size and so on), which may affect the extraction efficiency and selectivity, were explored. In addition, the procedure for method development and the parameters optimization strategies were also discussed.

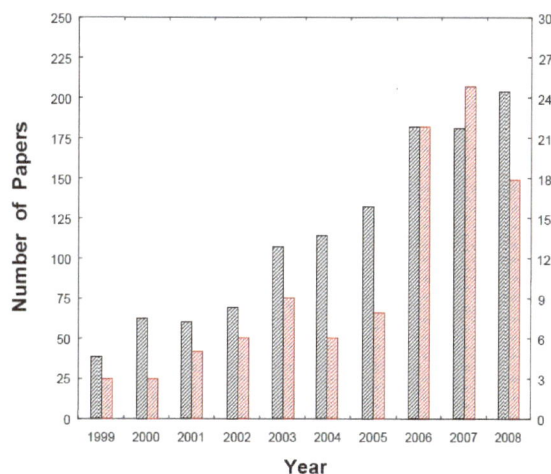

Figure 1: The annual growth of publications on pressurized liquid extraction (PLE) in various areas ( ) and in phytochemical analysis ( ) during the last decade based on the data from *ISI Web of Science*.

THE EFFECTS OF PARAMETERS ON PLE

A typical PLE process includes filling the extractor with solvent, heating (preheating if necessary) the extractor to a preset temperature, static (or dynamic) extraction, flushing the extract to the collection vial, and finally purging the extractor with nitrogen gas [4]. The parameters, such as the nature of solvent used, applied temperature and pressure, the time for extraction, particle size of samples and etc, would definitely influence the extraction efficiency and/or selectivity of PLE.

Solvent

Solvent is the key for affecting the extraction efficiency and/or selectivity of PLE [15-45]. In general, in order to extract the analytes of interest as many as possible from interfering compounds, the polarity of solvent should closely match that of target compounds, i.e. like dissolve like. Actually, more than 60% PLE in phytochemical analysis employed alcoholic solvents, especially methanol (Fig. **2**). The addition of water causes the plant material swelling thereby the solvent penetrates more easily into the solid matrix and increases extractability, hydrous methanol [18, 22, 36, 40, 46-51] and hydrous ethanol [16, 17, 19, 25, 34, 39, 52-63] have been commonly used as the optimum solvents. However, for those botanicals containing high content of polysaccharides such as starch, cellulose, gum or mucilage, hydrous alcohol (especially with the high percentage of water) are not suitable for PLE as it was observed that the powdered medicinal plants had a strong tendency to adsorb water which will eventually result in blockage of the system [25, 59, 64].

Water is also a commonly used solvent in PLE of plant materials [5]. Under ambient conditions, water is too polar to efficiently dissolve and extract most bioactive or marker compounds that are associated with botanicals. However, under subcritical status with temperature between 100 °C and 374 °C (the critical temperature) under high pressure (usually from 10 to 60 bar) in PLE system, the physicochemical properties of water change dramatically, which behaving like an organic solvent to extract a wide variety of phytochenical constituents from different matrices [65-74]. Furthermore, in order to disrupt the strong analyte-matrix interaction present naturally in the medicinal plants and improve the extraction efficiencies, the addition of surfactants such as Triton X-100 and SDS into water has also been proposed in phytochemical analysis [23, 75, 76].

Apart from extraction efficiency, selectivity is another important factor should be considered while choosing extraction solvent. As shown in Fig. **3**, although the extraction efficiency was not comparable to those of methanol and 70% methanol, acetonitrile was selected as the optimum solvent due to its high selectivity for the extraction of the four analytes [38].

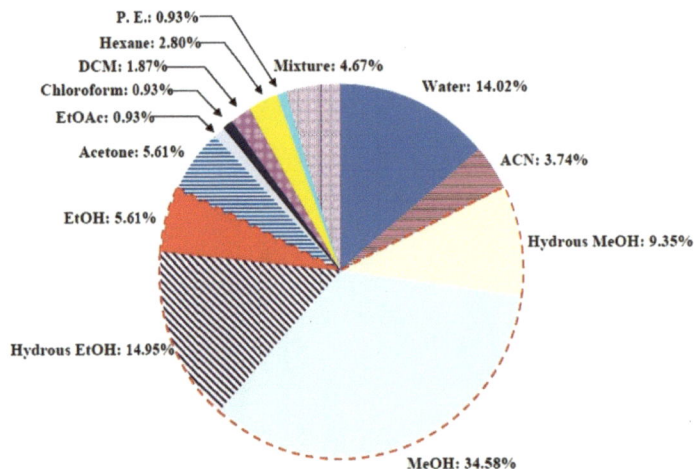

Figure 2: The types of solvent used for PLE in phytochemical analysis based on the data in Table **1-4**. ACN: acetonitrile; MeOH: methanol; EtOH: ethanol; EtOAc: ethyl acetate; DCM: dichloromethane; P. E.: petroleum ether.

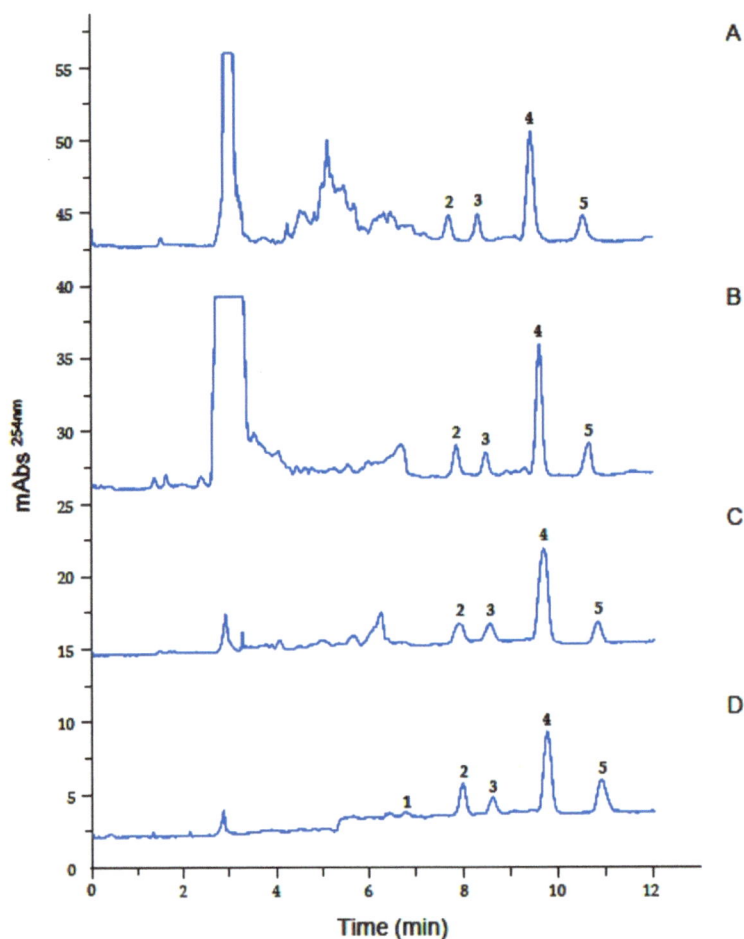

Figure 3: The comparison of (A) methanol, (B) 70% methanol, (C) ethanol and (D) acetonitrile as PLE solvents in CEC analysis of four tanshinones in *S. miltiorrhiza*. Conditions: 25 cm×100 μm CEC-Hypersil C18 column (3 μm particle size); BGE, 30 mM Tris-HCl (pH 8.5) with acetonitrile at the ratio of 1:3; electrokinetic injection, 10 kV for 5 seconds; temperature, 20 °C; voltage, 20kV; UV detection, 254 nm. **1**, dihydrotanshinone I; **2**, cryptotanshinone; **3**, tanshinone I; **4**, tanshinone IIA; **5**, physcion (internal standard). Reprinted from Ref. [38], with permission.

Temperature

Temperature is one of the most important factor affecting the extraction efficiency of PLE [16, 18-25, 27-30, 33, 36, 37, 40, 43, 44, 46, 51, 57-59, 61, 64-67, 70-72, 77-86]. Generally, increase of temperature will result in higher recoveries of the target compounds in plant matrices, which may attribute to increasing the soluble ability of solvents, speeding up the diffusion rate of analyte molecule and then accelerating the mass transfer, disrupting the strong solute-matrix interaction and decreasing viscosity and surface tension of solvent. Therefore, the extraction temperature applied in PLE was usually set above the normal boiling point of the solvents used [15, 21, 28-32, 37-39, 41, 43-45, 64, 66-71, 73, 74, 78-81, 83, 85, 87-112]. However, while the elevated extraction temperature increasing to a certain point, the stability of the analytes becomes an important issue as some of the marker compounds could decompose or degrade rapidly and thus results in lower recoveries [18-20, 22, 23, 36, 40, 42, 59, 61, 64, 71, 78]. For the extraction of glucosinolates such as epiprogoitrin, progoitrin, gluconapin and glucotropaeolin in *Isatis tinctoria*, thermal degradation of the analytes was observed at temperatures above 50 °C, where more than 60% of the glucosinolates was lost at 100 °C within 10 min [51]. A similar phenomenon was also found for the extraction of terpene trilactones such as ginkgolides and bilobalide from *Ginkgo biloba* leaves using PLE. Compared with the amount extracted at room temperature, a loss of about 25% and 85% of bilobalide was observed at 80 °C and 100 °C, respectively. Especially, no bilobalide could be detected when the temperature was up to 140 °C [72].

Fortunately, it was gratified that in the PLE process for the extraction of phytochemical constituents, decomposition or degradation of analytes did not always occur, even if for some thermal labile compounds. Z-ligustilide is a volatile and unstable compound, which can be changed to other phthalides at high temperature [113, 114]. Nevertheless, during PLE of bioactive compounds from *Angelica sinensis*, the temperature showed no obvious effect on the recovery of Z-ligustilide. The results showed that Z-ligustilide was stable at 110 °C within 25 min [30]. Coniferyl ferulate (CF), another biologically active compound in *Angelica sinensis*, is also unstable and readily hydrolyzed into ferulic acid. During supercritical fluid extraction (SFE), the extraction efficiency of CF within 240 min gradually decreased with the increase of temperature from 40 to 60 °C. But under PLE conditions (temperature, 100 °C and static extraction time, 10 min with 2 cycles), the recovery of CF was much higher than that obtained by SFE [97]. Actually, some complex natural compounds with chemically labile moieties such as an ester linkage and an o-diphenol could be extracted under a rather high temperature of 120 °C without degradation [62]. The possible reasons lied in the fact that the whole extraction of PLE performed under an inert atmosphere (N_2) and relatively short time [19, 30].

Besides the extraction efficiency, selectivity should also be considered for the optimization of temperature. For PLE extraction of secondary volatile metabolites in three common *Angelica* species, selectivity for the desired compounds was decreased though high temperature (> 80 °C) could increase the solubility and mass transfer [28]. As mentioned above, the physicochemical properties (especially dielectric constant) of water are very sensitive to temperature [115]. For PLE of polar antioxidants from rosemary extracts using water as the solvent, high polarity compounds (such as rosmarinic acid) were preferentially resolved and extracted at lower temperature (60 °C), while low polar compounds (such as carnosic acid) had better recovery at higher temperature (100 °C) [66].

Pressure

The main reason why high pressure is used during the process of PLE is to keep the solvent in a liquid state at elevated temperatures far above the boiling point, thereby improving analyte solubilities and the kinetics of their desorption from the matrices. Based on the data from Table **1-4**, nearly 80% applications of PLE for botanicals and medicinal plants were performed under the pressures between 1000 and 2000 psi, and 1500 psi was commonly used as the default value [23, 25, 26, 28, 30, 37, 41, 42, 45, 50, 52-54, 66, 73, 77, 82, 84-86, 89-96, 99, 102-106, 110, 112, 116-119].

Studies have demonstrated that pressure had little effect on the extraction efficiency of PLE [16, 18-21, 24, 36, 40, 58, 61, 82, 120]. Therefore, the applied pressure usually can be set at any levels as long as to maintain the extraction fluid as a liquid during PLE. However, a certain minimum pressure (threshold) may be required for the solvent molecules to overcome the surface barrier (such as leaf surface wax layer) and penetrate inside the sample particles in some cases. For the extraction of terpene trilactones including ginkgolides and bilobalide from leaves of *Ginkgo biloba* using PLE, the minimum pressure was approximately 50 atm [72]. It is notable that the elevated pressure during PLE process may also hinder the extraction, such as the isolation of caffeine from green tea leaves [67].

Extraction Time

In general, PLE can be carried out in two modes, static mode and dynamic mode. But static mode is widely used due to commercial success of Dionex ASE system. In this approach, static time is defined as the time of sample interacts with extraction solvent per cycle in extraction cell under defined PLE conditions. Generally, prolongation of static time can increase extraction efficiency of PLE. However, it should be noted that the extraction efficiency of some thermal unstable compounds could decrease significantly when static time increased to a certain level [22, 33, 38, 40, 59]. Most static extractions (nearly 90%) of PLE were achieved within 5-30 min which was much shorter than those of conventional extraction methods such as soxhlet extraction, reflux extraction, sonication and so on (Table **1, 2, 3** and **4**).

Actually, to produce a complete extraction, the effect of static time is always explored in conjunction with static cycles to maintain a favorable extraction equilibrium. According to Fick's law of diffusion, continuous interaction between sample matrix and fresh solvent could accelerate mass transfer. Therefore, increase of static cycles has been demonstrated a very efficacious approach to achieve a complete extraction, especially for sample types with a very high concentration of analyte [42, 61], or whole extraction process should be performed under low temperatures (≤70°C) [15, 35, 42, 51, 61, 63, 82, 119, 121-123]. Especially, dynamic mode, also commonly used in PLE [16, 34, 39, 43, 50, 56, 58, 64, 68-72, 74-76, 80, 81, 108, 110, 120], has no expected higher recovery and shorter extraction time comparing to static mode [34, 39, 43, 56, 58, 69-71, 75, 76]. A probable reason may be that the extractions were operated under either low extraction temperatures (e.g., ambient temperature) [34, 76] or low system pressures (e.g., 10-20 bar) [34, 39, 56, 58, 70, 71, 75, 76].

Miscellaneous

Particle size: For PLE, sample particle size has an important impact on the recoveries of target compounds, the larger the surface area of a sample, the more efficient extraction will occur. Consequently, the increased extraction efficiencies are often discovered with the particle size reduced [30, 31, 36, 37, 61, 84, 108]. The decrease of particle size from greater than 2.00 mm to less than 0.25 mm resulted in almost threefold increase of the extraction efficiency for the extraction of phenolic acids such as caffeic acid, ferulic acid, sinapic acid and isoferulic acid from black cohosh [36]. However, a dissimilar trend was observed for PLE of terpene trilactones from *Ginkgo biloba* leaves [72]. Since too fine sample particles would reduce the porosity or permeability of matrix and sequentially result in insufficient interaction between sample particles and solvent molecules, and approximately 20% decrease of the recoveries was observed when the sample particle size reduced from 42-60-mesh to ≤ 80-mesh. Therefore, in order to get an optimal extraction result, a suitable range of particle size should be decided.

Flow rate: The influence of flow rate on extraction efficiency of PLE is frequently overlooked as this factor is only considered in dynamic mode. The effect of flow rate on the extraction of three hydrolysable tannins, namely gallic acid, ellagic acid and corilagin from *Phyllanthus niruri* by PLE was investigated. An upward trend for the recoveries of the three marker compounds was observed with the reduction in flow rate (from 3.0 to 1.5 mL/min) [69].

METHOD DEVELOPMENT AND OPTIMIZATION

For effective optimization of the parameters mentioned above, a systematic approach is required. Many strategies involving univariate design, orthogonal design, and central composite design (CCD) have been used to optimize the parameters for achieving the best extraction efficiency and selectivity.

Procedure of Method Development

Based on our earlier works and other studies, the parameters predominant affecting the extraction efficiency of PLE for marker compounds in botanicals and herbal preparations were solvent and temperature. Particularly, the nature of solvent is crucial. Papagiannopoulos *et al.* [35] even considered that temperature and time are of marginal influence after choosing an optimal solvent composition. Therefore, the effects of solvent should be firstly evaluated. Since solvent is a non-continuous variable, univariate design can be selected as the optimal approach to explore this factor. Undoubtedly, optimization of extraction process should follow with an appropriate choice of temperature. When developing a new method, the system temperature should start at 100 °C, or 20 °C below the thermal degradation point of target analytes if they are known. The other factors could be fixed at their system default values (i.e. pressure, 1500 psi; extraction time, 5 min and flush volume, 60% for Dionex ASE® system) for solvent and temperature optimization. Actually, a satisfactory result can be obtained for most of analytes present in plants after the optimal solvent and temperature determined. However, for some special matrices (e.g. rigid seed, bark and leaf with surface wax layer), further optimization of the other parameters may be necessary in order to achieve an exhaustive extraction.

Univariate Design

Univariate design is the simplest and undoubtedly the most commonly used method for optimization of PLE parameters of phytochemicals from medicinal plants [18-24, 27-29, 31, 32, 34, 36-41, 43-46, 51, 52, 59, 67, 70, 72, 75, 76, 78, 81, 86, 98, 99, 108, 119]. The most attractive advantage of this approach is the simple operation, which requires only systematic alteration of one variable while keeping the others at a constant level. However, it is generally time consuming and labor intensive as only one factor could be decided at one time and thereby the variables have to be optimized one by one. Moreover, this method can not reflect the influence of mutual interactions. Actually, sometimes the interrelationships between different parameters are obvious. The interactions between temperature and flush volume and static time and flush volume have been discovered during the optimization of PLE for simultaneous extraction of *Z*-ligustilide, *Z*-butylidenephthalide and ferulic acid in *Angelica sinensis* [30]. In such case, another systematic optimization procedure such as orthogonal design could be considered.

Orthogonal Design

Orthogonal design, as a more systematic approach comparing to univariate analysis, has also been widely utilized for optimization of PLE parameters in phytochemical analysis [16, 59, 79, 102, 124]. Based on the scientific experimental design, a very limited number of experiments are performed for orthogonal design. Moreover, the significance of each variable and the influence order of factors on signal response could be confirmed by analysis of variance (ANOVA) [16, 79, 124] and range analysis [59, 79, 102], respectively. In the optimization of PLE procedure for extraction of ergosterol, nucleosides and their bases in *Cordyceps* by orthogonal test, higher extraction efficiency was obtained with the increase of temperature during the investigated range of levels (80-140 °C). Therefore, the further study of temperature (140-180 °C) should be performed and finally 160 °C was selected as the optimum [102]. Similarly, optimization of the extraction of isoflavonoids from *Pueraria lobata* by PLE using orthogonal design, total content of puerarin, daidzin and daizein increased with the elevation of temperature in the range of experiment (80-120 °C). Thus, higher temperatures (up to 180 °C) had to be further investigated [79]. These examples revealed that expanding the range of investigated levels is necessary in order to obtain the optimum values while the best result achieved at the end of the investigated level ranges.

Central Composite Design

Recently, a more powerful experimental design approach, i.e. central composite design (CCD) has been developed to optimize PLE parameters for extraction of marker compounds from plant materials [30, 33, 77, 83-85]. Since specifically developed to achieve a multivariate nonlinear regression and fit data into second-order polynomial model by means of statistical softwares (e.g., SAS and STATISTICA), this method could be utilized to confirm the optimum experimental levels more easily and more accurately. In order to obtain the highest recovery of Z-ligustilide, Z-butylidenephthalide and ferulic acid from *Angelica sinensis*, the most pronounced parameters, including temperature, static time and flush volume, discovered in the pilot experiments were further optimized by CCD [30]. Based on multiple regression analysis of the experimental data, the significance of each factor could be determined by Student's *t*-test and *P*-values, by which temperature was discovered as the most significant parameter affecting extraction efficiency. Furthermore, the optimal PLE conditions (temperature, 110 °C; static extraction time, 25 min, and flush volume, 10%) could also be easily obtained by the visual comparisons of the three-dimensional response surfaces mapped against experimental factors. Sure, a larger number of experiments need for CCD is the disadvantage [125]. Actually, extraction efficiency (response) of analytes to PLE parameters are usually linear [77]. Therefore, simple approaches such as univariate analysis and orthogonal design are sufficient to optimize PLE parameters.

APPLICATIONS IN SAMPLE PREPARATION FOR PHYTOCHEMICAL ANALYSIS

Flavonoids

Flavonoids, a major class of plant secondary metabolites, include flavonols, anthocyanins, proanthocyanidins (condensed tannins), and isoflavonoids. Owing to their multiple biological activities, PLE, as a new extraction technique, has been widely developed for sample preparation during analysis of flavonoids in various plant materials (Table **1**). Generally, alcohols (e.g., methanol and ethanol) and their mixtures with water are considered as optimum solvents [15-19, 21, 46-48, 52-57, 78, 79, 121, 122]. The effect of chemical additives on assay of isoflavones from soybean was investigated [126]. The results indicated that addition of dimethyl sulphoxide (DMSO) to aqueous ethanol (30:70, v/v) or aqueous acetonitrile (42:58, v/v) could obviously enhance total isoflavones recoveries. Besides, prior to extraction of target compounds, some low-polar reagents such as hexane were also utilized to remove the lipophilic impurities in sample matrices [47].

Table 1: Applications of PLE for the extractions of flavonoids from plant materials

Chemical constituents	Origin	Solvent	Temperature (°C)	Pressure (psi)	Time (min)	Other parameters	Optimization methods	Ref.
Biochanin A, formononetin, daidzein and genistein	*Trifolium arvense* L., *T. medium* L., *T. rubens* L., *T. pannonicum* L. and *T. pratense* L.	75% MeOH	125	1450	5 × 3 [a]	FV: 60%	UD	[46]
Casticine	*Vitex agnus-castus* L.	MeOH	70	1740	5 × 2			[15]
Six flavonoid glycosides (baicalin, dihydrobaicalin, lateriflorin and etc.) and three flavonoid aglycone (oroxylin A, baicalein and wogonin)	*Scutellaria lateriflora* L.	Water	85	1450	10 × 3		UD	[65]
Epimedin A, B, C, icariin and etc.	*Epimedium*	70% EtOH	120	1500	10	PS: 60-80 mesh ETs: 1	UD	[52-54]
Rutin	*Flos Lonicerae*	80% EtOH	100	1400	10	FV: 60%	UD	[17]
Euchrestaflavanone B, osajaxanthone, euchrestaflavanone C, alvaxanthone, macluraxanthone and 8-prenyltoxyloxanthone	*Maclura pomifera* Raf.	DCM	80	2000	5 × 3		UD	[87]
Rutin and isoquercitrin	*Sambucus nigra* L.	80% MeOH	100	870	20 × 3	FV: 60%	UD	[18]
Isoquercitrin and homoplantaginin	*Rosmarinus officinalis* L.	Water	100	1500	25	FV: 60%		[66]
Rutin, isoquercitrin, prunin, kaempferol-3-O-rutinoside and isorhamnetin-3-O-rutinoside	*Lysimachia clethroides* Duby	50% ACN	100	1500	25	FV: 70%	CCD	[77]
Ten isoflavones (genistin, daidzein, glycitin, ononin, formononetin and etc.)	Soybeans	90% MeOH	145	2030	5 × 2			[47]
Twelve isoflavones (daidzin, genistin, malonyldaidzin, malonylgenistin, daidzein and etc.)	Soybeans	DMSO: EtOH:Water (5:70:25, v/v/v) 70% EtOH	100 100	1000 1470	7 × 3 7 × 3	PS: < 0.825 mm	UD	[126] [19]
Puerarin, daidzin and daidzein	*Pueraria lobata*	95% EtOH	100	1400	10			[55]
Baicalein	*Radix Scutellariae*	20% EtOH	95	145-290	40	FR: 1.0 mL/min	UD	[56]
Catechin and epicatechin	tea leaves and grape seeds	MeOH	130	1470	5 × 2		UD	[78]
Eighteen anthocyanins (delphinidin-3-O-galactoside, delphinidin-3-O-glucoside, cyanidin-3-O-galactoside and etc.) and four flavonols (rutin, isoquercitrin, hyperoside and etc.)	*Calluna vulgaris* (L.) HULL, *Sambucus nigra* L. and *Vaccinium myrtillus* L.	80% MeOH	60-100	870-1000	(5-10) × 3	FV: 100%		[48]
Twenty-four anthocyanins	*Brassica oleracea* L. var. *capitata f. rubra*	Water/EtOH/FA (94/5/1, v/v/v)	99	725	7 [b]		CCD	[33]
Rotenone	*Derris elliptica* and *Derris malaccensis*	Chloroform	50	2000	30	FV: 60%	UD	[20]
Hyperoside, rutin and quercitrin	*Hypericum* species	MeOH	40	1450	5 × 4	FV: 60%		[121, 122]
Puerarin, daidzin and daizein	*Pueraria lobata* Willd. Ohwi	MeOH	140	1200	10	FV: 60%	OD	[79]
Rutin, hyperoside, isoquercitrin, quercitrin, quercetin and amentoflavone	*Hypericum perforatum* L.	MeOH or THF	100	2200	5 × 3	FV: 100%	UD	[21]
Quercetin, quercitrin, hyperoside and rutin	*Houttuynia cordata* Thunb	50% EtOH	70	1160	27	FR: 1.8 mL/min	OD	[16]
Daidzein, genistein, apigenin, biochanin A, kaempferol and coumestrol	*Matricaria recutita, Rosmarinus officinalis, Foeniculum vulgare,* and *Agrimonia eupatoria* L.	ACN	100	1450	15 × 2			[88]
Rutin, isoquercitrin, astragaline, cyanidin-3-sambubioside and cyanidin-3-glucoside	*Sambucus nigra* L.	80% EtOH	160	870	10	FV: 60%		[57]

Note: MeOH: methanol; EtOH: ethanol; DCM: dichloromethane; ACN: acetonitrile; DMSO: dimethyl sulphoxide; FA: formic acid; THF: tetrahydrofuran; FV: flush volume; PS: particle size; ETs: extraction times; FR: flow rate; UD: univariate design; OD: orthogonal design; CCD: central composite design.

[a] Time per cycle ×cycles.

[b] Five minites heat-up time included.

The effect of temperature on PLE is different based on the target compounds. An opposite trend for the recoveries of glycosides (e.g., baicalin) and aglycones (e.g., baicalein) from *Scutellaria lateriflora* was observed with increase of temperature [65]. Dawidowicz *et al.* [18] also explored the influence of temperature on PLE of rutin and isoquercitrin from *Sambucus nigra*. The yield of rutin began to decrease when the temperature was above 100 °C. A similar result was also obtained in the extraction of isoflavones from soybeans [19]. The possible reason was the degradation of the flavonol molecules at high temperatures rather than the hydrolysis of glycosides to aglycones [57].

Saponins

Saponins, one of the most widely distributed chemical groups in various plant species, can be divided into two types, triterpene and steroidal, based on the aglycone. PLE followed by HPLC coupled with UV [23, 58], DAD [22, 25, 80, 92], ELSD [24, 90, 93, 94], MS [49, 91, 127] or UPLC coupled with PDA [89] have been commonly utilized to determine saponins in some medicinal plants (Table **2**). Paul *et al.* [127] compared different extraction methods on isolation of steroidal glycoalkaloids from *Solanum xanthocarpum*. The results showed that PLE considerably enhanced extraction efficiency and reduced extraction time comparing to the conventional approaches such as ultrasonication and Soxhlet extraction. Several sample preparation techniques including PLE, ultrasonication, Soxhlet extraction and immersion were compared by Wan and his co-workers [92]. Likewise, PLE was considered as the method with the highest extraction efficiency and the best repeatability for the extraction of nine saponins from *Panax notoginseng*.

Table 2: Applications of PLE for the extractions of saponins, alkaloids and volatile oils from plant materials

Chemical constituents	Origin	Solvent	Temperature (°C)	Pressure (psi)	Time (min)	Other parameters	Optimization methods	Ref.
			Saponins					
Escin Ia, Ib, isoescin Ia and Ib	*Aesculus chinesis* Bunge	70% MeOH	120	N/A	7 × 2 [a]	FV: 60%	UD	[22]
Ginsenoside Rg$_1$, Re, Rb$_1$, Rc, and Rd	*Panax quinquefolium*	Water (1% Triton X-100)	120	1500	10		UD	[23]
Notoginsenoside R1, ginsenoside Rg1, Re, Rf, Rb1, Rg2, Rc, Rb2, Rb3, Rd and Rg3	*Panax notoginseng* (Burk.) F.H. Chen.	MeOH	150	1500	15	FV: 40% PS: 0.3-0.45 mm		[89-95]
Ginsenoside Rb1, Rb2, Rc, Rd, Re/Rg1	*Panax ginseng* and *Panax quinquefolium* L.	MeOH	140	362-435	20	FR: 1.0 mL/min PS: 0.5 mm		[80]
Ginsenoside Rb1, Rb2, Rg1, Rc and Rd	*Panax quinquefolium* L.	50% MeOH	120	1450	5 × 3			[49]
Solasonine, solamargine and β-2 solamargine	*Solanum xanthocarpum* Schrad. & Wendl.	MeOH	60	1450	10	FV: 60%		[127]
Jujuboside A and B	*Ziziphus jujube* Mill. var. *spinosa* (Bunge) Hu ex H. F. Chou	MeOH/EtOAc (95/5, v/v)	140	1200	15 × 2	FV: 40% PS: 40-60 mesh	UD	[24]
Gypenoside Rb1, Rb2, Rb3 and Rd	*Gynostemma pentaphyllum* (Thunb.) Makino	80% EtOH	100	215	180	FR: 10 mL/min SS[c]: 200 rpm		[58]
Segetoside B-I, K, L, vaccaroside A-H and vaccaroid A, B	*Vaccaria segetalis* Garcke, *Saponaria Vaccaria* L. and *Vaccaria pyramidata*	80% EtOH	150	1500	6 × 2	FV: 60%	UD	[25]
			Alkaloids					
Atropine Boldine Caffeine	*Atropa belladonna* L. *Peumus boldus* Mol. *Cola nitida* (Vent.) Schott et Endl.	DCM + 2ml dil. NH$_3$ DCM + 2ml dil. NH$_3$ MeOH	70	1740	5 × 3 5 × 3 5 × 2			[15]
Berberine, palmatine, jatrorrhizin and etc.	*Coptis chinensis* Franch	80% EtOH (with 0.50% HCl)	130	N/A	10	FV: 60% ETs: 2	UD & OD	[59]
Caffeine	Yunnan green tea	Water	150	580	10 × 5		UD	[67]
Pronuciferine, lotusine, nuciferine, liensinine, isoliensinine and neferine	*semen nelumbinis*	80% EtOH	100	1400	10	FV: 60%		[60]
Lycorine and galanthamine	*Narcissus jonquilla* 'Pipit'	MeOH (1% TA)	125	870	10	FV: 60%	UD	[27]
Hydrastine and berberine	*Hydrastis canadensis* L.	90% MeOH	100	1500	5 × 4			[50]

Table 2: cont....

Chemical constituents	Origin	Solvent	Temperature (°C)	Pressure (psi)	Time (min)	Other parameters	Optimization methods	Ref.
			Alkaloids					
Berberine and strychnine	*Strychnos nux-vomica* and *Rhizoma Coptidis*	MeOH	120	362-435	20	FR: 1 mL/min	UD	[81]
Ephedrine	*Ephedra sinica*	Water (0.4% SDS, w/v)	ambient	145-290	45-50	FR: 1.5 mL/min	UD	[76]
Tryptanthrin	*Isatis tinctoria* L.	MeOH	60	1740	5			[131]
			Volatile oils					
Eighteen secondary volatile metabolites (decursin, decursinol angelate, butylidene dihydrophthalide and etc.)	3 *Angelica* species	Hexane	80	1500	10 × 2	FV: 60%	UD	[28]
Twenty-eight essential oil components (thymol, *p*-cymene, *γ*-terpinene, carvacrol and etc.)	*Thymus vulgaris* L.	EtOAc	100	870	10	FV: 100%	UD	[29]
Ten volatile components (patchouli alcohol, pogostone, δ-guaiene, α-guaiene and etc.)	*Pogostemon cablin* (Blanco) Benth.	MeOH	80	1500	15	FV: 60% PS: 0.154 mm		[96]
Twelve volatile components (Z-ligustilide, E-butylidenephthalide, senkyunolide A, 6,7-dihydroxyligustilide and etc.)	*Angelica sinensis, Angelica acutiloba* and *Angelica gigas*	MeOH	100	1200	10 × 2	FV: 60% PS: 0.09-0.13 mm	UD	
E-, Z-butylidenephthalide and E-, Z-ligustilide	*Angelica sinensis*	MeOH	100	1200	10 × 2	FV: 60% PS: 0.09-0.13 mm		[97]
Z-ligustilide and Z-butylidenephthalide	*Angelica sinensis*	MeOH	110	1500	25	FV: 10% PS: 0.125-0.2 mm	CCD	[30]
β-caryophyllene, ar-curcumene, zingiberene, β-bisabolene, β-sesquiphellandrenendrene, and ar-, α-, β-turmerone	*Curcuma longa* L.	MeOH	140	1000	5	FV: 60% PS: 0.15-0.2 mm	UD	[98]
α-copaene, cyperene, β-selinene, β-cyperone and α-cyperone	*Cyperus rotundus* L.	MeOH	140	1000	10	FV: 60%	UD	[32]
Eleven sesquiterpenes (germacrene D, curzerene, γ-elemene, furanodienone, curcumol and etc.)	*Curcuma phaeocaulis, Curcuma wenyujin* and *Curcuma kwangsiensis*	MeOH	120	1500	5	FV: 60% PS: 0.2-0.3 mm	UD	[99]
Eleven sesquiterpenes (curcumenone, neocurdione, isocurcumenol, furanodiene, β-elemene and etc.)	*Curcuma phaeocaulis, Curcuma wenyujin* and *Curcuma kwangsiensis*	MeOH	100	1000	5	FV: 40% PS: 0.2-0.3 mm		[100, 101]
Thirty-one volatile components (1,2-dimethoxy-4-(2-propenyl)-benzene, 1,2-dimethoxy-4-(1-propenyl)-benzene, α-, β-asarone and etc.)	*Acorus tatarinowii* Schott.	Water	150	725	5	FR: 1 mL/min		[68]

Note: EtOAc: ethyl acetate; TA: tartaric acid; SDS: sodium dodecyl sulfate; SS: stirring speed; others: see Table 1.
[a]Time per cycle ×cycles.

Water, aqueous ethanol and methanol, as well as pure alcohol have been investigated as PLE solvents to extract saponins such as segetosides, vaccarosides and vaccaroids from cow cockle seed [25]. Under optimized conditions, acceptable recoveries could be obtained by means of whichever water, aqueous ethanol or methanol, but the chemical composition of the extracts was different (Fig. **4**). Moreover, a non-ionic surfactant solution as an alternative solvent system in PLE was reported for the extraction of ginsenosides in American ginseng [23]. The addition of Triton X-100 in water at a concentration above its critical micelle concentration (CMC) could enhance the recoveries of analytes.

Alkaloids

Alkaloids refer to the organic compounds normally with basic chemical properties and usually containing at least one nitrogen atom in a heterocyclic ring, originating chiefly from many vascular plants and some fungi. Many alkaloids exhibit remarkable bioactive activities, and therefore attract more and more attention of pharmacologists and chemists. Besides alcoholic solvents, acidic reagents were often utilized for the extraction of alkaloids in various plant materials. Recently, PLE combining with UPLC has been developed for qualitative and quantitative analysis of major alkaloids (e.g., berberine, palmatine and jatrorrhizin) in *Coptis chinensis* Franch, where 0.50% HCl existed in the aqueous ethanol or methanol was found to be beneficial to the yields of the three analytes [59]. Low concentration acid also could modify the selectivity of solvent to target alkaloids in PLE process for the extraction of two *Amaryllidaceae* alkaloids namely Lycorine and galanthamine from *Narcissus jonquilla* 'Pipit' [27]. In addition, pretreatment of sample matrices with ammonia could significantly increase PLE extraction efficiency of atropine, boldine and caffeine from *Atropa belladonna* L., *Peumus boldus* Mol. and *Cola nitida* (Vent.) Schott et Endl., respectively [15]. The addition of surfactants, e.g. SDS and Triton X-100, was also beneficial to PLE [76], which could disrupt the strong analyte-matrix interaction present naturally in plants and thus improve extraction efficiency (Table 2).

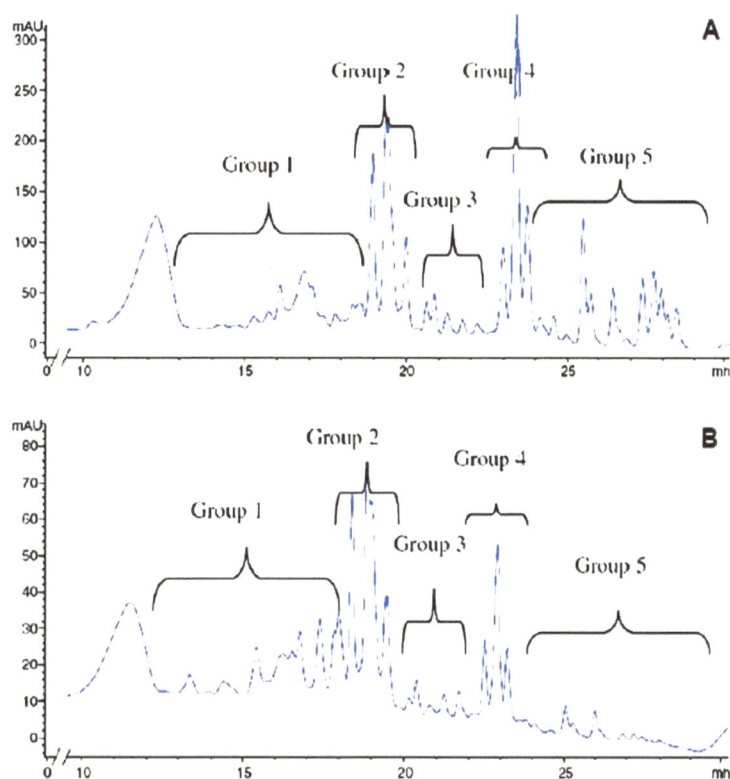

Figure 4: HPLC chromatograms of ground cow cockle seed extracts obtained using PLE with (A) 50% ethanol and (B) water at 125 °C for 15 min. Reprinted from Ref. [25], with permission.

4.4 Volatile Oils

Volatile oils (sometimes also called essential oils) are potentially diverse non-saponifiable lipids, which commonly contain terpenes, even though they may be minor constituents, and these can be monoterpene (C_{10}) or sesquiterpene (C_{15}) examples of unsaturated hydrocarbons or oxygen-containing terpenoids such as alcohols or carbonyl compounds. In last years, PLE has been successfully employed for the extraction of volatile components from various medicinal plants such as *Thymus vulgaris* L. [29], *Pogostemon cablin* (Blanco) Benth. [96], *Cyperus rotundus* L. [32], *Acorus tatarinowii Schott.* [68], and the species of *Angelica* [28, 30, 31, 97] and *Curcuma* [98-101] (Table 2). In these cases, methanol was proved as the most effective solvent for the extraction [30-32, 96-101]. As a polar reagent under ambient conditions, water, usually considered unsuitable for the extraction of lipophilic compounds, has also been attempted to extract some essential oil components in a traditional Chinese herb [68]. Especially, possible degradation phenomena were not observed during PLE of some thermally labile compounds, which might attribute to the inert (nitrogen) surroundings [30, 97]. Unfortunately, PLE has poor selectivity for the extraction of volatile components, and non-volatile ingredients could be co-extracted [29].

Phenols

Phenols are widely distributed in the plant kingdom and some of them possess germicidal, estrogenic or antioxidant property. Particularly, polyphenols, a major group of phenolic compounds with two or more hydroxy groups (-OH) bonded to the aromatic ring(s) in the same molecule, are usually unstable. So their extraction is very important. PLE was found to be a powerful, rapid and low solvent consumption method for extraction of phenolic compounds from various plant materials [82, 97, 110, 111] (Table 3). Even at room temperature, PLE showed comparable extraction efficiencies with heating under reflux for gastrodin and vanillyl alcohol in *Gastrodia elata* Blume [34]. Besides the commonly used hydroalcoholic mixtures, acetone and its aqueous mixture have also been demonstrated as felicitous solvents for the extraction of phenolic compounds [35, 61, 109]. It was reported that the use of acidified methanol (containing 0.2% of formic acid) could significantly improve the yields of phenolic compounds from *Mellissa officinalis* [128].

Table 3: Applications of PLE for the extractions of phenols, terpenoids and anthraquinones from plant materials

Chemical constituents	Origin	Solvent	Temperature (°C)	Pressure (psi)	Time (min)	Other parameters	Optimization methods	Ref.
Phenols								
3-isomangostin, gartanine, desoxygartanine, α-, β-mangostin and 9-hydroxycalabaxanthone	*Garcinia mangostana*	EtOH	100	N/A	5	FV: 60%		[107]
Eight phenolic compounds (apiin, malonyl-apiin, acetyl-apiin and etc.)	*Petroselinum crispum*	50% EtOH or 50% Acetone	70	1000	5 × 4 [a]	FV: 10% PS: < 0.425 mm	UD	[61]
Gastrodin and vanillyl alcohol	*Gastrodia elata* Blume	20% EtOH	ambient	145-290	40-50	FR: 1.5 mL/min	UD	[34]
Gallic acid, hydrolyzable and condensed tannins, myricetin derivatives, quercetin derivatives and kaempferol derivatives	*Ceratonia siliqua* L.	50% Acetone	60	N/A	5 × 2	FV: 50%		[35]
Sinapic, ferulic, coumarinic, caffeic, syringic, vanillic, and 4-hydroxybenzoic acids	*Majorana hortensis* L.	Acetone	150	2175	10 × 2			[109]
4-hydroxybenzoic acid, vanillic acid, 4-hydroxybenzaledehyde, vanillin, ferulic acid and ferulic aldehyde	*Curcuma longa* L.	MeOH	100	1500	5 × 4			[110]
Ten phenolic compounds (3,4-dihydroxybenzaldehyde, 4-hydroxybenzoic acid, vanillic acid, 4-hydroxybenzaldehyde and etc.)	*Vanilla planifolia*	EtOH	60	1500	5 × 2			[82]
Nine phenolic acids (gallic acid, chlorogenic acid, syringic acid, vanillic acid and etc.)	*Peucedanum alsaticum* L. and *Peucedanum cervaria* (L.) Lap.	MeOH	100	1450	10 × 5			[111]
Ferulic acid and coniferyl ferulate	*Angelica sinensis*	MeOH	100	1200	10 × 2	FV: 60%		[97]
Rosmarinic acid, caffeic acid, protocatechuic acid and protocatechuicaldehyde	*Mellissa officinalis*	MeOH (0.2% FA)	80	1450	5 × 2			[128]
Gallic acid, ellagic acid and corilagin	*Phyllanthus niruri* Linn.	Water	100	1450	60	FR: 1.5 mL/min	UD	[69]
Caffeic acid, ferulic acid, sinapic acid and isoferulic acid	*Cimicifuga racemosa*	60% MeOH	90	1000	5 × 2	FV: 50%	UD	[36]
Terpenoids								
Betulin	*Betula pendula* and *Betula pubescens*	EtOH	120	725	5 × 2	FV: 60%	CCD	[83]
Glycyrrhizin and 18β-glycyrrhetinic acid	*Radix glycyrrhizae*	MeOH	100	145-435	20	FR: 1.0 mL/min	UD	[64]
Glycyrrhizin	*Radix glycyrrhizae*	Water (0.4% SDS, w/v) or Water (1% Triton X-100, v/v)	ambient	145-290	45-50	FR: 1.5 mL/min	UD	[76]
Dihydrotanshinone I, cryptotanshinone, tanshinone I and tanshinone IIA	*Salvia miltiorrhiza*	ACN	120	1500	10	FV: 60% PS: 0.13-0.2 mm	UD	[38]
Tanshinone I and tanshinone IIA	*Salvia miltiorrhiza*	MeOH 30% EtOH	120 95	145-290 145-290	20 40	FR: 1.0 mL/min	UD	[39]
Ten diterpenoids (tanshinone I, tanshinone IIA, cryptotanshinone, dihydrotanshinone I and etc.)	*Salvia miltiorrhiza*	EtOH	100	1500	10	FV: 60%		[112]
Paclitaxel, baccatin III and 10-deacetylbaccatin III	*Taxus cuspidata* Sieb. et Zucc.	90% MeOH	150	1470	15	FV: 100%	UD	[40]
Caffeoyl esters of betulinic, morolic and oleanolic acid	*Oenothera biennis* L.	80% EtOH	120	1740	5 × 2		UD	[62]
Eight triterpenes (ganoderic acid A, ganoderic acid Y, ganoderic acid DM, ganoderol A and etc.)	*Ganoderma lucidum* and *Ganoderma sinense*	MeOH	100	1500	5	FV: 60%	UD	[41]
Mogroside IV, V, mogrol, 11-oxo-mogrol, siaminoside-1 and 11-oxo-mogroside V	*Siraitia grosvenorii* (Swingle) C. Jeffrey	Water	150	1700	30	FR: 0.7 mL/min	UD	[71]

Table 3: cont....

Chemical constituents	Origin	Solvent	Temperature (°C)	Pressure (psi)	Time (min)	Other parameters	Optimization methods	Ref.
Anthraquinones								
Rhein, emodin, aloe-emodin, chrysophanol and physcion,	*Rheum officinale* Baill	MeOH	140	1500	5	PS: 0.13-0.2 mm	UD	[37]
Hypericin and pseudohypericin	*Hypericum* species	MeOH	40	1450	5 × 4	FV: 60%		[121, 122]
Hypericin and pseudohypericin	*Hypericum perforatum* L.	Acetone	150	2200	5 × 3	FV: 100%	UD	[21]
Hypericin	*Hypericum perforatum* L.	MeOH	100	1500	10		UD	[26]
Damnacanthal	*Morinda citrifolia* L.	Water	170	580	200	FR: 2.4-4 mL/min	UD	[70]

Note: see Table 1 and Table 2.
[a] Time per cycle ×cycles.

Terpenoids

Terpenoids (sometimes called isoprenoids) are a large and diverse class of naturally-occurring organic chemicals similar to terpenes. Although terpenoids can be classified many types according to the number of isoprene units used, we only talk about diterpenoids and triterpenoids, two important types of termpenoids widely existed in various medicinal herbs with anti-bacterial, anti-neoplastic, and other pharmaceutical activities. *Salvia miltiorrhiza* is one of the most well-known traditional Chinese medicines for promoting blood circulation and removing stasis. The diterpenoids, such as tanshinone I, tanshinone IIA, cryptotanshinone and dihydrotanshinone I, are believed to be its major bioactive ingredients. Due to the importance of these compounds for quality control of *Salvia miltiorrhiza*, several analytical techniques including PLE coupled with CEC, ULPC and LC-ESI-MS have been developed for their qualitative and quantitative analysis in our laboratory [38, 112] and Ong's group [39]. Paclitaxel (generally known as Taxol), a complex nitrogen-containing diterpenoid, is famous for its significant anti-cancer activity. It is usually recognized as a thermal unstable compound and its extraction should be performed at room temperature. However, Kawamura and his co-workers successfully utilized PLE to extract paclitaxel and related compounds from *Taxus cuspidate* at 150 °C [40]. Recently, a PLE method, using water as the extraction solvent and several chromatographic support materials (Alumina, Celite and Silica gel) as filtration aids, was developed for the extraction of mogrosides from *Siraitia grosvenorii*, and the presence of these support materials was beneficial to the recovery of the target compounds [71]. The application of PLE for extraction of diterpenoids and triterpenoids was listed in Table 3.

Anthraquinones

Anthraquinone and its derivatives, including oxanthranol, anthranol, anthrone, the dimer of anthrone and so on, are the active substances of a series of medicinal preparations. In our previous work, a novel separation technique using PLE and capillary zone electrophoresis (CZE) was developed for simultaneous determination of five anthraquinones including aloe-emodin, emodin, chrysophanol, physcion, and rhein in *Rheum officinale* Baill [37]. Using subcritical water (temperature, 170 °C; pressure, 4 MPa), PLE provided a promising alternative method for extraction of damnacanthal from roots of *Morinda citrifolia* L. [70]. PLE has also been successfully applied to determine the anthraquinone derivatives, hypericin and pseudohypericin, in several species of *Hypericum* [21, 26, 121, 122] (Table 3).

Miscellaneous

Besides the major phytochemical compounds mentioned above, PLE has also been widely used for sample preparation during the analysis of phytochemicals, including nucleosides [102-106], saccharides [63, 73, 74], amino acids [65], fatty acids [24], vitamins [84, 117, 118, 123], carotenoids [42, 85], steroids [43], coumarins and furanocoumarins [44, 129, 130], lactones [42, 72, 119], glucosinolates [51], polyacetylenes [75], limonoid derivatives [45], and benzoxazinone derivatives [86, 116] in medicinal plants (Table **4**).

Table 4: Applications of PLE for the extractions of other bioactive or marker compounds from plant materials

Chemical constituents	Origin	Solvent	Temperature (°C)	Pressure (psi)	Time (min)	Other parameters	Optimization methods	Ref.
Nucleosides and bases (adenosine, cytidine, guanosine, adenine, cytosine, guanine and etc.)	*Cordyceps sinensis*	MeOH	160	1500	5		OD	[102-106]
Sucrose, raffinose, stachyose and verbascose	*Lupinus albus* and *Lupinus angustifolius*	48% EtOH	60	1450	5 × 5 [a]	FV: 150%		[63]

Table 4: cont....

Chemical constituents	Origin	Solvent	Temperature (°C)	Pressure (psi)	Time (min)	Other parameters	Optimization methods	Ref.
Polysaccharides	*Ganoderma lucidum, Trametes versicolor*, etc.	Water	120	1500	5 × 2			[73]
Stevioside and rebaudioside A	*Stevia rebaudiana* Bertoni	Water	100	160-190	15	FR: 1.5 mL/min	UD	[74]
Amino acids (GABA, glutamine and etc.)	*Scutellaria lateriflora* L.	Water	85	1450	10 × 3		UD	[65]
Nine fatty acids (lauric, myristic, palmitic, palmitoleic, stearic acid and etc.)	*Ziziphus jujube* Mill. var. *spinosa* (Bunge) Hu ex H. F. Chou	MeOH/EtOAc (95/5, v/v)	140	1200	15 × 2	FV: 40% PS: 40-60 mesh	UD	[24]
Fatty acids (palmitic, linoleic, stearic, lignoceric acid and etc.) and vitamin E	*Piper gaudichaudianum* Kunth	P. E. |EtOH	85 150	1500	10 20		CCD	[84, 117]
Vitamin E	*Corylus avellana* L.	Hexane (0.01% BHT)	60	1500	15	FV: 60%		[118]
δ-, (β+γ)- and α-tocopherol	almonds, sunflower seeds, hazelnuts and walnuts	ACN	50	1600	5 × 2			[123]
Eleven carotenoids (α-carotene, 13-*cis*-β-carotene, all-trans-β-carotene, 15-*cis*-β-carotene and etc.)	*Dunaliella salina*	EtOH	160	1500	17.5	FV: 60%	CCD	[85]
Six lactones (kavain, yangonin, dihydrokavain, methysticin, desmethoxyyangonin and dihydromethysticin)	*Piper methysticum* G. Forst.	Acetone MeOH	50 60	1500 2000	12 × 5 N/A	FV: 60%	UD	[119] [42]
Charantin	*Momordica charantia* L.	EtOH	120	1450	40	FR: 2.0 mL/min	UD	[43]
Eight furanocoumarins (umbelliferone, xanthotoxin, bergapten, isopimpinellin, hellopterin and etc.)	*Archangelica officinalis* Hoffm.	MeOH	100-130	870	10	FV: 60%	UD	[44]
Coumarins and furanocoumarins (umbelliferone and bergapten)	*Ammi majus* L.	Chloroform + MeOH	100	1015	5 × 3	FV: 33%		[129]
Eight furanocoumarins (xanthotoxin, bergapten, isopimpinellin, imperatorin, phellopterin and etc.)	*Pastinaca sativa*	P. E. + MeOH	100	870	N/A	FV: 60%		[130]
Five trilactones (ginkgolides A, B, C, J and bilobalide)	*Ginkgo biloba* L.	Water (0.2% HOAc, v/v)	ambient	1470	15 + 10 [b]	PS: 42-60 mesh FR: 1.5-2.0 mL/min	UD	[72]
Carotenoids (astaxanthin, β-carotene, lutein and etc.)	*Haematococcus pluvialis* and *Dunaliella salina*	Acetone	20	1500	5 × 3	FV: 60%		[42]
Nine glucosinolates (epiprogoitrin, sinigrin, progoitrin, gluconapin, sulfoglucobrassicin and etc.)	*Isatis tinctoria*	70% MeOH	50	1740	5 × 3	FV: 100% PS: 0.5 mm	UD	[51]
Three polyacetylene compounds	*Radix Codonopsis pilosula*	Water (0.01% Triton X-100)	95	145-290	40	FR: 1.0 mL/min	UD	[75]
Six limonoid derivatives (fraxinellone, obacunone, limonin and etc.)	*Dictamnus dasycarpus* L.	MeOH	150	1500	5	FV: 60%	UD	[45]
Benzoxazinone derivatives (HBOA, DIBOA, HMBOA, DIMBOA, BOA, MBOA and etc.)	wheat samples	MeOH (1% HOAc)	150	1500	5 × 3	FV: 60%	UD	[86, 116]

Note: P. E.: Petroleum ether; BHT: butylated hydroxytoluene; HOAc: acetic acid; others: see Table 1 and Table 2.
[a] Time per cycle ×cycles.
[b] Static extraction time (15 min) + dynamic extraction time (10 min).

CONCLUSIONS

As an ideal sample preparation technique, PLE can be almost applied for the extraction of all types of compounds present naturally in plant matrices. It usually shows better or at least comparable extraction efficiency to the conventional methods such as sonication, reflux extraction, Soxhlet extraction and microwave-assisted extraction. The advantages of PLE are low solvent consumption, short extraction time and high reproducibility and automation, which improve the wide application of PLE in phytochemical analysis.

ACKNOWLEDGEMENTS

This research was supported by grants from University of Macau (UL015/09-Y1 to S. P. Li).

REFERENCES

[1]   Li P, Li SP, Wang YT. Optimization of CZE for analysis of phytochemical bioactive compounds. Electrophoresis 2006; 27: 4808-19.

[2]   Romanik G, Gilgenast E, Przyjazny A, Kamiński M. Techniques of preparing plant material for chromatographic separation and analysis. J Biochem Biophys Methods 2007; 70: 253-61.

[3]   Raynie DE. Modern extraction techniques. Anal Chem 2006; 78: 3997-4004.

[4]   Zygmunt B, Namieśnik J. Preparation of samples of plant material for chromatographic analysis. J Chromatogr Sci 2003; 41: 109-16.

[5]   Ong ES, Cheong JS, Goh D. Pressurized hot water extraction of bioactive or marker compounds in botanicals and medicinal plant materials. J Chromatogr A 2006; 1112: 92-102.

[6]   Bjoërklund E, Nilsson T. Pressurised liquid extraction of persistent organic pollutants in environmental analysis. Trac-Trend Anal. Chem. 2000; 19: 434-445.

[7]   Ramos L, Kristenson EM, Brinkman UA. Current use of pressurised liquid extraction and subcritical water extraction in environmental analysis. J Chromatogr A 2002; 975: 3-29.

[8]   Fidalgo-Used N, Blanco-González E, Sanz-Medel A. Sample handling strategies for the determination of persistent trace organic contaminants from biota samples. Anal Chim Acta 2007; 590: 1-16.

[9]   Schantz MM. Pressurized liquid extraction in environmental analysis. Anal Bioanal Chem 2006; 386: 1043-7.

[10]   Wilkes JG, Conte ED, Kim Y, Holcomb M, Sutherland JB, Miller DW. Sample preparation for the analysis of flavors and off-flavors in foods. J. Chromatogr. A 2000; 880: 3-33.

[11]   Mendiola JA, Herrero M, Cifuentes A, Ibañez E. Use of compressed fluids for sample preparation: Food applications. J Chromatogr A 2007; 1152: 234-46.

[12]   Carabias-Martínez R, Rodríguez-Gonzalo E, Revilla-Ruiz P, Hernández-Méndez J. Pressurized liquid extraction in the analysis of food and biological samples. J Chromatogr A 2005; 1089: 1-17.

[13]   Beyer A, Biziuk M. Applications of sample preparation techniques in the analysis of pesticides and PCBs in food. Food Chem 2008; 108: 669-80.

[14]   Benthin B, Danz H, Hamburger M. Pressurized liquid extraction of medicinal plants. J Chromatogr A 1999; 837: 211-9.

[15]   Basalo C, Mohn T, Hamburger M. Are extraction methods in quantitative assays of pharmacopoeia monographs exhaustive? A comparison with pressurized liquid extraction. Planta Med 2006; 72: 1157-62.

[16]   Zhang Y, Li SF, Wu XW. Pressurized liquid extraction of flavonoids from Houttuynia cordata Thunb. Sep Purif Technol 2008; 58: 305-10.

[17]   Chen Y, Li SG, Lin XH, Luo HB, Li GW, Yao H. On-line screening and identification of radical scavenging compounds extracted from Flos Lonicerae by LC-DAD-TOF-MS. Chromatographia 2008; 68: 327-32.

[18]   Dawidowicz AL, Wianowska D, Gawdzik J, Smolarz DH. Optimization of ASE conditions for the HPLC determination of rutin and isoquercitrin in Sambucus nigra L. J Liq Chromatogr Relat Technol 2003; 26: 2381-97.

[19]   Rostagno MA, Palma M, Barroso CG. Pressurized liquid extraction of isoflavones from soybeans. Anal Chim Acta 2004; 522: 169-77.

[20]   Sae-Yun A, Ovatlarnporn C, Itharat A, Wiwattanapatapee R. Extraction of rotenone from Derris elliptica and Derris malaccensis by pressurized liquid extraction compared with maceration. J Chromatogr A 2006; 1125: 172-6.

[21]   Williams FB, Sander LC, Wise SA, Girard J. Development and evaluation of methods for determination of naphthodianthrones and flavonoids in St. John's wort. J Chromatogr A 2006; 1115: 93-102.

[22]   Chen J, Li W, Yang B, Guo X, Lee FS, Wang X. Determination of four major saponins in the seeds of Aesculus chinensis Bunge using accelerated solvent extraction followed by high-performance liquid chromatography and electrospray-time of flight mass spectrometry. Anal Chim Acta 2007; 596: 273-80.

[23]   Choi MP, Chan KK, Leung HW, Huie CW. Pressurized liquid extraction of active ingredients (ginsenosides) from medicinal plants using non-ionic surfactant solutions. J Chromatogr A 2003; 983: 153-62.

[24]   Zhao J, Li SP, Yang FQ, Li P, Wang YT. Simultaneous determination of saponins and fatty acids in Ziziphus jujuba (Suanzaoren) by high performance liquid chromatography-evaporative light scattering detection and pressurized liquid extraction. J Chromatogr A 2006; 1108: 188-94.

[25]   Güçlü-Üstündağ Ö, Balsevich J, Mazza G. Pressurized low polarity water extraction of saponins from cow cockle seed. J Food Eng 2007; 80: 619-30.

[26]   Anand R, Verma N, Gupta DK, Puri SC, Handa G, Sharma VK, Qazi GN. Comparison of extraction techniques for extraction of bioactive molecules from Hypericum perforatum L. plant. J Chromatogr Sci 2005; 43: 530-1.

[27] Mroczek T, Mazurek J. Pressurized liquid extraction and anticholinesterase activity-based thin-layer chromatography with bioautography of Amaryllidaceae alkaloids. Anal Chim Acta 2009; 633: 188-96.

[28] Cho SK, Abd El-Aty AM, Choi JH, Kim MR, Shim JH. Optimized conditions for the extraction of secondary volatile metabolites in Angelica roots by accelerated solvent extraction. J Pharm Biomed Anal 2007; 44: 1154-8.

[29] Dawidowicz AL, Rado E, Wianowska D, Mardarowicz M, Gawdzik J. Application of PLE for the determination of essential oil components from Thymus vulgaris L. Talanta 2008; 76: 878-84.

[30] Li P, Li SP, Lao SC, Fu CM, Kan KK, Wang YT. Optimization of pressurized liquid extraction for Z-ligustilide, Z-butylidenephthalide and ferulic acid in Angelica sinensis. J Pharm Biomed Anal 2006; 40: 1073-9.

[31] Lao SC, Li SP, Kan KK, Li P, Wan JB, Wang YT, Dong TT, Tsim KW. Identification and quantification of 13 components in Angelica sinensis (Danggui) by gas chromatography-mass spectrometry coupled with pressurized liquid extraction. Anal Chim Acta 2004; 526: 131-7.

[32] Tam CU, Yang FQ, Zhang QW, Guan J, Li SP. Optimization and comparison of three methods for extraction of volatile compounds from Cyperus rotundus evaluated by gas chromatography-mass spectrometry. J Pharm Biomed Anal 2007; 44: 444-9.

[33] Arapitsas P, Turner C. Pressurized solvent extraction and monolithic column-HPLC/DAD analysis of anthocyanins in red cabbage. Talanta 2008; 74: 1218-23.

[34] Ong ES, Heng MY, Tan SN, Hong Yong JW, Koh H, Teo CC, Hew CS. Determination of gastrodin and vanillyl alcohol in Gastrodia elata Blume by pressurized liquid extraction at room temperature. J Sep Sci 2007; 30: 2130-7.

[35] Papagiannopoulos M, Wollseifen HR, Mellenthin A, Haber B, Galensa R. Identification and quantification of polyphenols in carob fruits (Ceratonia siliqua L.) and derived products by HPLC-UV-ESI/MSn. J Agric Food Chem 2004; 52: 3784-91.

[36] Mukhopadhyay S, Luthria DL, Robbins RJ. Optimization of extraction process for phenolic acids from black cohosh (Cimicifuga racemosa) by pressurized liquid extraction. J Sci Food Agric 2006; 86: 156-62.

[37] Gong YX, Li SP, Wang YT, Li P, Yang FQ. Simultaneous determination of anthraquinones in Rhubarb by pressurized liquid extraction and capillary zone electrophoresis. Electrophoresis 2005; 26: 1778-82.

[38] Li P, Li SP, Yang FQ, Wang YT. Simultaneous determination of four tanshinones in salvia miltiorrhiza by pressurized liquid extraction and capillary electrochromatography. J Sep Sci 2007; 30: 900-5.

[39] Ong ES, Len SM. Evaluation of pressurized liquid extraction and pressurized hot water extraction for tanshinone I and IIA in Salvia miltiorrhiza using LC and LC-ESI-MS. J Chromatogr Sci 2004; 42: 211-6.

[40] Kawamura F, Kikuchi Y, Ohira T, Yatagai M. Accelerated solvent extraction of Paclitaxel and related compounds from the bark of Taxus cuspidata. J Nat Prod 1999; 62: 244-7.

[41] Zhao J, Zhang XQ, Li SP, Yang FQ, Wang YT, Ye WC. Quality evaluation of Ganoderma through simultaneous determination of nine triterpenes and sterols using pressurized liquid extraction and high performance liquid chromatography. J Sep Sci 2006; 29: 2609-15.

[42] Denery JR, Dragull K, Tang CS, Li QX. Pressurized fluid extraction of carotenoids from Haematococcus pluvialis and Dunaliella salina and kavalactones from Piper methysticum. Anal Chim Acta 2004; 501: 175-81.

[43] Pitipanapong J, Chitprasert S, Goto M, Jiratchariyakul W, Sasaki M, Shotipruk A. New approach for extraction of charantin from Momordica charantia with pressurized liquid extraction. Sep Purif Technol 2007; 52: 416-22.

[44] Waksmundzka-Hajnos M, Petruczynik A, Dragan A, Wianowska D, Dawidowicz AL. Effect of extraction method on the yield of furanocoumarins from fruits of Archangelica officinalis Hoffm. Phytochem Anal 2004; 15: 313-9.

[45] Jiang Y, Li SP, Chang HT, Wang YT, Tu PF. Pressurized liquid extraction followed by high-performance liquid chromatography for determination of seven active compounds in Cortex Dictamni. J Chromatogr A 2006; 1108: 268-72.

[46] Zgórka G. Pressurized liquid extraction versus other extraction techniques in micropreparative isolation of pharmacologically active isoflavones from Trifolium L. species. Talanta 2009; 79: 46-53.

[47] Klejdus B, Mikelová R, Petrlová J, Potěšil D, Adam V, Stiborová M, Hodek P, Vacek J, Kizek R, *et al.* Evaluation of isoflavone aglycon and glycoside distribution in soy plants and soybeans by fast column high-performance liquid chromatography coupled with a diode-array detector. J Agric Food Chem 2005; 53: 5848-52.

[48] Rieger G, Müller M, Guttenberger H, Bucar F. Influence of altitudinal variation on the content of phenolic compounds in wild populations of Calluna vulgaris, Sambucus nigra, and Vaccinium myrtillus. J Agric Food Chem 2008; 56: 9080-6.

[49] Ligor T, Ludwiczuk A, Wolski T, Buszewski B. Isolation and determination of ginsenosides in American ginseng leaves and root extracts by LC-MS. Anal Bioanal Chem 2005; 383: 1098-105.

[50] Schieffer GW, Pfeiffer K. Pressurized liquid extraction and multiple ultrasonically-assisted extractions of hydrastine and berberine from goldenseal (Hydrastis canadensis) with subsequent HPLC assay. J Liq Chromatogr Relat Technol 2001; 24: 2415-27.

[51] Mohn T, Cutting B, Ernst B, Hamburger M. Extraction and analysis of intact glucosinolates--a validated pressurized liquid extraction/liquid chromatography-mass spectrometry protocol for Isatis tinctoria, and qualitative analysis of other cruciferous plants. J Chromatogr A 2007; 1166: 142-51.

[52]   Chen XJ, Guo BL, Li SP, Zhang QW, Tu PF, Wang YT. Simultaneous determination of 15 flavonoids in Epimedium using pressurized liquid extraction and high-performance liquid chromatography. J Chromatogr A 2007; 1163: 96-104.

[53]   Chen XJ, Ji H, Wang YT, Li SP. Simultaneous determination of seven flavonoids in Epimedium using pressurized liquid extraction and capillary electrochromatography. J Sep Sci 2008; 31: 881-7.

[54]   Chen XJ, Ji H, Zhang QW, Tu PF, Wang YT, Guo BL, Li SP. A rapid method for simultaneous determination of 15 flavonoids in Epimedium using pressurized liquid extraction and ultra-performance liquid chromatography. J Pharm Biomed Anal 2008; 46: 226-35.

[55]   Lee MH, Lin CC. Comparison of techniques for extraction of isoflavones from the root of Radix Puerariae: Ultrasonic and pressurized solvent extractions. Food Chem 2007; 105: 223-8.

[56]   Ong ES, Len SM. Pressurized hot water extraction of berberine, baicalein and glycyrrhizin in medicinal plants. Anal Chim Acta 2003; 482: 81-9.

[57]   Dawidowicz AL, Wianowska D, Baraniak B. The antioxidant properties of alcoholic extracts from Sambucus nigra L. (antioxidant properties of extracts). LWT-Food Sci Technol 2006; 39: 308-15.

[58]   Chen CH, Huang TY, Lee MR, Hsu SL, Chang CMJ. Continuous pressurized fluid extraction of Gynostemma pentaphyllum and purification of gypenosides. Ind Eng Chem Res 2007; 46: 8138-43.

[59]   Chen J, Wang F, Liu J, Lee FS, Wang X, Yang H. Analysis of alkaloids in Coptis chinensis Franch by accelerated solvent extraction combined with ultra performance liquid chromatographic analysis with photodiode array and tandem mass spectrometry detections. Anal Chim Acta 2008; 613: 184-95.

[60]   Li WL, Chen JH, Yin YF, Wang XR, Lee FS. Analysis of alkaloids in Semen nelumbinis by accelerated solvent extraction-high performance liquid chromatography-diode array detection-electrospray ioninztion-time of flight-mass spectrometry. Fenxi Huaxue 2008; 36: 79-82.

[61]   Luthria DL. Influence of experimental conditions on the extraction of phenolic compounds from parsley (Petroselinum crispum) flakes using a pressurized liquid extractor. Food Chem 2008; 107: 745-52.

[62]   Zaugg J, Potterat O, Plescher A, Honermeier B, Hamburger M. Quantitative analysis of anti-inflammatory and radical scavenging triterpenoid esters in evening primrose seeds. J Agric Food Chem 2006; 54: 6623-8.

[63]   David B, Ingo S, Anne-Christin W. Highly automated and fast determination of raffinose family oligosaccharides in Lupinus seeds using pressurized liquid extraction and high-performance anion-exchange chromatography with pulsed amperometric detection. J. Sci. Food Agric. 2008; 88: 1949-1953.

[64]   Ong ES. Chemical assay of glycyrrhizin in medicinal plants by pressurized liquid extraction (PLE) with capillary zone electrophoresis (CZE). J Sep Sci 2002; 25: 825-31.

[65]   Bergeron C, Gafner S, Clausen E, Carrier DJ. Comparison of the chemical composition of extracts from Scutellaria lateriflora using accelerated solvent extraction and supercritical fluid extraction versus standard hot water or 70% ethanol extraction. J Agric Food Chem 2005; 53: 3076-80.

[66]   Herrero M, Arráez-Román D, Segura A, Kenndler E, Gius B, Raggi MA, Ibáñez E, Cifuentes A. Pressurized liquid extraction-capillary electrophoresis-mass spectrometry for the analysis of polar antioxidants in rosemary extracts. J Chromatogr A 2005; 1084: 54-62.

[67]   Dawidowicz AL, Wianowska D. PLE in the analysis of plant compounds. Part I. The application of PLE for HPLC analysis of caffeine in green tea leaves. J Pharm Biomed Anal 2005; 37: 1155-9.

[68]   Deng CH, Li N, Zhang XM. Rapid determination of essential oil in Acorus tatarinowii Schott. by pressurized hot water extraction followed by solid-phase microextraction and gas chromatography-mass spectrometry. J Chromatogr A 2004; 1059: 149-55.

[69]   Markom M, Hasan M, Daud WRW, Singh H, Jahim JM. Extraction of hydrolysable tannins from Phyllanthus niruri Linn.: Effects of solvents and extraction methods. Sep Purif Technol 2007; 52: 487-96.

[70]   Anekpankul T, Goto M, Sasaki M, Pavasant P, Shotipruk A. Extraction of anti-cancer damnacanthal from roots of Morinda citrifolia by subcritical water. Sep Purif Technol 2007; 55: 343-9.

[71]   Xia Y, Rivero-Huguet ME, Hughes BH, Marshall WD. Isolation of the sweet components from Siraitia grosvenorii. Food Chem 2008; 107: 1022-8.

[72]   Lang QY, Wai CM. Pressurized water extraction (PWE) of terpene trilactones from Ginkgo biloba leaves. Green Chem 2003; 5: 415-20.

[73]   Di X, Chan KK, Leung HW, Huie CW. Fingerprint profiling of acid hydrolyzates of polysaccharides extracted from the fruiting bodies and spores of Lingzhi by high-performance thin-layer chromatography. J Chromatogr A 2003; 1018: 85-95.

[74]   Teo CC, Tan SN, Hong Yong JW, Hew CS, Ong ES. Validation of green-solvent extraction combined with chromatographic chemical fingerprint to evaluate quality of Stevia rebaudiana Bertoni. J Sep Sci 2009; 32: 613-22.

[75]   Ong ES, Len SM. Evaluation of surfactant-assisted pressurized hot water extraction for marker compounds in Radix Codonopsis pilosula using liquid chromatography and liquid chromatography/electrospray ionization mass spectrometry. J Sep Sci 2003; 26: 1533-40.

[76]   Eng AT, Heng MY, Ong ES. Evaluation of surfactant assisted pressurized liquid extraction for the determination of glycyrrhizin and ephedrine in medicinal plants. Anal Chim Acta 2007; 583: 289-95.

[77]   Jiang Y, Li P, Li SP, Wang YT, Tu PF. Optimization of pressurized liquid extraction of five major flavanoids from Lysimachia clethroide. J Pharm Biomed Anal 2007; 43: 341-5.

[78]   Piñeiro Z, Palma M, Barroso CG. Determination of catechins by means of extraction with pressurized liquids. J Chromatogr A 2004; 1026: 19-23.

[79]   Wan JB, Xu C, Li SP, Kong LY, Wang YT. Studies on the pressurized solvent extraction of isoflavonoids from Pueraria lobata. Fenxi Huaxue 2005; 33: 1435-8.

[80]   Lee HK, Koh HL, Ong ES, Woo SO. Determination of ginsenosides in medicinal plants and health supplements by pressurized liquid extraction (PLE) with reversed phase high performance liquid chromatography. J Sep Sci 2002; 25: 160-6.

[81]   Ong ES, binte Apandi SN. Determination of berberine and strychnine in medicinal plants and herbal preparations by pressurized liquid extraction with capillary zone electrophoresis. Electrophoresis 2001; 22: 2723-9.

[82]   Sharma UK, Sharma N, Gupta AP, Kumar V, Sinha AK. RP-HPTLC densitometric determination and validation of vanillin and related phenolic compounds in accelerated solvent extract of Vanilla planifolia. J Sep Sci 2007; 30: 3174-80.

[83]   Co M, Koskela P, Eklund-Åkergren P, Srinivas K, King JW, Sjöberg PJ, Turner C. Pressurized liquid extraction of betulin and antioxidants from birch bark. Green Chem 2009; 11: 668-74.

[84]   Péres VF, Saffi J, Melecchi MI, Abad FC, Martinez MM, Oliveira EC, Jacques RA, Caramão EB. Optimization of pressurized liquid extraction of Piper gaudichaudianum Kunth leaves. J Chromatogr A 2006; 1105: 148-53.

[85]   Herrero M, Jaime L, Martín-Álvarez PJ, Cifuentes A, Ibáñez E. Optimization of the extraction of antioxidants from Dunaliella salina microalga by pressurized liquids. J Agric Food Chem 2006; 54: 5597-603.

[86]   Villagrasa M, Guillamón M, Eljarrat E, Barceló D. Determination of benzoxazinone derivatives in plants by combining pressurized liquid extraction-solid-phase extraction followed by liquid chromatography-electrospray mass spectrometry. J Agric Food Chem 2006; 54: 1001-8.

[87]   da Costa CT, Margolis SA, Benner J, B. A., Horton D. Comparison of methods for extraction of flavanones and xanthones from the root bark of the osage orange tree using liquid chromatography. J Chromatogr A 1999; 831: 167-78.

[88]   Bajer T, Adam M, Galla L, Ventura K. Comparison of various extraction techniques for isolation and determination of isoflavonoids in plants. J Sep Sci 2007; 30: 122-7.

[89]   Guan J, Lai CM, Li SP. A rapid method for the simultaneous determination of 11 saponins in Panax notoginseng using ultra performance liquid chromatography. J Pharm Biomed Anal 2007; 44: 996-1000.

[90]   Wan JB, Li P, Li SP, Wang YT, Dong TT, Tsim KW. Simultaneous determination of 11 saponins in Panax notoginseng using HPLC-ELSD and pressurized liquid extraction. J Sep Sci 2006; 29: 2190-6.

[91]   Qian ZM, Lu J, Gao QP, Li SP. Rapid method for simultaneous determination of flavonoid, saponins and polyacetylenes in folium ginseng and radix ginseng by pressurized liquid extraction and high-performance liquid chromatography coupled with diode array detection and mass spectrometry. J Chromatogr A 2009; 1216: 3825-30.

[92]   Wan JB, Lai CM, Li SP, Lee MY, Kong LY, Wang YT. Simultaneous determination of nine saponins from Panax notoginseng using HPLC and pressurized liquid extraction. J Pharm Biomed Anal 2006; 41: 274-9.

[93]   Wan JB, Li SP, Chen JM, Wang YT. Chemical characteristics of three medicinal plants of the Panax genus determined by HPLC-ELSD. J Sep Sci 2007; 30: 825-32.

[94]   Wan JB, Yang FQ, Li SP, Wang YT, Cui XM. Chemical characteristics for different parts of Panax notoginseng using pressurized liquid extraction and HPLC-ELSD. J Pharm Biomed Anal 2006; 41: 1596-601.

[95]   Wan JB, Zhang QW, Ye WC, Wang YT. Quantification and separation of protopanaxatriol and protopanaxadiol type saponins from Panax notoginseng with macroporous resins. Sep Purif Technol 2008; 60: 198-205.

[96]   Hu LF, Li SP, Cao H, Liu JJ, Gao JL, Yang FQ, Wang YT. GC-MS fingerprint of Pogostemon cablin in China. J Pharm Biomed Anal 2006; 42: 200-6.

[97]   Xie JJ, Lu J, Qian ZM, Yu Y, Duan JA, Li SP. Optimization and comparison of five methods for extraction of coniferyl ferulate from Angelica sinensis. Molecules 2009; 14: 555-65.

[98]   Qin NY, Yang FQ, Wang YT, Li SP. Quantitative determination of eight components in rhizome (Jianghuang) and tuberous root (Yujin) of Curcuma longa using pressurized liquid extraction and gas chromatography-mass spectrometry. J Pharm Biomed Anal 2007; 43: 486-92.

[99]   Yang FQ, Li SP, Chen Y, Lao SC, Wang YT, Dong TT, Tsim KW. Identification and quantitation of eleven sesquiterpenes in three species of Curcuma rhizomes by pressurized liquid extraction and gas chromatography - mass spectrometry. J Pharm Biomed Anal 2005; 39: 552-8.

[100]   Yang FQ, Wang YT, Li SP. Simultaneous determination of 11 characteristic components in three species of Curcuma rhizomes using pressurized liquid extraction and high-performance liquid chromatography. J Chromatogr A 2006; 1134: 226-31.

[101]   Yang FQ, Li SP, Zhao J, Lao SC, Wang YT. Optimization of GC-MS conditions based on resolution and stability of analytes for simultaneous determination of nine sesquiterpenoids in three species of Curcuma rhizomes. J Pharm Biomed Anal 2007; 43: 73-82.

[102]   Li SP, Li P, Lai CM, Gong YX, Kan KK, Dong TT, Tsim KW, Wang YT. Simultaneous determination of ergosterol, nucleosides and their bases from natural and cultured Cordyceps by pressurised liquid extraction and high-performance liquid chromatography. J Chromatogr A 2004; 1036: 239-43.

[103] Fan H, Li SP, Xiang JJ, Lai CM, Yang FQ, Gao JL, Wang YT. Qualitative and quantitative determination of nucleosides, bases and their analogues in natural and cultured Cordyceps by pressurized liquid extraction and high performance liquid chromatography-electrospray ionization tandem mass spectrometry (HPLC-ESI-MS/MS). Anal Chim Acta 2006; 567: 218-28.

[104] Yang FQ, Guan J, Li SP. Fast simultaneous determination of 14 nucleosides and nucleobases in cultured Cordyceps using ultra-performance liquid chromatography. Talanta 2007; 73: 269-73.

[105] Yang FQ, Li SP. Effects of sample preparation methods on the quantification of nucleosides in natural and cultured Cordyceps. J Pharm Biomed Anal 2008; 48: 231-5.

[106] Yang FQ, Li SP, Li P, Wang YT. Optimization of CEC for simultaneous determination of eleven nucleosides and nucleobases in Cordyceps using central composite design. Electrophoresis 2007; 28: 1681-8.

[107] Destandau E, Toribio A, Lafosse M, Pecher V, Lamy C, André P. Centrifugal partition chromatography directly interfaced with mass spectrometry for the fast screening and fractionation of major xanthones in Garcina mangostana. J Chromatogr A 2009; 1216: 1390-4.

[108] Ong ES, Woo SO, Yong YL. Pressurized liquid extraction of berberine and aristolochic acids in medicinal plants. J Chromatogr A 2000; 904: 57-64.

[109] Petr J, Vítková K, Ranc V, Znaleziona J, Maier V, Knob R, Ševčík J. Determination of some phenolic acids in Majorana hortensis by capillary electrophoresis with online electrokinetic preconcentration. J Agric Food Chem 2008; 56: 3940-4.

[110] Schieffer GW. Pressurized liquid extraction of curcuminoids and curcuminoid degradation products from turmeric (Curcuma longa) with subsequent HPLC assays. J Liq Chromatogr Relat Technol 2002; 25: 3033-44.

[111] Skalicka-Woźniak K, Głowniak K. Quantitative analysis of phenolic acids in extracts obtained from the fruits of Peucedanum alsaticum L. and Peucedanum cervaria (L.) lap. Chromatographia 2008; 68: s85-90.

[112] Li P, Xu G, Li SP, Wang YT, Fan TP, Zhao QS, Zhang QW. Optimizing ultra performance liquid chromatographic analysis of 10 diterpenoid compounds in Salvia miltiorrhiza using central composite design. J Agric Food Chem 2008; 56: 1164-71.

[113] Kobayashi M, Fujita M, Mitsuhashi H. Constituents of Cnidium officinale: occurrence of pregnenolone, coniferylferulate and hydroxyphthalides. Chem Pharm Bull 1987; 35: 1427-33.

[114] Kaouadji M, Pachtere FD, Pouget C, Chulia AJ, Lavaitte S. Three additional phthalide derivatives, an epoxymonomer and two dimmers from Ligusticum wallichii rhizomes. J Nat Prod 1986; 49: 872-7.

[115] Nerín C, Salafranca J, Aznar M, Batlle R. Critical review on recent developments in solventless techniques for extraction of analytes. Anal Bioanal Chem 2009; 393: 809-33.

[116] Bonnington L, Eljarrat E, Guillamón M, Eichhorn P, Taberner A, Barceló D. Development of a liquid chromatography-electrospray-tandem mass spectrometry method for the quantitative determination of benzoxazinone derivatives in plants. Anal Chem 2003; 75: 3128-36.

[117] Péres VF, Saffi J, Melecchi MI, Abad FC, de Assis Jacques R, Martinez MM, Oliveira EC, Caramão EB. Comparison of soxhlet, ultrasound-assisted and pressurized liquid extraction of terpenes, fatty acids and Vitamin E from Piper gaudichaudianum Kunth. J Chromatogr A 2006; 1105: 115-8.

[118] Sivakumar G, Bacchetta L, Gatti R, Zappa G. HPLC screening of natural vitamin E from mediterranean plant biofactories - a basic toot for pilot-scale bioreactors production of alpha-tocopherot. J Plant Physiol 2005; 162: 1280-3.

[119] Warburton E, Norris PL, Goenaga-Infante H. Comparison of the capabilities of accelerated solvent extraction and sonication as extraction techniques for the quantification of kavalactones in Piper methysticum (Kava) roots by high performance liquid chromatography with ultra violet detection. Phytochem Anal 2007; 18: 98-102.

[120] Ong ES. Extraction methods and chemical standardization of botanicals and herbal preparations. J Chromatogr B 2004; 812: 23-33.

[121] Smelcerovic A, Spiteller M, Zuehlke S. Comparison of methods for the exhaustive extraction of hypericins, flavonoids, and hyperforin from Hypericum perforatum L. J Agric Food Chem 2006; 54: 2750-3.

[122] Smelcerovic A, Verma V, Spiteller M, Ahmad SM, Puri SC, Qazi GN. Phytochemical analysis and genetic characterization of six Hypericum species from Serbia. Phytochemistry 2006; 67: 171-7.

[123] Delgado-Zamarreño MM, Bustamante-Rangel M, Sánchez-Pérez A, Carabias-Martínez R. Pressurized liquid extraction prior to liquid chromatography with electrochemical detection for the analysis of vitamin E isomers in seeds and nuts. J Chromatogr A 2004; 1056: 249-52.

[124] Chen PY, Tu YX, Wu CT, Jong TT, Chang CMJ. Continuous hot pressurized solvent extraction of 1,1-diphenyl-2-picrylhydrazyl free radical scavenging compounds from Taiwan yams (Dioscorea alata). J Agric Food Chem 2004; 52: 1945-9.

[125] Wsól V, Fell AF. Central composite design as a powerful optimisation technique for enantioresolution of the rac-11-dihydrooracin - the principal metabolite of the potential cytostatic drug oracin. J. Biochem. Biophys. Methods. 2002; 54: 377-390.

[126] Luthria DL, Biswas R, Natarajan S. Comparison of extraction solvents and techniques used for the assay of isoflavones from soybean. Food Chem 2007; 105: 325-33.

[127] Paul AT, Vir S, Bhutani KK. Liquid chromatography-mass spectrometry-based quantification of steroidal glycoalkaloids from Solanum xanthocarpum and effect of different extraction methods on their content. J Chromatogr A 2008; 1208: 141-6.

[128] Ziaková A, Brandšteterová E. Application of different preparation techniques for extraction of phenolic antioxidants from lemon balm (Melissa officinalis) before HPLC analysis. J Liq Chromatogr Relat Technol 2002; 25: 3017-32.

[129] Królicka A, Kartanowicz R, Wosiński SA, Szpitter A, Kamiński M, Łojkowska E. Induction of secondary metabolite production in transformed callus of Ammi majus L. grown after electromagnetic treatment of the culture medium. Enzyme Microb Technol 2006; 39: 1386-91.

[130] Waksmundzka-Hajnos M, Petruczynik A, Dragan A, Wianowska D, Dawidowicz AL, Sowa I. Influence of the extraction mode on the yield of some furanocoumarins from Pastinaca sativa fruits. J Chromatogr B 2004; 800: 181-7.

[131] Danz H, Baumann D, Hamburger M. Quantitative determination of the dual COX-2/5-LOX inhibitor tryptanthrin in Isatis tinctoria by ESI-LC-MS. Planta Med 2002; 68: 152-7.

INDEX

www.ingramcontent.com/pod-product-compliance
Lightning Source LLC
Chambersburg PA
CBHW041710210326
41598CB00007B/604